Lactobacillus Molecular Biology

From Genomics to Probiotics

Edited by

Åsa Ljungh
Faculty of Medicine
Lund University
Lund
Sweden

and

Torkel Wadström
Faculty of Medicine
Lund University
Lund
Sweden

 Caister Academic Press

Caister Academic Press
Norfolk, UK

www.caister.com

British Library Cataloguing-in-Publication Data
A catalogue record for this book is available from the British Library

ISBN: 978-1-904455-41-7

Description or mention of instrumentation, software, or other products in this book does not imply endorsement by the author or publisher. The author and publisher do not assume responsibility for the validity of any products or procedures mentioned or described in this book or for the consequences of their use.

Printed and bound in Great Britain

Contents

Contributors

Jenni Antikainen
Helsinki University Hospital Laboratory (HUSLAB)
Helsinki
Finland

jenni.antikainen@hus.fi

Denny Demeria
University of Alberta
Edmonton, Alberta
Canada

ddemeria@gmail.com

Lars Engstrand
Department of Microbiology
Tumor and Cell Biology
Karolinska Institutet
Stockholm
Sweden

lars.engstrand@smi.se

Julia Ewaschuk
University of Alberta
Edmonton, Alberta
Canada

jbe@ualberta.ca

Chandra Iyer
Department of Pathology
Texas Children's Hospital
Baylor College of Medicine
Houston, TX
USA

chandra@kibowbiotech.com

Cecilia Jernberg
Swedish Institute for Infectious Disease Control
Stockholm
Sweden

cecilia.jernberg@smi.se

John Keohane
Alimentary Pharmabiotic Centre
and Department of Medicine
Cork University Hospital
Cork
Ireland

keohanejohn@gmail.com

Timo K. Korhonen
General Microbiology
University of Helsinki
Helsinki
Finland

timo.korhonen@helsinki.fi

Veera Kuparinen
General Microbiology
University of Helsinki
Helsinki
Finland

veera.kuparinen@helsinki.fi

Åsa Ljungh
Faculty of Medicine
Lund University
Lund
Sweden

Asa.Ljungh@med.lu.se

Graciela L. Lorca
Department of Microbiology and Cell Science
University of Florida
Genetics Institute
Gainesville, FL
USA

glorca@ufl.edu

Karen Madsen
University of Alberta
Edmonton, Alberta
Canada

karen.madsen@ualberta.ca

Hans-Olof Nilsson
Faculty of Medicine
Lund University
Lund
Sweden

hans-olof.nilsson@med.lu.se

Elisabeth Norin
Department of Microbiology
Tumor and Cell Biology
Karolinska Institutet
Stockholm
Sweden

elisabeth.norin@ki.se

Paul W. O'Toole
Department of Microbiology
and Alimentary Pharmabiotic Centre
University College Cork
Cork
Ireland

pwotoole@ucc.ie

Jan-Peter Van Pijkeren
Department of Microbiology
and Alimentary Pharmabiotic Centre
University College Cork
Cork
Ireland

vanpijkeren@gmail.com

Bruno Pot
Laboratory 'Lactic Acid Bacteria and Mucosal Immunology'
Insitut Pasteur de Lille
Lille
France

bruno.pot@ibl.fr

Gregor Reid
Canadian Research and Development
Centre for Probiotics
Lawson Health Research Institute
London, Ontario
Canada

gregor@uwo.ca

Stefan Roos
Department of Microbiology
Swedish University of Agricultural Sciences
Uppsala
Sweden

stefan.roos@mikrob.slu.se

Kieran Ryan
Alimentary Pharmabiotic Centre
Cork University Hospital
Cork
Ireland

k.ryan@ucc.ie

Fergus Shanahan
Alimentary Pharmabiotic Centre
and Department of Medicine
Cork University Hospital
Cork
Ireland

F.shanahan@ucc.ie

Takahiro Toba
Department of Applied Biosciences
Faculty of Agriculture and Life Science
Hirosaki University
Hirosaki
Japan

ttakki@cc.hirosaki-u.ac.jp

Effie Tsakalidou
Agricultural University of Athens
Department of Food Science and Technology
Athens
Greece

et@aua.gr

Graciela Font de Valdez
Centro de Referencia para Lactobacilos
(CERELA-CONICET)
San Miguel de Tucumán
Argentina

gfont@cerela.org.ar

James Versalovic
Department of Pathology
Texas Children's Hospital
Baylor College of Medicine
Houston, TX
USA

jamesv@bcm.edu

Torkel Wadström
Faculty of Medicine
Lund University
Lund
Sweden

Torkel.Wadstrom@mmb.lu.se

Current Books of Interest

Microbial Toxins: Current Research and Future Trends	2009
Acanthamoeba	2009
Bacterial Secreted Proteins: Secretory Mechanisms and Role in Pathogenesis	2009
Lactobacillus Molecular Biology: From Genomics to Probiotics	2009
Mycobacterium: Genomics and Molecular Biology	2009
Real-Time PCR: Current Technology and Applications	2009
Clostridia: Molecular Biology in the Post-genomic Era	2009
Plant Pathogenic Bacteria: Genomics and Molecular Biology	2009
Microbial Production of Biopolymers and Polymer Precursors	2009
Plasmids: Current Research and Future Trends	2008
Pasteurellaceae: Biology, Genomics and Molecular Aspects	2008
Vibrio cholerae: Genomics and Molecular Biology	2008
Pathogenic Fungi: Insights in Molecular Biology	2008
Helicobacter pylori: Molecular Genetics and Cellular Biology	2008
Corynebacteria: Genomics and Molecular Biology	2008
Staphylococcus: Molecular Genetics	2008
Leishmania: After The Genome	2008
Archaea: New Models for Prokaryotic Biology	2008
RNA and the Regulation of Gene Expression	2008
Legionella Molecular Microbiology	2008
Molecular Oral Microbiology	2008
Epigenetics	2008
Animal Viruses: Molecular Biology	2008
Segmented Double-Stranded RNA Viruses	2008
Acinetobacter Molecular Biology	2008
Pseudomonas: Genomics and Molecular Biology	2008
Microbial Biodegradation: Genomics and Molecular Biology	2008
The Cyanobacteria: Molecular Biology, Genomics and Evolution	2008
Coronaviruses: Molecular and Cellular Biology	2007
Real-Time PCR in Microbiology: From Diagnosis to Characterisation	2007
Bacteriophage: Genetics and Molecular Biology	2007
Candida: Comparative and Functional Genomics	2007
Bacillus: Cellular and Molecular Biology	2007
AIDS Vaccine Development: Challenges and Opportunities	2007
Alpha Herpesviruses: Molecular and Cellular Biology	2007
Pathogenic *Treponema*: Molecular and Cellular Biology	2007
PCR Troubleshooting: The Essential Guide	2006
Influenza Virology: Current Topics	2006
Microbial Subversion of Immunity: Current Topics	2006
Cytomegaloviruses: Molecular Biology and Immunology	2006

Preface

Lactic acid bacteria (LAB) have a very long history of use in fermentation. At the beginning of the twentieth century, potential health-promoting effects associated with harbouring these bacteria were highlighted by Élie Metchnikoff. Since then, we have seen numerous commercial products with such postulated claims. However, the scientific basis for probiosis has been poor with respect to characterization of bacteria and their beneficial mechanisms, and declaration of species identity of strains, their numbers and their viability was often lacking in commercially available preparations. Furthermore, very few randomized case–control studies with follow-up analyses of colonization and persistence had been performed.

Thankfully, however, during recent decades, we have seen a breakthrough in LAB research. This volume, written by acknowledged experts in various fields provides authoritative up-to-date knowledge and understanding of most, if not all, aspects of LAB. The complex taxonomy of LAB, and their genome biology, are thoroughly reviewed. Since LAB form part of a complex intestinal microbiota, there is a pressing need for molecular approaches to study their composition, connection to health and disease, and changes induced by antibiotics and other treatments. LAB are exposed to acid as well as bile stress during passage through the gastrointestinal (GI) canal. This potentially influences expression of multiple genes, and changes the phenotypes of the corresponding LAB. The current knowledge of immunomodulatory effects and anti-cancer activities are presented. Finally, LAB in the GI tract and the vagina, the two most common areas for application of LAB, are reviewed.

We envisage that this book will be of great help and interest for all that are interested in ecology, non-antibiotic treatment of infections, and the use of LAB for specific purposes such as for vaccine delivery. It may also form a basis for rationalization of health claims, important for regulatory aspects in a growing number of countries.

History of Probiotics and Living Drugs

1

Åsa Ljungh and Torkel Wadström

The human gastrointestinal tract microbiota comprises trillions of organisms, and forms various complex ecosystems. It is the focus of major research from impact on health to various acute and chronic diseases, inflammatory conditions and cancer. We now face a 'boom' in interest in the human health, not the least our nutrition with stone age type of diets, antioxidants and various food phytochemicals.

Élie Metchnikoff and microbiologist colleagues in Russia, Bulgaria and other parts of Europe and the Middle East were involved in pioneer studies of sour milk and yoghurt microbiology, and described a number of organisms in natural fermentations, including the famous *Bacillus bulgaricus*, later renamed *Lactobacillus bulgaricus* and, more recently, *L. delbrückii* subspecies *bulgaricus*.

Metchnikoff wrote 'Essaies optimistes' in 1907 and discussed philosophy, religion, folklore and science of ageing, and microbial gut ecology during human life. This work is now regarded as the birth of probiotics, microbes ingested with the aim of promoting a good life and health. Lorendon M. Douglass wrote the volume *The Bacillus of Long Life* in 1911. Later, Rettger and colleagues at Yale showed that the *B. bulgaricus* could not survive in the human gut but well other lactic acid bacteria (LAB) such as *L. acidophilus* (1920), and clinical trials were performed first in patients with constipation with promising results (1935).

In the 1950s, Freter *et al.* showed that broad-spectrum antibiotics destroyed the mouse gut microflora, and enhanced the susceptibility of these animals to *Salmonella* and *Shigella* infections more than a 1000-fold. Collins (1978) showed that germ-free (GF) guinea pigs were killed by 10 cells (!) of *Salmonella enteritidis* strain, but only 10^9 cells killed a conventional animal with a complete gut microflora.

Pioneer studies in Japan in the 1940s and 1950s on the *L. casei* Yakult and various strains of some *Bifidobacteria* species claimed the role of microbes to protect the young mouse against gut infections after weanling. Some Japanese laboratories also early created a special *Bifidobacterium* journal (*Bioscience and Microflora*), and carried out a number of pioneer studies on various health effects associated with these s.c. probiotic bacteria. Moreover, later studies confirmed and expanded our knowledge that a number of these effects are associated with *in vivo* modulating effects of unique cell wall constituents, such as lipoteichoic acid (LTA) molecules and cell surface proteins (CSPs) of the *L. casei–paracasei* and also other species in the complex large LAB family, Bifidobacteria and other 'probiotic candidates' such as *Clostridium butyricum*. Many of the studies stress similar biological effects by living and killed and/or lysed microbes not accepted within the common probiotic definition in the EU and by FDA in the USA. Peptidoglycan and DNA of many pathogenic bacteria as well as LPS of most Gram-negative bacteria induce inflammation and coagulation markers in synergy upon gut translocation and septicaemia. Various probiotic LAB seem to counteract opening of gut epithelium tight junction (TJs) and to down-modulate these proinflammatory cascades (Amoureux *et al.*, 2005).

A number of published studies in the last few years indicate, however, that the *L casei/paracasei* group contains organisms with cell components (LTA, CSPs and maybe nucleic acid components) with anti-inflammatory properties (Chapter 7). The promising effect of *L. reuteri* on IBD induced by IL-10–/– *Helicobacter hepaticus* in a mouse model encourages studies on *other* LAB with other anti- to proinflammatory immunoproperties. The Nobel Prize in 2007 to the knock-out mouse, and the possibility to tailor-make models for specific human diseases such as atherosclerosis in the ApoE mouse is very encouraging for scientists who want to understand the mode of action of various LAB and other probiotic organisms including some *Bacillus* and *Clostridium* species. The induction of an IBD-like disease and colon cancer in the IL-$10^{-/-}$ mouse by an *Enterococcus faecalis* strain will now encourage more work on the gut microflora including organisms defined as possible probiotics (with *E. faecium* strains M74, SF68 in the literature). The recent pioneer studies by Gordon and colleagues in St. Louis, MO, on *specific* role of *specific Bacteroides* strains in the gut microflora and gut surface crosstalk is a very intriguing and promising area for future research (Wexler, 2007).

Infection with the protozoon *Babesia microti* can be treated in mice with *L. casei* strain ATCC 7469 live or dead bacteria with a very good outcome, at least partially due to stimulation of the innate immunity. Freund's complete adjuvant was used in the study, but is an example of several recent reports on probiotics modulating toll-like receptor (TLR) responses to various pathogens and inflammatory molecules, such as peptidoglycan and LPS fragments as well as microbial proinflammatory CpG DNA motifs.

This volume is an attempt to update us on probiotic lactic acid bacteria: from basic characteristics and biological effects to clinical application and use. The question about what we shall require from probiotic strains will be addressed.

References

Amoureux, M.-C., Rajapakse, N., Stipkovits, L., and Szathmary, S. (2005). Peptidoglycan and bacterial DNA induce inflammation and coagulation markers in synergy. Mediators of Inflammation 2, 118–120.

Fuller, R. ed. (1992). Probiotics. The Scientific Basis. (London: Chapman and Hall).

Metchnikoff, E. ed. (1907). The Prolongation Of Life. Optimistic Studies. London: William Heinemann.

Wexler, H.M. (2007). Bacteroides: the good, the bad and the nitty-gritty. Clin. Microbiol. Rev. *20*, 593–621.

Taxonomy and Metabolism of *Lactobacillus* 2

Bruno Pot and Effie Tsakalidou

Introduction

Today taxonomy is often based on a polyphasic approach (Vandamme *et al.*, 1996a), which involves genotypic and phenotypic methods. Pure historically phenotypic methods have dominated the identification and classification schemes of the lactic acid bacteria and the lactobacilli in particular. Today, 16S RNA sequencing has become the method of choice, not only because of its high degree of portability, but equally important, because of the availability of a large database of reference sequences. In a polyphasic approach, which involves both genotypic and phenotypic methods, one may encounter differences between, for example, the phylogenetic tree revealed by 16S RNA sequencing and the phenotypic groups based on fermentation profiles and metabolite production.

It would be wrong, however, to limit all classification schemes to the phylogenetic groupings based on 16S rRNA sequences exclusively. 16S rRNA suffers from considerable disadvantages, as will be shown below. A polyphasic approach that involves a balanced use of multiple genotypic and phenotypic methods will always yield a more balanced and reliable result. As sequencing becomes more and more accessible and cheaper, the role of a single molecule such as 16S rRNA, will tend to fade. The sequencing of a variety of individual (household) genes will diversify and broaden the taxonomic views, with the sequencing of the complete genome of many organisms as a feasible option within a few years.

Today, however, 16S rRNA in combination with DNA–DNA hybridizations is still the reference method. Therefore, the taxonomic discussion in this chapter will be based on a neighbour joining tree, obtained with the 16S rRNA of the *Lactobacillus* species known at the end of May 2008.

In the section 'Metabolism' below we describe the remarkable variation of catabolic activities within the genus *Lactobacillus*. This variation, together with the fact that lactobacilli are generally considered 'safe', has been the basis of their very broad use in food applications (Vankerckhoven *et al.*, 2008; Huys *et al.*, 2006). The active application of living microorganisms in foods requires proper labelling (Temmerman *et al.*, 2004) which makes proper identification, based on stable and reliable classification schemes, extremely important.

Metabolism

The first essential step in food fermentations is the catabolism of carbohydrates by the lactic acid bacteria. Lactic acid bacteria as a group exhibit an enormous capacity to degrade different carbohydrates and related compounds. The main end product is lactic acid (>50% of sugar carbon). It should be noticed, however, that lactic acid bacteria adapt to various conditions and change their metabolism accordingly. This may lead to significantly different end product patterns (Table 2.1).

The taxonomy of lactic acid bacteria for many decades heavily relied on the type of sugar fermentation. In order to deal with the large number of species being described, Orla-Jensen (1919, 1942, 1943) proposed a classification

Table 2.1 Historical subdivision of the genus *Lactobacillus* according to the type of fermentation

	I: Homofermentative		II: Heterofermentative	
Glucose fermented to lactic acid	≥ 85%		50%	
Formation of CO_2, acetic acid and ethanol	–		+	
CO_2 formed from glucose	–		+	
Thiamine required for growth	–		+	
Fructose diphosphate aldolase	+		–	
Orla-Jensen, 1919				
	'*Thermobacterium*'	'*Streptobacterium*'	'*Betabacterium*'	
	Obligate homofermentative	Facultative heterofermentative	Obligate heterofermentative	
van den Hamer, 1960				
Fructose-1,6-bisphosphate aldolase	+	+	–	
Glucose-6-phosphate dehydrogenase	–	+	+	
6-Phosphogluconate dehydrogenase	-	+	+	
Rogosa, 1970	**IA**	**IB**	**II**	
Growth at 45°C	+	d		
Growth at 15°C	–	d		
Ribose fermented	–	+	+	
CO_2 from gluconate	–	+	+	
Rogosa, 1974			**II**	**III**
Acidophilic			–	+
Ethanol tolerant			–	+
Most carbohydrates fermented			+	–

of lactic acid bacteria, which was based on morphology, temperature range of growth, nutritional characteristics, carbon sources utilization and agglutination effects. Orla-Jensen differentiated three major groups. The first group contained *Thermobacterium*, *Streptobacterium* and *Streptococcus*, which were all catalase negative and produce mainly lactic acid besides traces of other by-products. The second group contained *Betabacterium* and *Betacoccus*, which also lack catalase but as a rule formed detectable amounts of gas and other by-products, besides lactic acid. The third group consisting of *Microbacterium* and *Tetracoccus* show a positive catalase reaction.

In 1960, van den Hamer showed that representatives of *Betabacterium* did not posses fructose-1,6-bisphosphate aldolase, in contrast to *Thermobacterium* and

Table 2.1 *continued*

	I: Homofermentative		II: Heterofermentative	
Sharpe, 1979			**IIA**	**IIB**
Aerobic species	*Lactobacillus acidophilus*	*Lactobacillus casei*	*Lactobacillus brevis*	*Lactobacillus fructivorans*
	Lactobacillus delbrueckii	*Lactobacillus coryniformis*	*Lactobacillus buchneri*	*Lactobacillus hilgardii*
	Lactobacillus helveticus	*Lactobacillus curvatus*	*Lactobacillus confusus*	
	Lactobacillus jensenii	*Lactobacillus homohiochii*	*Lactobacillus fermentum*	
	Lactobacillus salivarius	*Lactobacillus plantarum*	*Lactobacillus viridescens*	
		Lactobacillus yamanshiensis		
Anaerobic species	*Lactobacillus ruminis*			
	Lactobacillus vitulinus			
Kandler and Weiss, 1986				
	Group I	Group II	Group III	
Hexose almost exclusively to lactic acid	+	+	–	
Hexose fermented to lactic-, acetic acid, ethanol, CO_2	–	–	+	
Lactic-, acetic-, formic acid, ethanol under glucose limitation	–	d	+	
Pentose phosphoketolase	–	+	+	
Gluconate fermented	–	+	+	

d, strain dependent; *L.*, *Lactobacillus*.

Streptobacterium. These findings supported the discrimination of three physiological groups: (i) the obligately homofermentative lactobacilli, lacking both glucose-6-phosphate dehydrogenase and 6-phosphogluconate dehydrogenase (*Thermobacterium*), (ii) the facultatively homofermentative lactobacilli, having both dehydrogenases, but degrading glucose preferably via the Embden–Meyerhof–Parnas pathway (*Streptobacterium*), and (iii) the obligately heterofermentative lactobacilli, lacking fructose-1,6-bisphosphate-aldolase (*Betabacterium*). *Thermobacterium*, *Streptobacterium* and *Betabacterium* were considered to be the three subgenera within the genus *Lactobacillus*.

The subdivision of lactobacilli into three major fermentation groups for taxonomic reasons was maintained until the late 1970s (Table 2.1). The development and application of advanced molecular techniques brought new insights in the taxonomy of the genus, which is meanwhile considered the most heterogeneous among the lactic acid bacteria, with currently 113 different species described (Table 2.2). However, for practical rea-

Table 2.2 List of *Lactobacillus* species with some details on their original description and their current taxonomic status. The table also contains some information on their respective phylogenetic position, metabolism and peptidoglycan type

Number	Genus	Species	Subspecies	References
1	*Lactobacillus*	*acetotolerans*		Entani *et al.* 1986
2	*Lactobacillus*	*acidifarinae*		Vancanneyt *et al.* 2005b
3	*Lactobacillus*	*acidipiscis*		Tanasupawat *et al.* 2000
4	*Lactobacillus*	*acidophilus*		(Moro 1900) Hansen and Mocquot 1970; Johnson *et al.* 1980
5	*Lactobacillus*	*agilis*		Weiss *et al.* 1981,1982
6	*Lactobacillus*	*algidus*		Kato *et al.* 2000
7	*Lactobacillus*	*alimentarius*		(ex Reuter 1970) Reuter 1983a,b
8	*Lactobacillus*	*amylolyticus*		Bohak *et al.* 1998, 1999
9	*Lactobacillus*	*amylophilus*		Nakamura and Crowell 1979,1981
10	*Lactobacillus*	*amylotrophicus*		Naser *et al.* 2006c
11	*Lactobacillus*	*amylovorus*		Nakamura 1981
12	*Lactobacillus*	*animalis*		Dent and Williams 1982, 1983
13	*Lactobacillus*	*antri*		Roos *et al.* 2005
14	*Lactobacillus*	*apodemi*		Osawa *et al.* 2006
	Lactobacillus	*"arizonensis"*		Swezey *et al.* 2000
15	*Lactobacillus*	*aviarius*		Fujisawa *et al.* 1984,1985
		aviarius	*araffinosus*	Fujisawa *et al.* 1984,1986
		aviarius	*aviarius*	Fujisawa *et al.* 1984,1986
	Lactobacillus	*"backi"*		Bohack *et al.*, 2006
	Lactobacillus	*"bavaricus"*		Stetter and Stetter 1980
	Lactobacillus	*"bifermentans"*		(ex Pette and van Beynum 1943) Kandler *et al.* 1983b,c
16	*Lactobacillus*	*brevis*		(Orla-Jensen 1919) Bergey *et al.* 1934
17	*Lactobacillus*	*buchneri*		(Henneberg 1903) Bergey *et al.* 1923
	Lactobacillus	*"bulgaricus"*		(Orla-Jensen 1919) Rogosa and Hansen 1971
18	*Lactobacillus*	*camelliae*		Tanasupawat *et al.* 2007
	"Lactobacillus"	*"carnis"*		Shaw and Harding 1985, 1986
19	*Lactobacillus*	*casei*		(Orla-Jensen 1916) Hansen and Lessel 1971
	Lactobacillus	*"casei"*	*"alactosus"*	Mills and Lessel 1973
	Lactobacillus	*casei*	*casei*	(Orla-Jensen 1916) Hansen and Lessel 1971
	Lactobacillus	*"casei"*	*"pseudoplantarum"*	Abo-Elnaga and Kandler 1965b
	Lactobacillus	*"casei"*	*"rhamnosus"*	Hansen 1968
	Lactobacillus	*"casei"*	*"tolerans"*	Abo-Elnaga and Kandler 1965b
20	*Lactobacillus*	*catenaformis*		(Eggerth 1935) Moore and Holdeman 1970
	Lactobacillus	*"cellobiosus"*		Rogosa *et al.* 1953
21	*Lactobacillus*	*ceti*		Vela *et al.* 2008
22	*Lactobacillus*	*coleohominis*		Nikolaitchouk *et al.* 2001
23	*Lactobacillus*	*composti*		Endo and Okada 2007a
24	*Lactobacillus*	*collinoides*		Carr and Davies 1972

Current name (reference)	Phylogenetic group	Metabolism type §	Mol% G + C content	Peptidoglycan type	Lactic acid type
	acidophilus	B	35–37	Lys-D-Asp	DL
	buchneri	C	51	NA	DL
	salivarius	B	38–41	Lys-D-Asp	L
	acidophilus	A	34–37	Lys-D-Asp	DL
	salivarius	B	43–44	DAP	L
	salivarius	B	36–37	DAP	L
	plantarum	B	36–37	Lys-D-Asp	L-DL
	acidophilus	A	39	Lys-D-Asp	DL
	acidophilus	A	44–46	Lys-D-Asp	L
	acidophilus	A	43.5	NA	L
	acidophilus	A	40–41	Lys-D-Asp	DL
	salivarius	A	41–44	Lys-D-Asp	L
	reuteri	C	44–45	Lys-D-Asp	DL
	salivarius	B	38.5	L-Lys-D-Asp	L
Lactobacillus plantarum (Kostinek *et al.* 2005)	plantarum	B	48	NA	DL
	salivarius	A	39–43	Lys-D-Asp	DL
	salivarius	A	39–43	Lys-D-Asp	DL(D <15%)
	salivarius	A	39–43	Lys-D-Asp	DL
Not yet validated	NA	NA	NA	NA	NA
Lactobacillus sakei (Kagermeier-Callaway and Lauer 1995)	casei	B	41–43	Lys-D-Asp	L
	casei	B	45	Lys-D-Asp	DL
	buchneri	C	44–47	Lys-D-Asp	DL
	buchneri	C	44–46	Lys-D-Asp	DL
Lactobacillus delbrueckii subsp. *bulgaricus* (Weiss *et al.* 1983b)	acidophilus	A	49–51	Lys-D-Asp	D
	casei	A	51.9	Lys-D-Asp	L
Carnobacterium maltaromaticus (Collins *et al.* 1987)					
	casei	B	45–47	Lys-D-Asp	L
Lactobacillus paracasei subsp. *paracasei* (Collins *et al.* 1989b)	casei	B	45–47	Lys-D-Asp	L
See text for further explanation	casei	B	45–47	Lys-D-Asp	L
Lactobacillus paracasei subsp. *paracasei* (Collins *et al.* 1989b)	casei	B	45–47	Lys-D-Asp	L
Lactobacillus rhamnosus (Collins *et al.* 1989b)	casei	B	45–47	Lys-D-Asp	L
Lactobacillus paracasei subsp. *tolerans* (Collins *et al.* 1989b)	casei	B	45–47	Lys-D-Asp	L
	vitulinus-catenaformls	A	31–33	Lys-Ala	D
Lactobacillus fermentum (Dellaglio *et al.* 2004a)	reuteri	C	53	Orn-D-Asp	L or DL
	salivarius	B	NA	Lys-L-Ser	L
	reuteri	B	NA	mDAP	DL
	perolens	C	46	Lys-D-Asp	DL
	plantarum	B	48	no mDAP	DL

Table 2.2 *continued*

Number	Genus	Species	Subspecies	References
25	*Lactobacillus*	*concavus*		Tong and Dong 2005
	"Lactobacillus"	*confusus*		(Holzapfel and Kandler 1969) Sharpe *et al.* 1972
26	*Lactobacillus*	*coryniformis*		Abo-Elnaga and Kandler 1965b
		coryniformis	*coryniformis*	Abo-Elnaga and Kandler 1965b
		coryniformis	*torquens*	Abo-Elnaga and Kandler 1965b
27	*Lactobacillus*	*crispatus*		(Brygoo and Aladame 1953) Moore and Holdeman 1970
28	*Lactobacillus*	*curvatus*		(Troili-Petersson 1903) Abo-Elnaga and Kandler 1965b emend. Klein *et al.* 1996
		curvatus	*curvatus*	Torriani *et al.* 1996
		curvatus	*"melibiosus"*	Torriani *et al.* 1996
	Lactobacillus	*"cypricasei"*		Lawson *et al.* 2001a
29	*Lactobacillus*	*delbrueckii*		(Leichmann 1896) Beijerinck 1901
		delbrueckii	*bulgaricus*	(Orla-Jensen 1919) Weiss *et al.* 1983b,1984
		delbrueckii	*delbrueckii*	(Leichmann 1896) Beijerinck 1901
		delbrueckii	*indicus*	Dellaglio *et al.* 2005
		delbrueckii	*lactis*	(Orla-Jensen 1919) Weiss *et al.* 1983b,1984
30	*Lactobacillus*	*diolivorans*		Krooneman *et al.* 2002
	Lactobacillus	*"disidiosus"*		Vaughn, *et al.* 1949
	"Lactobacillus"	*divergens*		Holzapfel and Gerber 1983,1984
	Lactobacillus	*"durianis"*		Leisner *et al.* 2002
31	*Lactobacillus*	*equi*		Morotomi *et al.* 2002
32	*Lactobacillus*	*farciminis*		(ex Reuter 1970) Reuter 1983a,b
33	*Lactobacillus*	*farraginis*		Endo and Okada 2007b
	Lactobacillus	*"ferintoshensis"*		Simpson *et al.*2001, 2002
34	*Lactobacillus*	*fermentum*		Beijerinck 1901 emend. Dellaglio *et al.* 2004a
35	*Lactobacillus*	*fornicalis*		Dicks *et al.* 2000
36	*Lactobacillus*	*fructivorans*		Charlton *et al.* 1934
	"Lactobacillus"	*fructosus*		Kodama 1956
37	*Lactobacillus*	*frumenti*		Müller *et al.* 2000
38	*Lactobacillus*	*fuchuensis*		Sakala *et al.* 2002
39	*Lactobacillus*	*gallinarum*		Fujisawa *et al.* 1992
40	*Lactobacillus*	*gasseri*		Lauer and Kandler 1980
41	*Lactobacillus*	*gastricus*		Roos *et al.* 2005
42	*Lactobacillus*	*ghanensis*		Nielsen *et al.* 2007
43	*Lactobacillus*	*graminis*		Beck *et al.* 1988, 1989
	"Lactobacillus"	*"halotolerans"*		Kandler *et al.* 1983a,c
44	*Lactobacillus*	*hammesii*		Valcheva *et al.* 2005
45	*Lactobacillus*	*hamsteri*		Mitsuoka and Fujisawa 1987,1988
46	*Lactobacillus*	*harbinensis*		Miyamoto *et al.* 2005,2006
47	*Lactobacillus*	*hayakitensis*		Morita *et al.* 2007

Current name (reference)	Phylogenetic group	Metabolism type §	Mol% G + C content	Peptidoglycan type	Lactic acid type
	perolens	A	46–47	mDAP	DL (D 5%)
Weissella confusa (Collins *et al.* 1993)					
	casei	B	45	Lys-D-Asp	DL
	casei	B	45	Lys-D-Asp	DL (L<15%)
	casei	B	45	Lys-D-Asp	D
	acidophilus	A	35–38	Lys-D-Asp	DL
	casei	B	42–44	Lys-D-Asp	DL
	casei	B	42–44	Lys-D-Asp	DL
Lactobacillus sakei subsp. *carnosus* (Koort *et al.* 2004)	casei	B	42–44	Lys-D-Asp	DL
Lactobacillus acidipiscis (Naser *et al.* 2006b)	salivarius	B	40	NA	L
	acidophilus	A	49–51	Lys-D-Asp	D
	acidophilus	A	49–51	Lys-D-Asp	D
	acidophilus	A	49–51	Lys-D-Asp	D
	acidophilus	A	NA	NA	D
	acidophilus	A	49–51	Lys-D-Asp	D
	buchneri	C	NA	NA	NA
Lactobacillus kefiri (Marshall *et al.* 1984).	buchneri	C	NA	NA	NA
Carnobacterium divergens (Collins *et al.*, 1987)					
Lactobacillus vaccinostercus (Dellaglio *et al.* 2006)	reuteri	C	43	NA	DL
	salivarius	A	38–39	NA	DL
	plantarum	A	34–36	Lys-D-Asp	L (D<15%)
	buchneri	B	40–41	no mDAP	DL
Lactobacillus parabuchneri (Vancanneyt *et al.* 2005a)	buchneri	C	43	NA	DL
	reuteri	C	52–54	Orn-D-Asp	DL
	acidophilus	B	37	NA	DL
	buchneri	C	38–41	Lys-D-Asp	DL
Leuconostoc fructosum (Antunes *et al.* 2002)					
	reuteri	C	43–44	Lys-D-Asp	L
	casei	B	41–42	NA	L (D<40%)
	acidophilus	A	36–37	Lys-D-Asp	DL
	acidophilus	A	33–35	Lys-D-Asp	DL
	reuteri	C	41–42	L-Orn-D-Asp	DL
	salivarius	A	38	mDAP	DL
	casei	B	41–43	Lys-D-Asp	DL
Lactobacillus viridescens subsp. *halotolerans; Weissella halotolerans* (Collins *et al.*1993,1994)					
	buchneri	B	NA	L-Lys-D-Asp	DL
	acidophilus	B	33–35	Lys-D-Asp	DL
	perolens	B	53–54	NA	L
	salivarius	A	34	Lys-Asp	L

Table 2.2 *continued*

Number	Genus	Species	Subspecies	References
48	*Lactobacillus*	*helveticus*		(Orla-Jensen 1919) Bergey *et al.* 1925
	Lactobacillus	*"heterohiochii"*		Kitahara *et al.* 1957a,b; Momose *et al.* 1974
49	*Lactobacillus*	*hilgardii*		Douglas and Cruess 1936
50	*Lactobacillus*	*homohiochii*		Kitahara *et al.* 1957a,b
51	*Lactobacillus*	*iners*		Falsen *et al.* 1999
52	*Lactobacillus*	*ingluviei*		Baele *et al.* 2003
53	*Lactobacillus*	*intestinalis*		(ex Hemme 1974) Fujisawa *et al.* 1990
54	*Lactobacillus*	*jensenii*		Gasser *et al.* 1970
55	*Lactobacillus*	*johnsonii*		Fujisawa *et al.* 1992
	Lactobacillus	*"jugurtii"*		Orla-Jensen, 1919
56	*Lactobacillus*	*kalixensis*		Roos *et al.* 2005
	"Lactobacillus"	*kandleri*		Holzapfel and van Wyk 1982,1983
57	*Lactobacillus*	*kefiranofaciens*		Fujisawa *et al.* 1988 emend. Vancanneyt *et al.* 2004
		kefiranofaciens	*kefiranofaciens*	Fujisawa *et al.* 1988
		kefiranofaciens	*kefirgranum*	(Takizawa *et al.* 1994) Vancanneyt *et al.* 2004
	Lactobacillus	*"kefirgranum"*		Takizawa *et al.* 1994
58	*Lactobacillus*	*kefiri*		Kandler and Kunath 1983a,b
59	*Lactobacillus*	*kimchii*		Yoon *et al.* 2000
60	*Lactobacillus*	*kitasatonis*		Mukai *et al.* 2003
61	*Lactobacillus*	*kunkeei*		Edwards *et al.* 1998a,b
	Lactobacillus	*"lactis"*		(Orla-Jensen 1919) Bergey *et al.* 1934
	Lactobacillus	*"leichmannii"*		(Henneberg 1903) Bergey *et al.* 1923
62	*Lactobacillus*	*lindneri*		(ex Henneberg 1901) Back *et al.* 1996, 1997
63	*Lactobacillus*	*malefermentans*		(ex Russell and Walker 1953) Farrow *et al.* 1988,1989
64	*Lactobacillus*	*mali*		(Carr and Davies 1970) Kaneuchi *et al.* 1988
		mali	*mali*	(Carr and Davies 1970) Kaneuchi *et al.* 1988
		mali	*yamanashiensis*	Nonomura 1983, Kaneuchi *et al.* 1988
	"Lactobacillus"	*"maltaromicus"*		Miller *et al.* 1974
65	*Lactobacillus*	*manihotivorans*		Morlon-Guyot *et al.* 1998
66	*Lactobacillus*	*mindensis*		Ehrmann *et al.* 2003
	"Lactobacillus"	*"minor"*		Kandler *et al.* 1983a,c
	"Lactobacillus"	*minutus*		(Hauduroy *et al.* 1937) Moore and Holdeman 1972; Olsen *et al.* 1991
67	*Lactobacillus*	*mucosae*		Roos *et al.* 2000
68	*Lactobacillus*	*murinus*		Hemme *et al.* 1980, 1982
69	*Lactobacillus*	*nagelii*		Edwards *et al.* 2000
70	*Lactobacillus*	*namurensis*		Scheilrinck *et al.* 2007
71	*Lactobacillus*	*nantensis*		Valcheva *et al.* 2006
72	*Lactobacillus*	*oligofermentans*		Koort *et al.* 2005a,b

Current name (reference)	Phylogenetic group	Metabolism type §	Mol% G + C content	Peptidoglycan type	Lactic acid type
	acidophilus	A	38–40	Lys-D-Asp	DL
Lactobacillus fructivorans (Weiss *et al.* 1983a)	buchneri	C	38–40	Lys-D-Asp	DL
	buchneri	C	39–41	Lys-D-Asp	DL
	buchneri	B	35–38	Lys-D-Asp	DL
	acidophilus	A	34–35	Lys-D-Asp	L
	reuteri	C	49–50	NA	NA
	acidophilus	B	33–35	Lys-D-Asp	DL
	acidophilus	B	35–37	Lys-D-Asp	D
	acidophilus	A	33–35	Lys-D-Asp	DL
Lactobacillus helveticus (Simonds *et al.* 1971)	acidophilus	A	NA	NA	NA
	acidophilus	A	35–36	Lys-D-Asp	DL
Weissella kandleri (Collins *et al.* 1993,1994)					
	acidophilus	A	34–38	NA	DL
	acidophilus	A	34–38	NA	DL
	acidophilus	A	34–38	NA	DL
Lactobacillus kefiranofaciens subsp. *kefiranofaciens* (Vancanneyt *et al.* 2004)	acidophilus	A	34–38	no mDAP	DL
	buchneri	C	41–42	Lys-D-Asp	DL
	plantarum	B	35	NA	DL
	acidophilus	B	37–40	NA	DL
	buchneri	C	NA	Lys-D-Asp	L
Lactobacillus delbrueckii subsp. *lactis* (Weiss *et al.* 1983b,1984)	acidophilus	A	50	NA	NA
Lactobacillus delbrueckii subsp. *lactis* (Weiss *et al.* 1983b,1984)	acidophilus	A	51	NA	NA
	buchneri	C	35	Lys-D-Asp	DL
	plantarum	C	41–42	Lys-D-Asp	NA
	salivarius	A	32–34	DAP	L
	salivarius	A	32–34	DAP	L
	salivarius	A	32–34	mDAP	L
Carnobacterium piscicola (Mora *et al.* 2003) *Carnobacterium maltaromaticus* (Mora *et al.* 2003)					
	casei	A	48–49	NA	L
	plantarum	A	37–38	Lys-D-Asp	DL
Lactobacillus viridescens subsp. minor; Weissella minor (Collins *et al.* 1993,1994)					
Atopobium minutum (Collins and Wallbanks 1992, 1993)					
	reuteri	C	46–47	Orn-D-Asp	DL
	salivarius	B	43–44	Lys-D-Asp	L
	salivarius	A	NA	NA	DL
	buchneri	C	52	NA	DL
	plantarum	B	38.6	NA	DL
	reuteri	C	35.3–39.9	NA	DL(D 30%)

Table 2.2 *continued*

Number	Genus	Species	Subspecies	References
73	*Lactobacillus*	*oris*		Farrow and Collins 1988
74	*Lactobacillus*	*panis*		Wiese *et al.* 1996
75	*Lactobacillus*	*pantheris*		Liu and Dong 2002
76	*Lactobacillus*	*parabrevis*		Vancanneyt *et al.* 2006b
77	*Lactobacillus*	*parabuchneri*		Farrow *et al.* 1988, 1989
		parabuchneri	*parabuchneri*	Farrow *et al.* 1988, 1989
		parabuchneri	*ferintoshensis*	Vancanneyt *et al.*, 2005a
78	*Lactobacillus*	*paracasei*		Collins *et al.* 1989b
		paracasei	*paracasei*	Collins *et al.* 1989b
		paracasei	*tolerans*	(Abo-Elnaga and Kandler 1965b) Collins *et al.* 1989b
79	*Lactobacillus*	*paracollinoides*		Suzuki *et al.* 2004
80	*Lactobacillus*	*parafarraginis*		Endo and Okada 2007b
81	*Lactobacillus*	*parakefiri*		Takizawa *et al.* 1994
82	*Lactobacillus*	*paralimentarius*		Cai *et al.* 1999
83	*Lactobacillus*	*paraplantarum*		Curk *et al.* 1996
	Lactobacillus	*"pastorianus"*		Van Laer, 1892
84	*Lactobacillus*	*pentosus*		(ex Fred *et al.* 1921) Zanoni *et al.* 1987
85	*Lactobacillus*	*perolens*		Back *et al.* 1999, 2000
	"Lactobacillus"	*"piscicola"*		Hiu *et al.* 1984
86	*Lactobacillus*	*plantarum*		(Orla-Jensen 1919) Bergey *et al.* 1923
		plantarum	*argentoratensis*	Bringel *et al.* 2005
		plantarum	*plantarum*	(Orla-Jensen 1919) Bergey *et al.* 1923; Bringel *et al.* 2005
87	*Lactobacillus*	*pontis*		Vogel *et al.* 1994
88	*Lactobacillus*	*psittaci*		Lawson *et al.* 2001b
89	*Lactobacillus*	*rennini*		Chenoll *et al.* 2006a
90	*Lactobacillus*	*reuteri*		Kandler *et al.* 1980, 1982
91	*Lactobacillus*	*rhamnosus*		(Hansen 1968) Collins *et al.* 1989b
	"Lactobacillus"	*rimae*		Olsen *et al.* 1991
92	*Lactobacillus*	*rogosae*		Holdeman and Moore 1974
93	*Lactobacillus*	*rossiae*		Corsetti *et al.* 2005
94	*Lactobacillus*	*ruminis*		Sharpe *et al.* 1973
95	*Lactobacillus*	*saerimneri*		Pedersen and Roos 2004
96	*Lactobacillus*	*sakei*		Katagiri *et al.* 1934 emend. Klein *et al.* 1996
		sakei	*carnosus*	Torriani *et al.* 1996; Koort *et al.* 2004
		sakei	*sakei*	Katagiri *et al.* 1934 emend. Klein *et al.* 1996
97	*Lactobacillus*	*salivarius*		Rogosa *et al.* 1953
		salivarius	*salicinius*	Rogosa *et al.* 1953 emend. Li *et al.* 2006
		salivarius	*salivarius*	Rogosa *et al.* 1953 emend. Li *et al.* 2006
98	*Lactobacillus*	*sanfranciscensis*		(ex Kline and Sugihara 1971) Weiss and Schillinger 1984a,b
99	*Lactobacillus*	*satsumensis*		Endo and Okada 2005
100	*Lactobacillus*	*secaliphilus*		Ehrmann *et al.* 2007
101	*Lactobacillus*	*sharpeae*		Weiss *et al.* 1981,1982

Current name (reference)	Phylogenetic group	Metabolism type §	Mol% G + C content	Peptidoglycan type	Lactic acid type
	reuteri	C	49–51	Orn-D-Asp	DL
	reuteri	C	49–51	Lys-D-Asp	DL
	casei	A	52–53	NA	D
	buchneri	C	49	NA	DL
	buchneri	C	44	Lys-D-Asp	NA
	buchneri	C	44	Lys-D-Asp	NA
	buchneri	C	43	NA	DL
	casei	B	45–47	Lys-D-Asp	L
	casei	B	45–47	Lys-D-Asp	L
	casei	B	45–47	Lys-D-Asp	L
	plantarum	C	44–45	NA	D
	buchneri	B	40	no mDAP	DL (D< 70%)
	buchneri	C	41–42	NA	L
	plantarum	B	37–38	NA	NA
	plantarum	B	44–45	DAP	DL
Lactobacillus paracollinoides (Ehrmann and Vogel 2005)		C	NA	NA	NA
	plantarum	B	46–47	DAP	DL
	perolens	B	49–53	Lys-D-Asp	L
Carnobacterium piscicola (Collins *et al.,* 1987), *Carnobacterium maltaromaticus* (Mora *et al.* 2003)					
	plantarum	B	44–46	DAP	DL
	plantarum	B	44–46	NA	DL
	plantarum	B	44–46	DAP	DL
	reuteri	C	53–56	Orn-D-Asp	DL
	acidophilus	C	NA	NA	NA
	casei	B	NA	L-Lys-D-Asp	DL
	reuteri	C	40–42	Lys-D-Asp	DL
	casei	B	45–47	Lys-D-Asp	L
Atopobium rimae (Collins and Wallbanks 1992,1993)					
Taxonomic status unclear due to lack of type strain (Felis *et al.* 2004)	NA	NA	NA	NA	NA
	reuteri	C	44–45	Lys-Ser-Ala,	DL
	salivarius	A	44–47	DAP	L
	salivarius	A	42–43	DAP	DL
	casei	B			
	casei	B	42–44	NA	DL
	casei	B	42–44	NA	DL
	salivarius	A	34–36	Lys-D-Asp	L
	salivarius	A	NA	NA	NA
	salivarius	A	34–36	Lys-D-Asp	L
	buchneri	C	36–38	Lys-Ala	DL
	salivarius	A	39–41	DAP	L
	reuteri	B	48	L-Lys-D-Asp	L (D 5%)
	casei	A	53	DAP	L

Table 2.2 *continued*

Number	Genus	Species	Subspecies	References
102	*Lactobacillus*	*siliginis*		Aslam *et al.* 2006
	Lactobacillus	*"sobrius"*		Konstantinov *et al.* 2006
103	*Lactobacillus*	*spicheri*		Meroth *et al.* 2004a,b
104	*Lactobacillus*	*suebicus*		Kleynmans *et al.* 1989
	Lactobacillus	*"suntoryeus"*		Cachat and Priest 2005
105	*Lactobacillus*	*thailandensis*		Tanasupawat *et al.* 2007
	Lactobacillus	*"thermotolerans"*		Niamsup *et al.* 2003
	Lactobacillus	*"trichodes"*		Fornachon *et al.* 1949
	Lactobacillus	*"tucceti"*		Chenoll *et al.* 2006b
	"Lactobacillus"	*uli*		Olsen *et al.* 1991
106	*Lactobacillus*	*ultunensis*		Roos *et al.* 2005
107	*Lactobacillus*	*vaccinostercus*		Okada *et al.* 1979; Kozaki and Okada 1983 emend. Dellaglio *et al.*2006
108	*Lactobacillus*	*vaginalis*		Embley *et al.* 1989
109	*Lactobacillus*	*versmoldensis*		Kröckel *et al.* 2003
110	*Lactobacillus*	*vini*		Rodas *et al.* 2006
	"Lactobacillus"	*viridescens*		Niven and Evans 1957
111	*Lactobacillus*	*vitulinus*		Sharpe *et al.* 1973
	"Lactobacillus"	*"xylosus"*		Kitahara 1938
	Lactobacillus	*"yamanashiensis"*		Nonomura 1983
112	*Lactobacillus*	*zeae*		(ex Kuznetsov 1959) Dicks *et al.* 1996
113	*Lactobacillus*	*zymae*		Vancanneyt *et al.* 2005b

§Type of glucose fermentation as defined by Hammes and Vogel (1995) and Hammes and Hertel (2003): A = homofermentative, B = facultatively heterofermentative, C = obligately heterofermentative, NA = not available.

sons the genus today is still considered divided in the same three major groups, namely group I (obligately homofermentative lactobacilli), group II (facultatively homofermentative) and group III (obligately heterofermentative) (Tables 2.1 and 2.2). In addition, the accumulated knowledge on their sugar fermentation patterns created a solid basis on which further research was carried out, including other metabolic properties of the lactobacilli, such as proteolytic and lipolytic activities, which are equally important in food applications. These aspects are further discussed below.

Carbon sources metabolism in lactobacilli

Lactose fermentation

Lactose fermentation is by far the most studied disaccharide metabolism in lactic acid bacteria, since it is the major carbohydrate of milk. As shown for *Lactobacillus casei*, lactose is taken up via the phosphoenolpyruvate-dependent phosphotransferase system (PTS) and enters the cytoplasm as lactose phosphate (Chassy and Alpert, 1989). Lactose phosphate is cleaved by phospho-β-D-galactosidase (P-β-gal) to yield glucose and galactose-6-phosphate. Glucose is

Current name (reference)	Phylogenetic group	Metabolism type §	Mol% G + C content	Peptidoglycan type	Lactic acid type
	reuteri	C	44.5	L-Lys-D-Glu-L-Ala	NA
Lactobacillus amylovorus (Viljanen *et al.* 2008)	acidophilus	B	35–36	NA	DL
	buchneri	B	55	Lys-D-Asp	DL
	reuteri	C	40–41	DAP	DL
Lactobacillus helveticus (Naser *et al.* 2006a)	acidophilus	A	NA	NA	NA
	casei	A	49–50	no mDAP	DL
Lactobacillus ingluviei (Felis *et al.* 2006)	reuteri	C	49–51	no mDAP; Lys-Asp	DL
Lactobacillus fructivorans (Weiss *et al.* 1983a)	buchneri	C	NA	NA	NA
not yet validated	plantarum	A	ND	L-Lys-Gly-D-Asp	DL
Olsenella uli (Dewhirst *et al.* 2001)					
	acidophilus	A	35–36	Lys-D-Asp	DL
	reuteri	C	36–37	DAP	NA
	reuteri	C	38–41	Orn-D-Asp	NA
	plantarum	A	40–41	NA	L
	salivarius	B	39.4	L-Lys-D-Asp	DL
Weissella viridescens (Collins *et al.* 1993,1994)					
	vitulinus-catenaformis	A	34–37	mDAP	D
Lactococcus lactis subsp. *lactis* (Schleifer *et al.*, 1985)					
Lactoabcillus mali (Kaneuchi *et al.*, 1988)	salivarius	A	32–34	mDAP	L
	casei	B	48–49	Lys-D-Asp	L
	buchneri	C	53–54	NA	DL

phosphorylated by glucokinase and metabolized through either the glycolytic pathway or the pentose phosphate pathway. Galactose-6-phosphate is metabolized through the tagatose-6-phosphate pathway (Bisset and Anderson, 1974), while the Leloir pathway is used by galactose-fermenting lactic acid bacteria, which transport galactose with a permease and which lack the galactose-PTS (Konings *et al.*, 1989). The enzyme systems of lactose-PTS and P-β-gal are generally inducible, and repressed by glucose (Kandler, 1983). An equally common way for lactic acid bacteria to metabolize lactose is by means of a lactose carrier (permease) and subsequent cleavage by β-galactosidase (β-gal) to yield glucose and galactose, which may again enter the two major pathways (McKay *et al.*, 1970; Bhowmik and Marth, 1990). Some of the thermophilic lactobacilli, such as *Lactobacillus delbrueckii* subsp. *bulgaricus*, *L. delbrueckii* subsp. *lactis* and *Lactobacillus acidophilus*, only metabolize the glucose moiety after transport of lactose and cleavage by β-gal, while galactose is excreted into the medium (Hickey *et al.*, 1986; Hutkins and Morris, 1987).

Glucose fermentation

For glucose fermentation two major pathways occur in lactic acid bacteria. The Embden–Meyerhof–Parnas pathway (glycolysis) is used by all lactic acid bacteria except leuconostocs, group III lactobacilli (obligately heterofermenative species), oenococci and weissellas. It is characterized by the formation of fructose-1,6-diphosphate (FDP), which is split by the

FDP aldolase into dihydroxyacetone-phosphate (DHAP) and glycerinaldehyde-3-phosphate (GAP). GAP (and DHAP via GAP) is then converted to pyruvate in a metabolic sequence including substrate level phosphorylation. One mole of glucose results in 2 moles of lactic acid and a net gain of 2 ATP. The glycolysis pathway is used by the *homofermentative* lactic acid bacteria. The other main fermentation pathway is the pentose phosphate pathway. The key step is the phosphoketolase split of xylulose-5-phoshate to glycerinaldehyde-3-phosphate (GAP) and acetyl-phosphate. GAP is then converted to lactate, while acetyl-phosphate to acetate and ethanol. This pathway is used by the heterofermentative lactic acid bacteria. Heterolactic fermentation gives 1 mole each of lactic acid, ethanol and CO_2 and 1 ATP per mole of glucose. It should be noted that glycolysis may lead to a heterolactic fermentation (meaning significant amounts of other end-products besides lactic acid) under certain conditions, and some lactic acid bacteria, regarded as homofermentative, use the pentose phosphate pathway when metabolizing certain substrates (Axsellson, 1998).

Among lactic acid bacteria, those found in sourdough fermentations belong mainly to the heterofermentative lactobacilli, which catabolize glucose via the pentose phosphate pathway. Under micro-aerophilic conditions, both oxygen and fructose can be used as electron acceptors. This gives rise to the formation of additional metabolites such as acetate and mannitol (Hammes and Gänzle, 1998).

Maltose fermentation

In sourdough, maltose is the most abundant fermentable carbohydrate, and hence maltose catabolism is a key process during fermentation. Microbial associations of maltose-positive and maltose-negative lactic acid bacteria strains are typical for sourdoughs dominated by *Lactobacillus sanfranciscensis* (Gobbetti, 1998). In *L. sanfranciscensis*, *Lactobacillus reuteri* and *Lactobacillus fermentum* a constitutive intracellular maltose phosphorylase catalyses the phosphorolytic cleavage of maltose, yielding glucose 1-phosphate and glucose (Vogel *et al.*, 1994). Glucose 1-phosphate is then converted by phosphoglucomutase to glucose 6-phosphate, which is further metabolized via the pentose phosphate pathway (Hammes *et al.*, 1996; Vogel *et al.*, 1999). On the other hand, hexokinase activity, which catalyses the conversion of glucose to glucose-6-phosphate, is virtually absent in cells growing exponentially in maltose-containing media, and thus the non-phosphorylated glucose becomes excreted in the medium in a molar ratio with maltose of about 1:1 (Stolz *et al.*, 1993; Gobbeti *et al.*, 1994). It has been shown, however, that no glucose accumulation occurred in the fermentation broth, and no maltose phosphorylase activity could be detected in cell extracts prepared from cells grown in the presence of both maltose and fructose, suggesting that in the presence of both maltose and fructose in the medium, induction of hexokinase activity does occur (De Vuyst *et al.*, 2003). Similarly, in experiments performed with growing cells of *L. sanfranciscensis*, no significant accumulation of glucose was observed in the medium as that reported for resting cells of *L. sanfranciscensis*, *L. reuteri*, and *Lactobacillus pontis* (Neubauer *et al.*, 1994; Stolz *et al.*, 1995a, b). It is also believed that hexokinase activity is induced in the presence of glucose or fructose in the medium (Stolz *et al.*, 1996).

Fructose fermentation

L. sanfranciscensis and *L. pontis* are able to use fructose as carbon source; however, in the presence of maltose they use it mainly as an electron acceptor and fructose is reduced to mannitol (Stolz *et al.*, 1995a; Hammes *et al.*, 1999; Wolfrum and Vogel, 1999). According to Röcken and Voysey (1995), oxygen was proved to be the preferred hydrogen acceptor for the *L. sanfranciscensis* strains. When oxygen is depleted, fructose is used as an electron acceptor (Gobbetti *et al.*, 1995; Stolz *et al.*, 1995a). Through the reduction of fructose to mannitol, extra ATP is produced via the acetate kinase reaction, and thus maltose-fructose co-metabolism yields shorter lag phase and higher growth rate and biomass production. It has been shown that at a molar ratio of 4:1 (fructose–maltose), acetic acid is the main product (Martinez-Anaya *et al.*, 1994; Gobbetti *et al.*, 1995, 2000; Stolz *et al.*, 1995a). *L. sanfranciscensis* converts stoichiometrically fructose to mannitol, while *L. pontis* produces small amounts of lactic acid and ethanol (Hammes *et al.*, 1996).

Pentose fermentation

As far as pentose fermentation is concerned (Kandler, 1983; Posthuma *et al.*, 2002), despite some strain and species differences, group II and group III lactobacilli are pentose positive. In general, specific permeases are used to transport the sugars into the cell. The pentoses are then phosphorylated and converted by epimerases or isomerases to ribulose 5-phosphate or xylulose 5-phosphate, respectively, which can be metabolized by the lower half of the pentose phosphate pathway (Kandler, 1983).

Citrate fermentation

Citrate, which is present in many raw materials such as milk, vegetables, etc., can also serve as energy source for lactic acid bacteria. It is generally accepted that next to carbohydrates, citrate metabolism plays an important role in food fermentations. The ability of lactic acid bacteria to metabolize citrate is invariably linked to endogenous plasmid that contains the gene encoding the transporter, which is responsible for citrate uptake from the medium (Hugenholtz, 1993). Within the cell, citrate is initially converted by the citrate lyase to acetate and oxaloacetate. Oxaloacetate is then decarboxylated to pyruvate. According to the intracellular enzyme pool, pyruvate may be then converted (i) to acetyl CoA (via the pyruvate dehydrogenase complex), which leads to acetate (via the acetate kinase) and acetaldehyde/ethanol formation (via the alcohol dehydrogenase), (ii) to formate (via the pyruvate formate lyase), (iii) to alpha-acetolactate (via the acetolactate synthase), which leads to acetoin (acetolactate decarboxylate), and diacetyl and 2,3-butanediol (via the diacetyl/acetoin reductase), and finally (iv) to lactate (via the lactate dehydrogenase). The energy is mostly generated from the conversion of acetyl CoA to acetate, meaning that citrate acts as electron acceptor, resulting in a higher production of acetate and ATP probably via the acetate kinase pathway. Additional energy is produced during the initial breakdown of citrate into pyruvate (Hugenholtz, 1993). Furthermore, recent studies performed with *Lactococcus lactis* subsp. *lactis* biovar diacetylactis (Hugenholtz, 1993; Hugenholtz *et al.*, 1993) and *Leuconostoc oenos* (Marty-Teysset *et al.*, 1996) indicated that the uptake of citrate is coupled to the generation of a proton-motive force, which was shown to be strong enough to drive the additional ATP synthesis. Some of the products of citrate catabolism, such as diacetyl, acetaldehyde and acetoin, have very distinct aroma properties and influence significantly the quality of fermented foods. For instance, diacetyl determines the aromatic properties of fresh cheese, fermented milk, cream and butter (De Figureoa *et al.*, 1998). The breakdown of citrate results as well in the production of carbon dioxide, which can add to the texture of some fermented dairy products (Kimoto *et al.*, 1999).

It has been shown that several strains of *L. sanfranciscensis* are able to use citrate as electron acceptor in the presence of maltose (Stolz *et al.*, 1995a). On the other hand, co-metabolism of maltose and citrate has not been observed for *L. pontis* (Hammes *et al.*, 1996). According to Gobbetti and Corsetti (1996), during co-metabolism of maltose and citrate, lactic acid and acetic acid are initially produced, but when citrate is exhausted, lactic acid and ethanol are the main products. In all cases, maltose serves as carbon source, while citrate as electron acceptor. The production of small amounts of succinate from citrate has been also observed, indicating the presence of citrate lyase, malate dehydrogenate, fumarase and succinate dehydrogenase. A putative citric acid cycle (PCAC) for *L. casei* was recently generated, utilizing the genome sequence and metabolic flux analyses (Diaz-Muniz *et al.*, 2006). Although it was possible to construct a unique PCAC for *L. casei*, its full functionality was unknown. Therefore, the *L. casei* PCAC was evaluated utilizing end product analyses of citric acid catabolism during growth in modified chemically defined media (mCDM), and Cheddar cheese extract (CCE). Results suggest that under energy source excess and limitation in mCDM this micro-organism produces mainly L-lactic acid and acetic acid, respectively. Both organic acids were produced in CCE. Additional end products include D-lactic acid, acetoin, formic acid, ethanol, and diacetyl. Production of succinic acid, malic acid, and butanendiol was not observed. It is thus concluded that under conditions similar to those present in ripening cheese, citric acid is converted to acetic acid, L/D-lactic acid, acetoin, diacetyl, ethanol, and formic acid. The PCAC suggests that conversion of the citric acid-derived pyruvic acid into acetic acid, instead of lactic acid, may yield two ATPs per molecule of citric acid.

Lactobacilli usually dominate the lactic acid bacteria microflora in naturally fermented sausages. The growth and metabolism of lactic acid bacteria is affected by the presence of oxygen. Usually the carbohydrates are metabolized via glycolysis. However, under certain conditions, the heterofermentative pathway is activated, resulting in undesirable flavour components, i.e. acetate (Jessen, 1995). In the presence of oxygen, metabolites other than those found in anaerobic conditions may be observed. Despite the formation of hydrogen peroxide, which may be formed during the aerobic metabolism of glucose, the yield of lactic acid, acetic acid, acetoin and ethanol are affected. *Lactobacillus plantarum*, which under anaerobic conditions mainly forms lactic acid from glucose, shows a dramatic increase in the production of acetic acid under aerobic conditions, together with small amounts of acetoin (Kröckel, 1995).

Proteolytic system of lactobacilli

Proteolysis

Proteolysis is considered the most complex of the three primary events during food fermentations, the other two being carbohydrate fermentation and lipolysis. It is a general belief that lactic acid bacteria have limited abilities to synthesize amino acids, which are essential for their growth, and most raw food materials contain insufficient amounts of free amino acids and low molecular mass peptides to sustain growth (Law and Kolstadt, 1983; Thomas and Pritchard, 1987). Although they are considered as weak proteolytic bacteria compared with other groups of microorganisms, it has be shown that lactic acid bacteria posses a complex proteolytic system capable of hydrolysing food proteins to peptides and amino acids (Kunji *et al.*, 1996; Mierau *et al.*, 1997). Furthermore, it is generally accepted that their proteolytic system contributes to the degradation of food protein and hence to the texture, taste and aroma of fermented products (McSweeney and Sousa, 2000).

The most extensively studied proteolytic system is that of *Lactococcus lactis*, and it serves as a model for all lactic acid bacteria. The second best unravelled proteolytic systems are those of *Lactobacillus* species, most notably *Lactobacillus helveticus*, *Lactobacillus bulgaricus* and *L. casei*. An extracellular membrane-anchored serine proteinase (PrtP) is an essential component of this system. PrtP exists in at least two variants with somewhat different specificities in the degradation of milk casein. The gene encoding PrtP has been cloned and sequenced for a number of *Lactobacillus paracasei* (Holck and Nes, 1992) and *L. bulgaricus* (Gilbert *et al.*, 1996) strains. The *L. paracasei* enzyme shows more than 95% similarity to the lactococcal one, while the *L. delbrueckii* proteinase shows up to 40% identity over the first 820 residues when compared to the lactococcal enzymes; however the C-terminal part does not share any homology. Studies have indicated that *L. helveticus* may contain two proteinases with different substrate specificities (Gilbert *et al.*, 1997), while a cell envelope-associated proteinase gene (prtH) was identified in *L. helveticus* CNRZ32, with a deduced amino acid sequence having significant identity (45%) to that of the lactococcal PrtP proteinases (Pederson *et al.*, 1999).

The majority of sourdough lactic acid bacteria does not exhibit cell wall-associated proteinase activity (Pepe *et al.*, 2003; Vermeulen *et al.*, 2005). Generally, a comparable extent of protein degradation is observed in wheat sourdough and in chemically acidified dough (Thiele *et al.*, 2004; Loponen *et al.*, 2004). However, several strains of sourdough lactic acid bacteria strains exhibiting proteolytic activity were characterized (Gobbetti *et al.*, 1996a; Di Cagno *et al.*, 2002; Pepe *et al.*, 2003) and a contribution of selected lactic acid bacteria to proteolysis could be demonstrated by analysis of the degradation of albumins, globulins, and gliadins in wheat sourdoughs (Di Cagno *et al.*, 2002; Pepe *et al.*,

2003; Zotta *et al.*, 2006). The analysis of peptide and amino acid levels in wheat sourdoughs indicate that *L. sanfranciscensis* preferably utilizes peptides during growth in sourdough (Thiele *et al.*, 2004). Comparable to *Lactococcus lactis* and *L. plantarum*, *L. sanfranciscensis* expresses transport systems for oligo- and dipeptides (Vermeulen *et al.*, 2005) and peptides are hydrolysed by intracellular peptidases, several of which have been characterized at the biochemical or genetic level (Gobbetti *et al.*, 1996a; Gallo *et al.*, 2005; Vermeulen *et al.*, 2005). Analysis of the regulation of peptide uptake systems and peptidases during growth of *L. sanfranciscensis* has shown that genes coding for the peptide uptake systems for dipeptides (DtpT) (Foucaud *et al.*, 1995) and an oligopeptide transport system (Opp) were expressed during exponential growth in sourdough and their expression was reduced in stationary phase cells or when the peptide supply in dough was increased (Vermeulen *et al.*, 2005).

Amino acid transport

In lactococci, the products of the initial casein degradation (amino acids and peptides) are transported into the cell by transport systems specific for amino acids (Konings *et al.*, 1989), two di- and tripeptides (DtpT and DtpP) (Foucaud *et al.*, 1995) and an oligopeptide transport system (Opp) accepting four to eight residue peptides (Tynkkynen *et al.*, 1993). However, little information is available on the transport of casein breakdown products in lactobacilli. Results suggest that the amino acid transport systems in *L. helveticus* are similar to the lactococcal (Nakajima *et al.*, 1998). The gene coding for a branched chain amino acid carrier (*brnC*) of *L. delbreuckii* subsp. *lactis* has been cloned and sequenced (Stucky *et al.*, 1995), and it is driven by the proton motive force. For the same organism the genes coding for an aromatic acid and a dipeptide transporter, *aroP* and *dppE*, respectively, have also been cloned and sequenced (Kunji *et al.*, 1996). For *L. helveticus*, a homologue of DtpT is specified by a sequence located downstream of pepN (Christensen *et al.*, 1995), and transport experiments have shown that substrates, typical for the lactococcal DtpT, are indeed transported by this organism. Experiments also indicate that an oligopeptide transport system is present in *L. helveticus* (Nakajiama *et al.*, 1998). Inside the cell, several peptidases with a wide range of specificity complete the degradation (Christensen *et al.*, 1999).

Amino acid metabolism

In addition to proteolysis, bacterial amino acid metabolism contributes to flavour formation during fermentation. In recent years, it has become clear that a number of enzymes are involved in the conversion of amino acids to flavour components. Indeed, the genome sequence analysis of several species of lactic acid bacteria provided insight into the metabolic pathways for amino acid conversion (van Kranenburg *et al.*, 2002). These enzymes may catalyse reactions such as transamination, deamination, decarboxylation, and cleavage of the amino acid side chain. Branched-chain amino acids can be transaminated to ketoacids, which then undergo either spontaneous degradation or they are enzymatically converted to the corresponding aldehydes or carboxylic acids (Smit *et al.*, 2000). Amino acid transamination is a key step in the amino acid conversion to aroma compounds by lactic acid bacteria. Indeed, in lactic acid bacteria catabolism of ArAAs, BcAAs and Met is essentially initiated by a transamination reaction since the degradation occurs only in presence of an alpha-ketoacid which is used as amino group acceptor. This was demonstrated in mesophilic lactobacilli such as *L. paracasei*, *L. casei*, *L. plantarum*, *Lactobacillus rhamnosus* (Gummalla and Braodbent, 1996; Tammam *et al.*, 2000) and also in thermophilic lactobacilli such as *L. helveticus*, *L. delbrueckii* subsp. *lactis*, and *L. delbrueckii* subsp. *bulgaricus* (Gummalla and Broadbent, 1999). α-Ketoglutarate serves as amino acceptor in the transamination reaction of leucine, phenylalanine and other amino acids, and the addition of α-ketoglutarate strongly increases amino acid conversion of *Lactobacillus sakei* and *L. plantarum* (Yvon *et al.*, 1998; Larrouture *et al.*, 2000). Lactic acid bacteria exhibit glutamate dehydrogenase activity in a strain specific manner. The enzyme catalyses the NAD(P)H-dependent recycling of glutamate to a-ketoglutarate, and consequently increases the flux through the transaminase reaction (Gänzle *et al.*, 2007). The catabolism of leucine and phenylalanine was analysed in detail with strains of *L. sakei* and *L. plantarum* (Groot and de Bont, 1998; Larrouture *et al.*, 2000), valine and isoleucine are degraded by comparable metabolic pathways. Cystathionine lyase (Cxl) is a key enzyme in the metabolism of methionine and cysteine in lactic acid bacteria Cystathionine-γ-lyase was purified and characterized from *L. fermentum* and *L. reuteri* (De Angelis *et al.*, 2002). Vermeulen *et al.* (2003) reported that *L. fermentum*, *L. reuteri*, *L. pontis*, *Lactobacillus panis* and *Lactobacillus mindensis* but not *L. sanfranciscensis*, *L. plantarum* and *Lactobacillus brevis* expressed genes coding for Cxl. Cxl activity of lactic acid bacteria contributes to the flavour development during cheese ripening (Weimer *et al.*, 1999), and a cysteine uptake system was shown to be essential for oxygen tolerance in *L. fermentum* (Turner *et al.*, 1999) but a possible functional role of cysteine and methionine metabolism in sourdough remains to be determined.

Lipolytic system of lactobacilli

Lipolysis is among the principal events occurring during cheese ripening. Free fatty acids can be further converted to methyl ketones, lactones, thioesters, keto and hydroxy acids, which contribute in addition to the free fatty acids to the flavour of the ripened product, while the volatile short chain fatty acids are responsible for the rancid flavour of milk (El Soda *et al.*, 1995). The main lipolytic agents in cheese include the indigenous milk lipoprotein lipase, but also the lipases and esterases produced by the starter and non-starter bacteria, and depending on the cheese variety enzyme preparations added during manufacturing.

To hydrolyse milk fat in milk and cheese, lactic acid bacteria possess esterolytic and lipolytic enzymes capable of hydrolysing a range of esters of FFA, tri-, di, and monoacylglyceride substrates (Fox and Wallace, 1997). Despite the presence of these enzymes, lactic acid bacteria, especially *Lactococcus* and *Lactobacillus* spp. are generally considered to be weakly lipolytic in comparison to species such as *Pseudomonas*, *Acinetobacter* and *Flavobacterium* (Fox *et al.*, 1993). However, because of their presence in cheese at high numbers over an extended ripening period, lactic acid bacteria are considered likely to be responsible for the liberation of significant levels of FFA. To date, lipases/esterases of lactic acid bacteria appear to be exclusively intracellular and a number have been identified and characterized (Chich *et al.*, 1997; Castillo *et al.*, 1999; Liu *et al.*, 2001). El-Soda *et al.* (1986) found intracellular esterolytic activities against substrates up to C5:0 in *L. helveticus*, *L. delbrueckii* subsp. *bulgaricus*, *L. delbrueckii* subsp. *lactis* and *L. acidophilus*, with *L. delbrueckii* subsp. *lactis* and *L. acidophilus* displaying the highest activities. Khalid and Marth (1990) reported the quantitative estimation of the lipolytic activity of *L. casei*, *L. plantarum* and *L. helveticus* towards milk fat, olive oil and tributanoic acid emulsions. The three emulsions were hydrolysed by the lactobacilli with the exception of one *L. casei* strain, which failed to hydrolyse olive oil. According to Lee and Lee (1990), esterolytic and lipolytic enzymes were produced by cell lysis of *L. casei* subsp. *casei* LLG, while *L. fermentum* contains a cell surface-associated esterase specific for C4:0, which can hydrolyse β-naphthyl esters of fatty acids from C2:0 to C10:0 (Gobbetti *et al.*, 1997). Gobbetti *et al.* (1996b) reported the purification of an intracellular lipase from a *L. plantarum* strain isolated from Cheddar cheese, with a molecular mass of 65 kDa, and pH and temperature optima of 7.5 and 35°C, respectively. The enzyme was relatively heat stable to a temperature of 65°C but was irreversibly inactivated on heating to 75°C for 2 min. Hydrolysis of triaclyglycerides indicated that the enzyme had highest activity on tributanoic acid, less activity on tridodecanoic and trihexadecanoic acids and no activity on tri-cis-9-octadecenoic acid.

Lipids are only a minor component of wheat and rye flours but have a significant effect on bread quality. *L. sanfranciscensis* is auxotroph for unsaturated fatty acids (Sugihara and Kline, 1975). Unsaturated fatty acids are subject to autoxidation during flour storage, and are oxidized by cereal lipoxygenase activity during dough mixing (Laignelet and Dumas, 1984). (E)-2-Nonanal and other aldehydes resulting from lipid oxidation are key aroma compounds in wheat and rye bread that impart a 'fatty', 'metallic' or 'green' flavour (Hansen and Schieberle, 2005) and the concentrations of these flavour compounds are significantly reduced during sourdough fermentation (Czerny and Schieberle, 2002). The SC-ADH activity of lactobacilli contributes to the reduction of these flavour compounds during sourdough fermentation, yeasts additionally may exhibit ADH activity in dough.

The taxonomy of the genus *Lactobacillus*

Some taxonomic background

The genus *Lactobacillus* belongs to the large group of lactic acid bacteria, which are all Gram-positive non-sporing cocci, coccobacilli or rods, having a DNA base composition of less than 50 mol% G + C. As mentioned before they lack catalase and need a fermentable carbohydrate for growth. The lactic acid bacteria in the broad sense comprise genera such as *Aerococcus, Alloiococcus, Atopobium, Carnobacterium, Enterococcus, Lactobacillus, Lactococcus, Leuconostoc, Oenococcus, Paralactobacillus, Pediococcus, Streptococcus, Tetragenococcus, Vagococcus* and *Weissella*. The genus *Bifidobacterium, Gardnerella, Scardovia* and *Parascardovia* are often also included in this collection, although phylogenetically they belong to the Actinobacteria subdivision (PHYLUM 3), of the Gram-positive Eubacteria (the Firmicutes), comprising also *Propionibacterium, Brevibacterium* and the microbacteria. The latter taxa are only very distantly related to the genuine lactic acid bacteria.

The genus *Lactobacillus* belongs phylogenetically to the phylum *Firmicutes* (Garrity *et al.*, 2004). The family *Lactobacillaceae* comprises the main family in the order *Lactobacillales* which itself belongs to the class *Bacilli*. From the other members of the family mentioned above, the genera *Paralactobacillus* and *Pediococcus* are most noteworthy since species of these genera tend to intermingle phylogenetically with the variety of species of the genus *Lactobacillus*.

This classification was mainly build on the results of 16S rRNA sequence analysis (*Taxonomic Outline of the Procaryotes;* Garrity *et al.*, 2004), and as mentioned before does not necessarily reflect the metabolic diversity discussed above. Equally essential as the 16S rRNA sequencing technique to this scheme, is the species concept (*The ad hoc committee for the re-evaluation of the species definition in bacteriology;* Stackebrandt *et al.*, 2002). Although the species concept remains subject of animated debates among taxonomists (Rosselló-Mora and Amann, 2001; Rosselló-Mora, 2003; Gevers *et al.*, 2005), it remains the formal unit of bacterial classification and is defined as 'a monophyletic and genomically coherent cluster of individual organisms that show a high degree of overall similarity with respect to many independent characteristics', and 'is diagnosable by a discriminative phenotypic property' (Rosselló-Mora and Amann, 2001). This phylophenetic concept requires that besides genetic evidence, some phenotypic characteristics will discriminate a possible new species from its closest phylogenetic neighbours. When phenotypic variation in a species is considerable, a species may be further subdivided into subspecies, based on this phenotypic variation; subspecies can but need not be supported by genetic determinants (Rosselló-Mora and Amann, 2001).

In practice, a species is defined by two main genotypic criteria: strains with a total DNA similarity of 70% or higher (relative binding in a hybrid DNA reassociation experiment) and a difference in the melting temperature (ΔT_m) equal to or lower than 5°C, will be considered to belong to a single species; in addition the 16S rRNA gene sequence similarity should not differ more than 3%. As mentioned above, phenotypic features should be sought that confirm the proposed groupings.

DNA–DNA similarity measures are thus still considered the gold standard technique for the delineation of bacterial species. Since this technique is laborious, it is very unpractical for the 113 species of the genus *Lactobacillus*. Therefore, the closest phylogenetic species are often identified by a (partial) 16S rRNA sequencing, which after comparison with the large collection of 16S rRNA sequences available in public databases, allows the identification of the most closely related species for which DNA–DNA hybridizations will need to be set up.

For several reasons, 16S rRNA gene sequencing can never be used as the sole method for species delineation (Stackebrandt and Goebel, 1994).

1 A first limit might be that often a single representative strain per species is used for the 16S rRNA analysis, lacking the possibility to position a new isolate in the biological diversity of the species considered.
2 The use of partial sequences is also making the result of 16S rRNA sequences less reliable.
3 The many sequencing errors present in the reference sequences (often from the early days of sequencing), will also influence the final tree.
4 Sequence alignment, essential for sequence similarity calculation, is a highly subjective business. Not only does it rely on a wealth of algorithms, but often manual editing is necessary to 'improve' the result obtained. Critical are the ease one allows the software to create gaps, and the 'cost' defined to extend these gaps.
5 Also, in the pairwise calculation of the sequence similarities, gaps can be included or not and phylogenetic corrections can be applied or not.
6 The cluster algorithm chosen, as well as the selection of reference sequences included, will also affect the shape of the final tree.
7 Finally, as a conserved taxonomic marker, 16S rRNA is not really suitable to study small differences between closely related species.

For these reasons, 16S rRNA sequencing will be useful to frame a new isolate in a well-known phylogenetic scheme, but may not solve the real identification or classification problem. A polyphasic approach (Vandamme *et al.*, 1996a), taking into consideration a variety of information sources, should result in a more reliable identification or classification.

In view of the wide use of lactic acid bacteria in food applications, identification and classification, however, are very important. The discussion whether evolutionary deductions should automatically reflect on nomenclatural designation (Dellaglio *et al.*, 2004b) is a very relevant one. Nomenclature is essential for proper food labelling and will allow producers to communicate in a formal way about the bacteria they add to foods. Safety aspects, for example, have also been linked to species definition (the QPS principle; http://www.efsa.europa.eu/EFSA/efsa_locale-1178620753812_1178620759439.htm). One could therefore support the automatic link between evolution and speciation (de Quieroz and Gauthier, 1992; Woese, 1998; Cantino 1999), but its automatic translation into nomenclatural designations can be questioned for practical and other reasons.

As an example we could use the *Lactobacillus casei* case. The taxonomic controversy has been going on for quite some time. *Lactobacillus casei* was described by Orla-Jensen in 1919. Hansen and Lessel (1971) choose strain ATCC 393 as the neotype strain based on a limited number of phenotypic traits. DNA–DNA hybridization experiments (Dellaglio *et al.*, 1975) showed that strain ATCC 393^{T} had high DNA similarity with the former type strain of '*Lactobacterium zeae*' (Kuznetsov, 1959) and as such was shown not the best neotype strain for *L. casei*. Using DNA–DNA hybridization experiments Collins *et al.* (1989b) confirmed the separate position of strain ATCC 393^{T} and transferred all other *L. casei* strains to a new species *L. paracasei*.

In a first request for opinion, Dellaglio *et al.* (1991) proposed strain *Lactobacillus casei* ATCC 334 (not investigated by Collins *et al.*, 1989b), as an alternative neotype strain of *L. casei* in place of ATCC 393^{T} and requested the rejection of the name *L. paracasei*. This request was denied by the Judicial Commission of the International Committee on Systematic Bacteriology (Wayne, 1994), as this would create a precedent in replacing a type strain that was officially described and which was still readily available (Wayne, 1994). In 1996, Dicks *et al.* reclassified *L. casei* subsp. '*casei*' ATCC 393^{T} and '*Lactobacterium zeae*' ATCC 15820 as *L. zeae* nom. rev., and designated strain ATCC 334 as the neotype of *L. casei* subsp. '*casei*'. They formerly rejected the name *L. paracasei*. Since this species, however, was validly published, the situation was unclear, as both proposals were standing and a new opinion of the Judicial Commission was required. The proposal of Dicks *et al.* (1996) was supported by Mori *et al.* (1997) and by the Subcommittee on the taxonomy of *Bifidobacterium, Lactobacillus* and related organisms (Biavati, 2001; Klein 2001). Consequently, Dellaglio *et al.* reformulated their earlier request for opinion in 2002. In 2005, however, at the occasion of the IUMS meeting (International Union of Microbiological Societies) in San Francisco, the Judicial Commission of the International Committee on Systematic Bacteriology rejected again the proposal for the designation of a new type strain for *L. casei*, even if arguments were raised on the industrial importance of this species and its implication in patents, production processes, food product labels, etc. As a consequence, *L. casei*, used in numerous food applications, today is restricted to only a few strains, including ATCC 393^T and NCFB 173. Almost all other '*L. casei*' strains, many of which are used in food and feed fermentations, or are part of probiotic formulations, are to be named *L. paracasei* subsp. *paracasei*.

These ever-changing names (other examples are the yoghurt bacteria *Lactobacillus 'bulgaricus'* and *Streptococcus 'thermophilus'*) should not determine technological or safety criteria of foods. Taxonomist need and will continue to publish new insights in taxonomy, based on new methods and new taxa, but one should take care not to impact too much on the nomenclature, which for obvious reasons needs stability (Principle 1 of the *Bacteriological Code*; Dellaglio *et al.*, 2004b). This stability, even without frequent name switches, will be difficult given the ever increasing number of species and taxonomic entities being described. Fig. 2.1 illustrates the increase of the number of *Lactobacillus* species described since 1900. This observation warrants that proper classification schemes are handled, and proper taxonomic rules as well as common sense, are necessary to fit this new wealth of information.

Taxonomic techniques used with lactobacilli

As discussed above, classification systems tend to be 'dynamic' and as a consequence the *Lactobacillus* taxonomy, based on biochemical and physiological criteria, differs considerably of the schemes revealed by molecular methods (Stiles and Holzapfel, 1997).

Morphological techniques

Lactobacilli, being rods, are difficult to recognize by simple morphological tests. Moreover, growth conditions and growth stage of the bacterial cells may seriously affect their morphology and many heterofermentative lactobacilli may exhibit a coccobacillus morphology, which is difficult to discriminate from, for example, leuconostocs. Other morphological criteria may include the presence of endospores or inclusion bodies, the number and type of flagella, and the size.

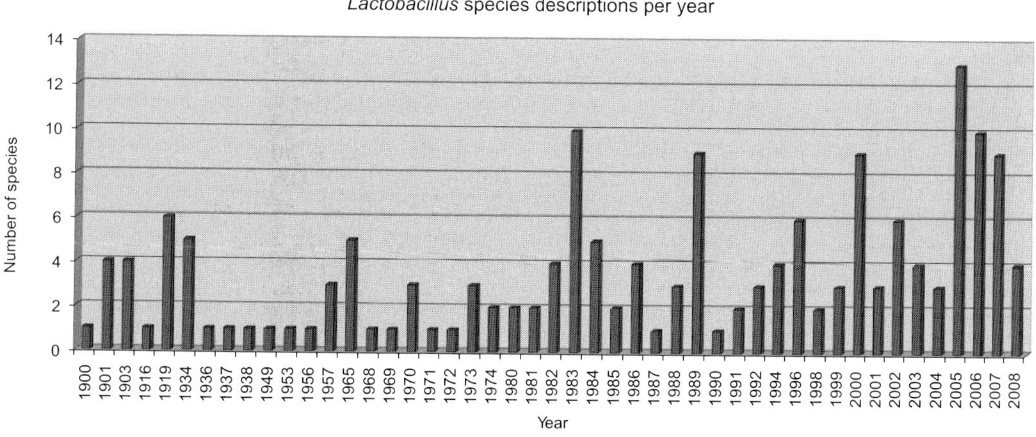

Figure 2.1 Number of *Lactobacillus* species descriptions since 1900. The value for the year 2008 is registered in May.

Phenotypic techniques

As mentioned before, phenotypic characters are important in species description. They also have practical importance: discrimination between *Lactobacillus* and *Carnobacterium*, for example, can easily be achieved since the latter cannot grow at pH 4.5 or on acetate agar but can grow at pH 9.0 (Hammes *et al.*, 1991). For discriminating the 113 species of *Lactobacillus*, simple physiological tests are not sufficient and should be supplemented with additional analyses of fermentation products (Hammes *et al.*, 1991) and genotypic data.

The taxonomic use of fermentation patterns, extensively discussed above, suffers from a number of serious disadvantages such as interlaboratory variation, strain to strain variation and long processing times, which generally result in a rather low reliability of the classification/identification obtained.

In order to improve reproducibility, highly standardized, commercially available systems can be used, such as API 20 STREP, API 50CH (bioMérieux, France), Diatabs (Rosco, Denmark) or BIOLOG GP MicroPlate (BIOLOG Inc., USA). When a database has to be prepared, e.g. for identification purposes, these systems are preferred, as they often also offer relevant reference databases.

Alternatively, immunological techniques allowed the use of, for example, fructose-1,6-biphosphate aldolase (London and Chace, 1976; London *et al.*, 1975; London and Kline, 1973), malic enzymes (London *et al.*, 1971a,b), and glyceraldehyde-3-phosphate dehydrogenase (London and Chace, 1983) as evolutionary probes.

Phenotypic tests also include, for example, the registration of the electrophoretic mobility of lactic acid dehydrogenases (LDH; Gasser, 1970) or the electrophoretic separation of whole cell proteins (sodium dodecyl sulphate polyacrylamide gel electrophoresis; SDS-PAGE). The first technique has only historical importance and was used in the *L. acidophilus* group (see below) (Fujisawa *et al.*, 1992).

SDS-PAGE, however, is still used today, as it is reliable for most lactic acid bacteria on the species and subspecies level (Schleifer and Stackebrandt, 1983; Dicks and van Vuuren, 1987; Pot *et al.*, 1992, 1993, 1994a,b, 1996; Dicks, 1995; Vandamme *et al.*, 1996b). The major disadvantage of the technique is the fact that it yields only discriminative information on the species level, requesting a certain degree of pre-identification. This problem has been overcome by the creation of a database of digitized and normalized protein patterns for a wide selection of reference strains (Vandamme *et al.*, 1996a).

The instability of several phenotypic properties of lactic acid bacteria can probably be linked to the presence of plasmids. This phenomenon has hampered the use of phenotypic tests in the early taxonomy of the lactic acid bacteria.

Cell wall composition

The presence or absence of *meso*-diaminopimelic acid in the cell wall was one of the key parameters in earlier identification of lactobacilli (Hammes *et al.*, 1991, Kandler and Weiss, 1986; Hammes and Hertel, 2003). It involves a relatively simple thin-layer plate chromatographic procedure, which can be done on a large number of strains (Kandler and Weiss, 1986).

In Table 2.2, the peptidoglycan type (Schleifer and Kandler, 1972; Schleifer and Stackebrandt, 1983) is listed for all *Lactobacillus* species, where available. Peptidoglycan types differ in the amino acid sequence of the peptide moiety and the cross-linkage type. For many species these differences are stable on the species level. The Lys-D-Asp type is the predominant type within the genus *Lactobacillus* (Table 2.2) but might, for example, be useful to differentiate the genus *Weissella* from other lactic acid bacteria (Björkroth and Holzapfel, 2003), and *L. reuteri* from *L. fermentum* (Kandler and Weiss, 1986). The analysis, however, requires the preparation of purified cell walls and is therefore time-consuming.

Serology

Some *Lactobacillus* species have been characterized by serology. In total seven serological groups, labelled A–G, have been described (Sharpe, 1955, 1970, 1979, 1981; Kandler and Weiss, 1986). Unlike for streptococci, these studies did not contribute significantly to the identification and the classification of lactobacilli.

Chemotaxonomic markers

Some chemotaxonomic markers have been found in *Lactobacillus*, but their taxonomic role has been rather limited. Menaquinones (Mk), predominantly with eight and nine isoprene units (Mk-8 and Mk-9), have been found in *Lactobacillus 'yamanashiensis'* (now *Lactobacillus mali*) only (Collins and Jones, 1981).

Fatty acid methyl ester (FAME) analysis has been used in lactic acid bacteria (Schleifer *et al.*, 1985; Pompei *et al.*, 1992, Collins *et al.*, 1986, 1989a; Schmitt *et al.*, 1989; Shaw and Harding, 1989; Björkroth and Holzapfel, 2003). The major disadvantage of FAME analysis is the sensitivity of the method to growth conditions, and growth temperature in particular.

Genotypic techniques

DNA–DNA hybridizations and DNA base content

As mentioned above, DNA–DNA hybridizations are extremely important in the species delineation (Wayne *et al.*, 1987). Also, the DNA base content is part of every species definition (see also Table 2.2), and for lactic acid bacteria it is generally below 50%. Some *Lactobacillus* species, however, have values up to 55% (Table 2.2). In practice, the G + C content should not vary more than 5% within a species and 10% within a genus (Schleifer and Stackebrandt, 1983). For *Lactobacillus* this limit is clearly exceeded (Axelsson, 2004), indicating the necessity of subdivision.

DNA–DNA hybridization has been applied extensively in the study of the taxonomy of *Lactobacillus* (Simonds *et al.*, 1971; Dellaglio *et al.*, 1973, 1975; Vescovo *et al.*, 1979; Johnson *et al.*, 1980; Lauer *et al.*, 1980; Dellaglio and Torriani, 1986). As mentioned above, hybridizations are often limited to the type strains of a small set of closely related species, often revealed after (partial) 16S rRNA sequencing (e.g. *Lactobacillus concavus*: Tong and Dong, 2005). Problems arise with DNA–DNA hybridization when strains are phenotypically identical but reassociation values around or below the 70% limit are obtained (Vandamme *et al.*, 1996a).

Comparative analysis of 16S/23S rRNA sequences

As indicated above, 16S rRNA sequencing today is one of the most powerful tools for finding phylogenetic relationships among bacteria. Through its universal presence and its conserved nature, this molecule has excellent properties as molecular clock and was shown to be extremely reliable to study phylogeny. As a consequence, large databases for sequence comparison have been produced, which fortified the usefulness of the technique. Since the 16S rRNA encoding gene is a relatively small molecule, it can be sequenced directly from PCR amplicons obtained using universal 16S primers. Preferably full 16S rRNA sequences are obtained (about 1500 base pairs), although public databases contain relatively large numbers of partial 16S rRNA sequences. Once sequences are obtained, they can be aligned and similarity can be calculated. As argued above, both processes can be performed using a variety of algorithms and settings, implemented in a variety of computer programmes. The results will slightly differ, but overall stability of individual branches of the tree can be tested, for example by bootstrap analysis. Another way to reconstruct phylogeny is based on parsimony, which means that evolution is assumed to have reached the current situation by the shortest possible route. Although this principle can be questioned, it has provided a stable basis for phylogenetic analysis over the years.

Following the report of the Ad Hoc Committee for the Re-Evaluation of the Species Definition in Bacteriology, most species descriptions of the lactic acid bacteria include a 16S rRNA gene sequence (partial or full). As a conserved housekeeping gene, it is useful as a phylogenetic marker to reveal relationships between more distantly related genera of LAB or to reveal misidentifications on the genus level.

The Neighbour joining tree provided in Fig. 2.2 represents all *Lactobacillus* species listed in Table 2.2. Parameters used for the creation of the tree are mentioned in the legend of Fig. 2.2. The tree shows considerable 16S rRNA heterogeneity, confirming the large spread of the G + C content found in the genus and announcing future subdivision of the genus. Fig. 2.2 clearly shows different phylogenetic subgroups, which

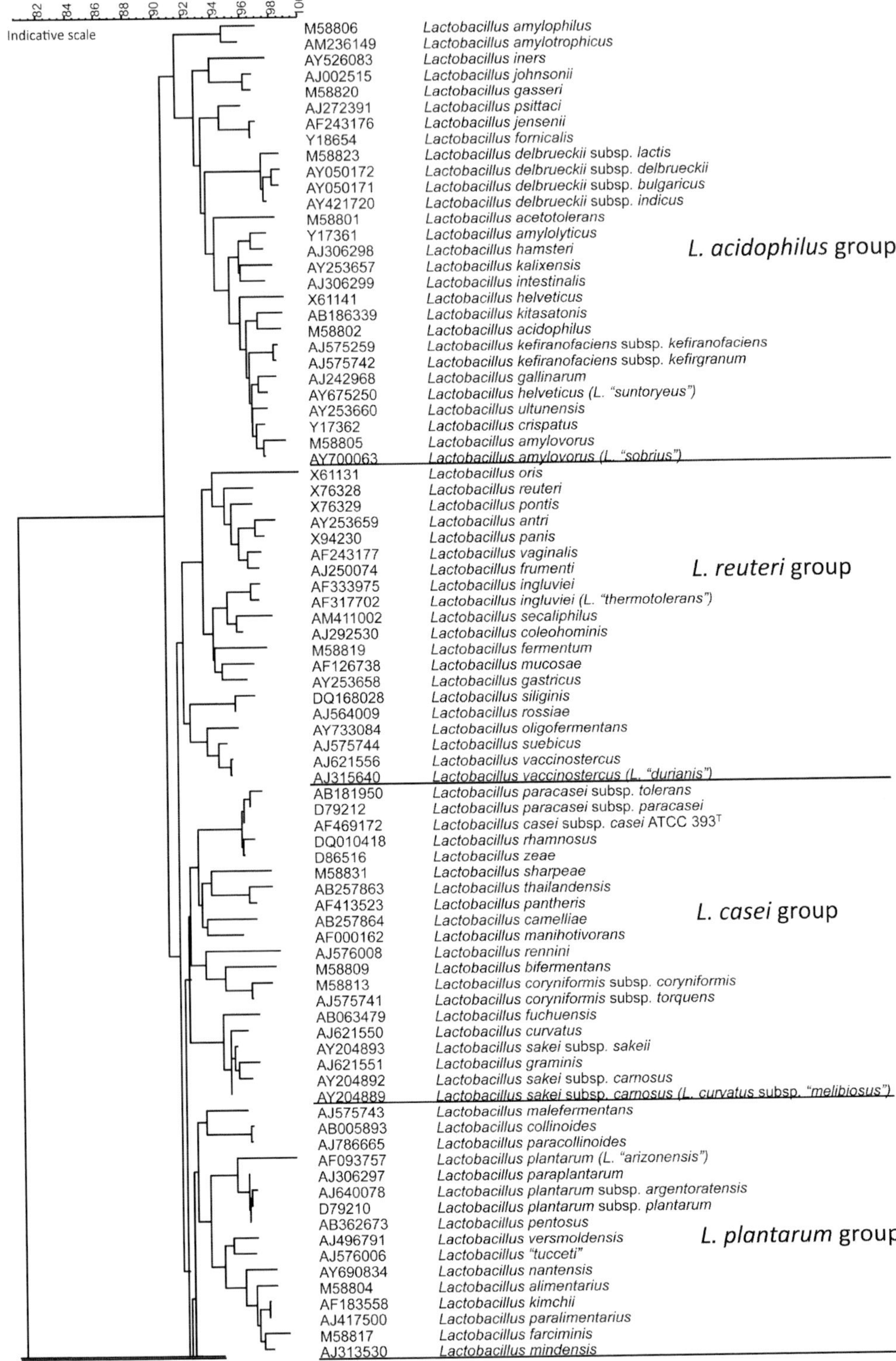
82 84 86 88 90 92 94 96 98 100
Indicative scale
M58806 Lactobacillus amylophilus
AM236149 Lactobacillus amylotrophicus
AY526083 Lactobacillus iners
AJ002515 Lactobacillus johnsonii
M58820 Lactobacillus gasseri
AJ272391 Lactobacillus psittaci
AF243176 Lactobacillus jensenii
Y18654 Lactobacillus fornicalis
M58823 Lactobacillus delbrueckii subsp. lactis
AY050172 Lactobacillus delbrueckii subsp. delbrueckii
AY050171 Lactobacillus delbrueckii subsp. bulgaricus
AY421720 Lactobacillus delbrueckii subsp. indicus
M58801 Lactobacillus acetotolerans
Y17361 Lactobacillus amylolyticus
AJ306298 Lactobacillus hamsteri
AY253657 Lactobacillus kalixensis
AJ306299 Lactobacillus intestinalis
X61141 Lactobacillus helveticus
AB186339 Lactobacillus kitasatonis
M58802 Lactobacillus acidophilus
AJ575259 Lactobacillus kefiranofaciens subsp. kefiranofaciens
AJ575742 Lactobacillus kefiranofaciens subsp. kefirgranum
AJ242968 Lactobacillus gallinarum
AY675250 Lactobacillus helveticus (L. "suntoryeus")
AY253660 Lactobacillus ultunensis
Y17362 Lactobacillus crispatus
M58805 Lactobacillus amylovorus
AY700063 Lactobacillus amylovorus (L. "sobrius")
L. acidophilus group
X61131 Lactobacillus oris
X76328 Lactobacillus reuteri
X76329 Lactobacillus pontis
AY253659 Lactobacillus antri
X94230 Lactobacillus panis
AF243177 Lactobacillus vaginalis
AJ250074 Lactobacillus frumenti
AF333975 Lactobacillus ingluviei
AF317702 Lactobacillus ingluviei (L. "thermotolerans")
AM411002 Lactobacillus secaliphilus
AJ292530 Lactobacillus coleohominis
M58819 Lactobacillus fermentum
AF126738 Lactobacillus mucosae
AY253658 Lactobacillus gastricus
DQ168028 Lactobacillus siliginis
AJ564009 Lactobacillus rossiae
AY733084 Lactobacillus oligofermentans
AJ575744 Lactobacillus suebicus
AJ621556 Lactobacillus vaccinostercus
AJ315640 Lactobacillus vaccinostercus (L. "durianis")
L. reuteri group
AB181950 Lactobacillus paracasei subsp. tolerans
D79212 Lactobacillus paracasei subsp. paracasei
AF469172 Lactobacillus casei subsp. casei ATCC 393T
DQ010418 Lactobacillus rhamnosus
D86516 Lactobacillus zeae
M58831 Lactobacillus sharpeae
AB257863 Lactobacillus thailandensis
AF413523 Lactobacillus pantheris
AB257864 Lactobacillus camelliae
AF000162 Lactobacillus manihotivorans
AJ576008 Lactobacillus rennini
M58809 Lactobacillus bifermentans
M58813 Lactobacillus coryniformis subsp. coryniformis
AJ575741 Lactobacillus coryniformis subsp. torquens
AB063479 Lactobacillus fuchuensis
AJ621550 Lactobacillus curvatus
AY204893 Lactobacillus sakei subsp. sakeii
AJ621551 Lactobacillus graminis
AY204892 Lactobacillus sakei subsp. carnosus
AY204889 Lactobacillus sakei subsp. carnosus (L. curvatus subsp. "melibiosus")
L. casei group
AJ575743 Lactobacillus malefermentans
AB005893 Lactobacillus collinoides
AJ786665 Lactobacillus paracollinoides
AF093757 Lactobacillus plantarum (L. "arizonensis")
AJ306297 Lactobacillus paraplantarum
AJ640078 Lactobacillus plantarum subsp. argentoratensis
D79210 Lactobacillus plantarum subsp. plantarum
AB362673 Lactobacillus pentosus
AJ496791 Lactobacillus versmoldensis
AJ576006 Lactobacillus "tucceti"
AY690834 Lactobacillus nantensis
M58804 Lactobacillus alimentarius
AF183558 Lactobacillus kimchii
AJ417500 Lactobacillus paralimentarius
M58817 Lactobacillus farciminis
AJ313530 Lactobacillus mindensis
L. plantarum group

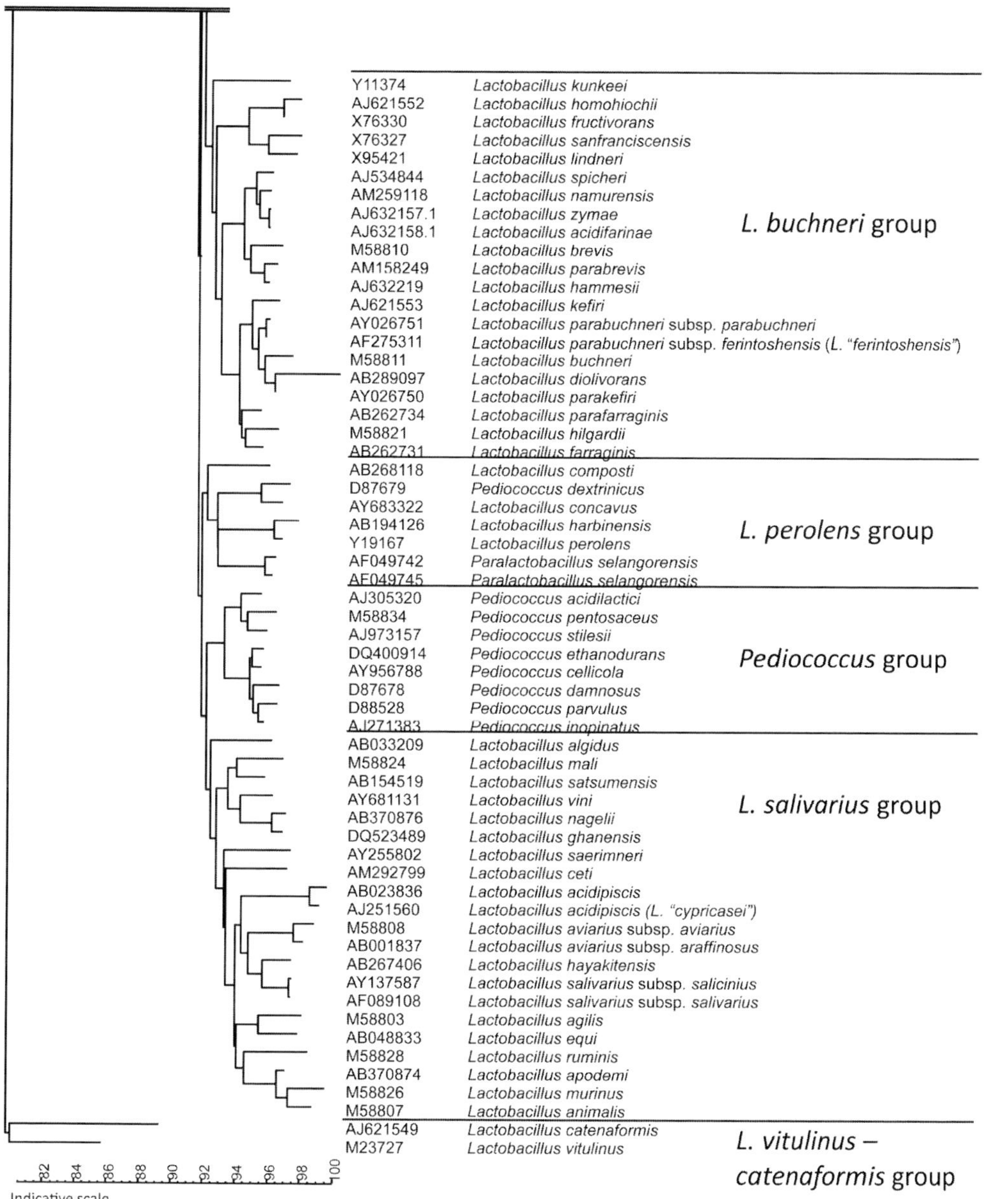

Figure 2.2 Phylogenetic structure of the genus *Lactobacillus*. The similarity axis indicates similarity in %; total distance is to be calculated as an accumulation of horizontal distances between members. 16S ribosomal DNA sequences (partial sequences only used when full sequence was not available) were obtained from the NCBI. The tree is a Neighbour Joining tree calculated on a pairwise similarity matrix obtained from aligned sequences. The BioNumerics software (Applied Maths NV, Belgium) was used with the following settings: Pairwise similarity calculation: Open gap penalty 100%; Unit gap penalty 0%, fast algorithm with Minimum match sequence = 2 and Maximum number of gaps = 9. The multiple alignment was prepared from a UPGMA tree topology calculated from this similarity matrix, using the same settings, but without the fast algorithm. Manual corrections were limited. The multiple aligned sequences were then grouped phylogenetically using the neighbour Joining algorithm with a gap penalty of 0%, and no Jukes and Cantor or Kimura2 correction.

may be the basis for this division and which will be discussed in detail below.

Historically, early 16S rRNA sequencing contributions for the lactic acid bacteria were published by Yang and Woese (1989), Collins *et al.* (1989a, 1990, 1991), Martinez-Murcia and Collins (1990), Williams *et al.* (1990), Wallbanks *et al.* (1990), Bentley *et al.* (1991).

As a major result, one should conclude that the generic description of various species of *Lactobacillus* does not fit with the phylogenetic structure (Collins *et al.*, 1991). The most striking conclusions are: (i) the heterogeneity of the lactobacilli, (ii) the rather close relationships between some lactobacilli and some pediococci, (iii) the separate position of the carnobacteria, more distantly related from the lactobacilli and more closely grouped with the enterococci and with *Vagococcus* (Pot, 2007).

Obligately heterofermentative as well as the obligately homofermentative lactobacilli are scattered over various evolutionary branches. A number of branches, however, do show a common type of carbohydrate fermentation, which could be used for reclassification, supported by evolutionary relationships (Hammes *et al.*, 1991; see also Table 2.2).

As mentioned earlier, one of the disadvantages of the 16S rRNA sequencing technique is that the molecule is too conserved to provide resolution at the species and subspecies levels (De Parasis and Roth, 1990; Weisburg *et al.*, 1991; Fox *et al.*, 1992; Martinez-Murcia *et al.*, 1992; Stackebrandt and Goebel, 1994). Consequently, different species of the lactic acid bacteria have been found to share (nearly) identical 16S sequences (Yoon *et al.*, 2000; Švec *et al.*, 2001; Björkroth *et al.*, 2002; Leisner *et al.*, 2002; Kim *et al.*, 2003; Cachat and Priest, 2005). Some of the *Lactobacillus* species have been renamed in the mean time (see examples in Fig. 2.2), for other species there are sufficient phenotypic and genotypic differences to maintain a separate species status. As for G + C content, 16S rRNA similarity values can thus only be used as an excluding criteria, and other genotypic methods, such as DNA–DNA hybridizations, but also multilocus sequence analysis, may need to be used.

Sequencing of housekeeping genes

According to the Ad Hoc Committee for the Re-Evaluation of the Species Definition in Bacteriology (Stackebrandt *et al.*, 2002), sequencing of housekeeping genes is a promising tool for phylogenetic studies. It was recommended that a multilocus sequence typing (MLST) approach be used, with the sequencing of at least five genes located in diverse chromosomal loci and widely distributed among taxa. For taxa as *Neisseria*, the MLST technique has been very useful, for other taxa such as *Helicobacter* it was not very conclusive (http://pubmlst.org/). Traditional MLST schemes have been proposed for streptococci and enterococci (http://efaecium.mlst.net/, Enright and Spratt, 1998; Enright *et al.*, 2001; Homan *et al.*, 2002; King *et al.*, 2002; Jones *et al.*, 2003).

In *Lactobacillus*, sequencing of a more limited number of selected genes has proven to be quite promising. The *recA*, *cpn*60, *tuf* and *slp* genes have been used in *Lactobacillus* (Felis *et al.*, 2001, Torriani *et al.*, 2001; Dellaglio *et al.*, 2004b; Bringel *et al.*, 2005; Cachat and Priest, 2005) as well as the phenylalanyl-tRNA synthase alpha subunit (*pheS*) and the μ RNA polymerase alpha subunit (*rpoA*) (Vancanneyt *et al.*, 2006b; Naser *et al.*, 2007). The alpha subunit of ATP synthase (*atpA*), *rpoA* and *pheS* have also been used successfully in *Enterococcus* (Dahlhof *et al.*, 2000; Naser *et al.*, 2005) and the genes *gyrA*, *gyrB*, *sodA* and *parC* in *Streptococcus* (Kawamura *et al.*, 2005). The most interesting aspect of these evolutions is the comparison of the phylogenetic trees obtained with different housekeeping genes, combinations of housekeeping genes or the comparison with trees obtained from 16S or 23S ribosomal RNA sequences.

Typing of lactobacilli

A large number of typing techniques, revealing relationships at the subspecific or strain level, have been used with lactobacilli. These techniques include DNA restriction fragment length polymorphism (RFLP) (Yavuz *et al.*, 2004), ribotyping (Björkroth and Korkeala, 1997; Björkroth *et al.*, 2000, 2002; Suzuki *et al.*, 2004; Kostinek *et al.*, 2005) or involve the use of specific oligonucleotide probes or primers which have been designed to allow quick identification (Andrighetto *et al.*, 1998; Torriani *et al.*, 1999; Lick *et al.*, 2000).

A variant of these techniques is the amplified ribosomal DNA restriction analysis (ARDRA) technique, allowing the construction of a library of reference strains for identification. Some examples can be found in Giraffa *et al.* (1998) and Ventura *et al.* (2001).

Polymerase chain reaction (PCR)-based techniques have also been very successful. The target sequence can be a repetitive extragenic palindromic (REP) sequence (ERIC-PCR, GTG5-PCR, Box-PCR) (De Urraza *et al.*, 2000; Gevers *et al.*, 2001; Masco *et al.*, 2003; Kostinek *et al.*, 2005; Švec *et al.*, 2005a,b) or a random sequence (random amplified polymorphic DNA; RAPD) (Moschetti *et al.*, 1997; Torriani, *et al.*, 1999, Dellaglio *et al.*, 2005; Valcheva *et al.*, 2005). These techniques all have in common that they are easy to perform, without extensive laboratory equipment. The disadvantage, however, is a differing degree of reproducibility, with RAPD-PCR to be the least reproducible and REP-PCR allowing the construction of reference databases within a single laboratory.

More reliable is the amplified fragment length polymorphism (AFLP) fingerprinting technique, which also gives information at the species level. This technique has been used in lactic acid bacteria by Dellaglio *et al.* (2005), Valcheva *et al.* (2005) and Vancanneyt *et al.* (2005b). A fluorescent variant has been described (fAFLP; Vancanneyt *et al.*, 2005a). AFLP fingerprinting has also been used to trace the origin of LAB that are dominant during a fermentation process (Nokuthula *et al.*, 2000).

Not extensively used in lactic acid bacteria is the pulsed-field gel electrophoresis (PFGE) technique. In clinical microbiology this method is considered as the gold standard for bacterial typing and epidemiology. Examples in lactics can be found in Guopeng and Holley (1999), Orrhage *et al.* (2000), Blaiotta *et al.* (2001), Somers *et al.* (2001), Coppola *et al.* (2003) and Dalgaard *et al.* (2003).

Methodologies under development

Identification using microarrays will increase as more genome sequences become available (Ye *et al.*, 2001). These arrays may involve functionality genes besides the above mentioned taxonomic markers. It can be anticipated that the combination of multiple markers, in the line of polyphasic taxonomic thinking, will yield more reliable classification and identification schemes. The fact that functionality genes may be read as well from the arrays may yield a 'functional identification', which covers criteria other than those covered by the traditional taxonomic identification. For application in the food industry this would clearly solve the problem of the ever-changing bacterial nomenclature, which often renders the labelling of, for example, fermented foods problematic, and would enable labelling of individual food products with a description of those aspects/risks which are important and relevant for that product.

The announced sequencing of non pathogenic lactic acid bacteria (Canchaya *et al.*, 2006; Makarova *et al.*, 2006; Makarova and Koonin, 2007) and the availability of high-throughput sequencing platforms will speed up the knowledge gathering process from lactic acid bacteria. The availability of large numbers of chromosome sequences may also shed some light on the importance of, for example, lateral gene transfers and the possible role (or not) of species barriers therein.

The application of DNA microarrays has also extended into the fields of environmental microbiology and microbial ecology, and acts as a substitute for existing molecular tools (Guschin *et al.*, 1997). Whereas opportunities to use this technology to address important questions in microbial ecology are abundant (Cho and Tiedje, 2001), several practical limitations slow its implementation (Jin-Woo *et al.*, 2005).

The *Lactobacillus* phylogeny

The phylogenetic position of *Lactobacillus*, according to the *Taxonomic Outline of the Prokaryotes* (Garrity *et al.*, 2004), is within the phylum *Firmicutes*, class *Bacilli*, order *Lactobacillales*, family *Lactobacillaceae*. Its closest neighbours are the genera *Leuconostoc*, *Oenococcus* and *Weissella* (Hammes and Hertel, 2003; Pot, 2007). Most species of the genus *Pediococcus* and *Paralactobacillus selangorensis*, however, fall within the phylogenetic group of the 'lactobacilli' (Fig. 2.2).

Collins *et al.* (1991) originally discriminated three phylogenetic groups, the *L. delbrueckii* group, the *L. casei* – *Pediococcus* group and the *Leuconostoc* group. Later they reclassified some *Lactobacillus* species in the genera *Leuconostoc* and *Weissella* reducing the heterogeneity of the genus *Lactobacillus* (Collins *et al.*, 1993). The *L. delbrueckii* group was renamed as *L. acidophilus* group by Schleifer and Ludwig (1995) and the heterogeneous *L. casei–Pediococcus* group was split into smaller subgroups: the *Lactobacillus buchneri* group, the *L. casei* group, the *L. plantarum* group, the *L. reuteri* group and the *L. salivarius* group. According to the dendrogram presented in Fig. 2.2 a few additional groups can be discriminated.

In the meantime, some species of the genus have also been transferred to genera as *Carnobacterium* and *Atopobium*. It is very likely that the genus will be further subdivided into multiple new genera, restricting the name *Lactobacillus* to the group of species around *L. delbrueckii*, the type species. As mentioned before, this taxonomic heterogeneity is nicely illustrated by the DNA G + C content which ranges from 32 to 54 mol%, far too broad for a genus (Schleifer and Stackebrandt, 1983).

How many genera will need to be created, could be a matter of long debate. As mentioned before, the parameters and algorithms used for phylogenetic tree construction as well as the number of 16S rRNA sequences used, will heavily determine the result of the analysis. More diverse sequences will result in many more gaps in the aligned sequences and therefore give rise to different levels of similarity. Phylogenetic positions of species, represented by a single sequence, will therefore easily change positions when included in a dendrogram representing the family *Lactobacillaceae*, the genus *Lactobacillus* or the subgroup *L. acidophilus*. Dellaglio and Felis (2005) also noted that 'the addition of novel species dramatically changes the phylogenetic structure of the genus, even in a short time', which can also explain why in the taxonomic literature many different groupings and dendrograms will be encountered for the genus *Lactobacillus*. This phenomenon will clearly render the decision to create new genera for 'recognized phylogenetic groups' rather difficult. The lack of clear phenotypic markers that allow to recognize these 'phylogenetic groups' will render the task extremely difficult. It can be hoped that the gradual availability of other sequences than 16S rRNA sequences (e.g. protein coding sequences from the *recA*, *cpn60*, *tuf*, *slp*, *pheS*, and *rpoA* genes) will help to delineate 'consensus' groups that can be considered for the definition of new genera.

Given this 'inevitable' variation, the dendrogram presented in Fig. 2.2 should therefore be considered as one representation of a single momentary registration of the species of the genus *Lactobacillus* available in May 2008. Identification of new sequences or classification of potential new species should, for the same reason, not rely on the comparison of 16S rRNA sequences only.

In this chapter we consider a number of phylogenetic subgroups of the genus *Lactobacillus* which may have some historical meaning, which will facilitate description, but which may not necessarily represent fixed phenotypic behaviour or ecological links.

The *Lactobacillus acidophilus* group

The *L. acidophilus* group (Schleifer and Ludwig, 1995), also referred to as the *L. delbrueckii* group (Hammes and Hertel, 2003; Dellaglio and Felis, 2005), is to be considered a very important group from historical point of view, with a subdivision supported by a lot of taxonomic papers. Although *L. delbrueckii* is the type species of the genus *Lactobacillus*, the name '*L. acidophilus* group' has a much larger historical meaning.

L. acidophilus today is an industrially important species, since many strains are known to have probiotic properties. In the past the species was studied intensively for its taxonomic heterogeneity (Moro, 1900; Orla-Jensen *et al.*, 1936; Hansen and Mocquot, 1970; Gasser, 1970; Sharpe, 1970; Gasser and Janvier, 1980; Johnson *et al.*, 1980; Lauer *et al.*, 1980; Sarra *et al.*, 1980; Pot *et al.*, 1993; Gancheva *et al.*, 1999). Based on the results of DNA–DNA hybridizations *L. acidophilus* was finally split into *L. acidophilus sensu stricto* (DNA group A1), *Lactobacillus crispatus* (DNA group A2), *Lactobacillus amylovorus* (DNA group A3), *Lactobacillus gallinarum* (DNA group A4), *Lactobacillus gasseri* (DNA group B1) and *Lactobacillus johnsonii* (DNA group B2). These six species are difficult to differentiate based on simple phenotypic tests. Characteristics which can be used for discrimination are given by Fujisawa *et al.* (1992). Electrophoretic analysis of soluble cellular proteins, whole cell proteins or lactate dehydrogenases, detailed cell wall studies, 16S rDNA sequencing and polyphasic studies including RADP-PCR and AFLP also allow the discrimination of these species (Gancheva *et al.*, 1999).

L. delbrueckii is the oldest species of the genus and is therefore the type species. It is now represented by four subspecies: *L. delbrueckii* subsp. *bulgaricus* (formerly *Lactobacillus* '*bulgaricus*'), *Lactobacillus delbrueckii* subsp. *delbrueckii* (formerly *Lactobacillus* '*fermentum*' var. *delbrücki*, Leichmann, 1896), *Lactobacillus delbrueckii* subsp. *lactis* (formerly *Lactobacillus* '*lactis*' and *Lactobacillus* '*leichmannii*', Weiss *et al.*, 1983b), and *Lactobacillus delbrueckii* subsp. *indicus* (Dellaglio *et al.* 2005). The four subspecies share at least 78% DNA similarity (Dellaglio *et al.* 2005). The G + C content of the species *Lactobacillus delbrueckii* is at least 10% higher than the other species in the subgroup (49–51 mol%, see Table 2.2) which is one of the few ways to discriminate this species from, for example, *Lactobacillus jensenii*.

Besides the species mentioned above, the group contains also *Lactobacillus acetotolerans*, *Lactobacillus amylophilus*, *Lactobacillus amylolyticus*, *Lactobacillus amylotrophicus*, *Lactobacillus fornicalis*, *Lactobacillus hamsteri*, *L. helveticus*, *Lactobacillus iners*, *Lactobacillus intestinalis*, *Lactobacillus jensenii*, *Lactobacillus kalixensis*, *Lactobacillus kefiranofaciens* with the subsp. *kefiranofaciens* and *kifirgranum*, *Lactobacillus kitasatoris*, *Lactobacillus psittaci*, *Lactobacillus ultunensis*, and the former species *Lactobacillus* '*sobrius*,' now *L. amylovorus* and *Lactobacillus* '*suntoryeus*', now *L. helveticus*. Table 2.2 contains the literature references towards the original descriptions of these species. *Lactobacillus* '*jugurti*' was shown to be synonymous with *L. helveticus* by Simonds *et al.* (1971) and Dellaglio *et al.* (1973).

Interestingly, the *L. acidophilus* group contains almost exclusively obligately homofermentative lactobacilli (Table 2.2). Since *L. acidophilus* and *L. delbrueckii* will remain the core of the genus *Lactobacillus*, even after extensive restructuring of the genus, this obligately homofermentative character may become an important property of the redefined genus.

Lactobacillus animalis, *Lactobacillus aviarius*, *Lactobacillus farciminis*, *Lactobacillus mali*, *Lactobacillus ruminis*, *Lactobacillus salivarius*, *Lactobacillus sharpeae* and *Lactobacillus vitulinus*, are also homofermentative (Hammes *et al.*, 1991), but these species are phylogenetically spread over different phylogenetic subgroups (see below).

The *Lactobacillus reuteri* group

Lactobacillus fermentum, one of the best-known species of this group, is also one of the first formal *Lactobacillus* species described (Beijerinck, 1901). Like many of these older species, *L. fermentum* became broader and broader over the years and was subject of major taxonomic changes. In 1968, for instance, Hansen designated a reference strain for the *L. fermentum* biotype IIb, which in 1980 became the type strain of *L. reuteri* (Kandler and Weiss, 1980). Besides DNA–DNA hybridization results (Vescovo *et al.*, 1979; Sarra *et al.*, 1980; Collins *et al.*, 1991) both species can be discriminated by SDS-PAGE of whole-cell proteins, by differences in peptidoglycan type or by determinations of the DNA base compositions (Kandler and Weiss, 1986). Molecular techniques remain the most reliable method for identification (Klein *et al.*, 1998; Weiss *et al.* 2005).

Lactobacillus cellobiosus was shown to be identical to *L. fermentum* by DNA–DNA homology (Vescovo *et al.*, 1979). The closest phylogenetic neighbours of *L. fermentum* are *Lactobacillus* '*thermotolerans*' and *Lactobacillus ingluviei*. *Lactobacillus ingluviei* was isolated from the intestinal tract of pigeons (Baele *et al.* 2003) and in the same year, *L. thermotolerans* was described for heterofermentative and thermotolerant strains isolated from chicken faeces. Felis *et al.* in 2006 showed, in a polyphasic study involving 16S rRNA, the hsp60 gene and recA gene sequencing, DNA:DNA homology, G + C contents and biochemical characteristics, that both taxa should constitute a single species, *L. ingluviei* (first described species). Euzéby, however, pointed out several reasons why both species should be kept separate (http://www.bacterio.cict.fr/bacdico/ll/thermotolerans.html). Maybe two subspecies should be created to accommodate the phylogenetic similarity as well as the physiological/ecological diversity.

The *Lactobacillus reuteri* group contains at least 6 species isolated from sourdough exclusively (*Lactobacillus frumenti*, *Lactobacillus pontis* and *L. panis* around *L. reuteri*, and *Lactobacillus rossiae*, *Lactobacillus siliginis* and *Lactobacillus secaliphilus* around *Lactobacillus fermentum*). It should be noted that *L. reuteri* strains can also be a dominant member of sourdough populations (Dal Bello *et al.* 2005). This concentration of species, adapted to the particular niche of sourdough, may be an interesting challenge for a comparative genomic study that tries to link ecological specification to functional characteristics embedded in the genetic potential of the strains. This study may be especially interested in the *L. reuteri* group since a considerable subgroup of species were isolated from human or porcine origin. *Lactobacillus oris* was already isolated in the 1950s from human oral cavities, although originally classified as *Lactobacillus 'brevis'*. Based on nucleic acid and biochemical data Farrow and Collins (1988) proposed the name *L. oris*. *Lactobacillus vaginalis* (isolated from vaginal swabs from patients with trichomoniasis) phenotypically resembles *L. fermentum* and *L. reuteri* although phylogenetically it is most closely related to *Lactobacillus frumenti*, although it lacks the potential to survive in the sourdough ecosystem. *Lactobacillus gastricus* strains (group G1, isolated from human stomach mucosa; Roos *et al.*, 2005) showed the highest levels of 16S rRNA sequence similarity to *Lactobacillus mucosae* and *L. ingluviei*. *Lactobacillus antri* (group G2, also isolated from human stomach mucosa) showed the highest levels of similarity to *Lactobacillus oris*, *L. panis* and *L. frumenti*. DNA–DNA hybridizations confirmed the separate species status of *Lactobacillus antri*. The groups G3 and G4 from human stomach mucosa have been assigned to the species *Lactobacillus kalixensis* and *Lactobacillus ultunensis* respectively. These species are further discussed below. *Lactobacillus coleohominis* was also derived from human sources (Nikolaitchouk *et al.*, 2001). Two strains were isolated from the vaginas of healthy women, while two other strains were recovered from urine and from a human cervix. *Lactobacillus mucosae*, finally, was derived from the small intestine of pigs, and was found in an attempt to isolate *L. reuteri* strains carrying the *mub* gene (coding for the mucus binding protein, thought to be important in adhesion to pig intestine; Roos *et al.*, 2000).

The close association of *Lactobacillus* species and the intestinal tracts and mucous membranes of humans and animals, has often been shown to be beneficial (Wibowo *et al.*, 1985, Aguirre and Collins, 1993; Sanders, 1994; Roos *et al.*, 2000). Some studies have compared so-called 'probiotic' strains (strains that beneficially affect the host) to strains derived from lesions (sepsis, endocarditis, etc.; Vancanneyt *et al.*, 2006a; Vankerckhoven *et al.*, 2008).

The third and last subcluster of the *L. reuteri* group contains four species. *Lactobacillus oligofermentans* was described for strains associated with spoilage of modified atmosphere-packaged poultry products (Koort *et al.* 2005a,b). *Lactobacillus suebicus* (Kleynmans *et al.*, 1989) is an obligately heterofermentative *Lactobacillus* species isolated from fruit mashes. Collins *et al.* (1991) showed that the closest neighbour of *L. suebicus* is *Lactobacillus vaccinostercus*. Both species are the only two heterofermentative lactobacilli which also have a meso-diaminopimelic acid direct type murein. *L. vaccinostercus* was originally described for a strain isolated from cow (vaccinus) dung (stercus) by Okada *et al.* (1979). Leisner *et al.* (2002) described a new species, *Lactobacillus 'durianis'*, for the predominant micro-organisms in a Malaysian acid-fermented condiment, tempoyak, prepared from the durian fruit. In their comparative study, however, they did not include *Lactobacillus vaccinostercus*. Dellaglio *et al.* (2006) formally questioned the taxonomic status of the newly described species, as highly similar 16S rRNA gene sequences and high homologies between partial recA and hsp60 gene sequences suggested a close relationship with *Lactobacillus vaccinostercus*.

It should be noted that this third subgroup is phylogenetically more distantly removed from the *L. fermentum* and the *L. reuteri* core groups (Fig. 2.2), so their future inclusion in a single genus together with these subgroups might be uncertain.

The *Lactobacillus casei* group

As mentioned above, the '*Lactobacillus casei*' case has attracted a lot of attention over the last 20 years. This is, without any doubt, linked to the considerable economic importance of the *Lactobacillus (para)casei* species, which is used in many food and feed applications such as dairy products and silage, and has a proven record in human and animal health. The name *Lactobacillus casei* in the strictest sense, is now reserved for strains resembling strain NCFB 161 (= ATCC 393), the neotype strain (Hansen and Lessel, 1971), or strain NCFB 173, which showed 80% DNA relatedness with the former. Most other strains are to be considered as *L. paracasei*, which is the official name to be used in, for example, production protocols and on packaging labels. In Fig. 2.2, the position of the *L. casei* type strain (ATCC 393) is very close to the representative strains of the two subspecies of *L. paracasei*. The species *L. zeae* and *L. rhamnosus* in this version of the dendrogram occupy a somewhat separate position, probably due to a shorter available sequence length. From this dendrogram it is clear that 16S rRNA sequence analysis might not be the method of choice to assign *L. (para)casei* strains to one or the other species.

As for the rest of the *L. casei* group, early ribosomal RNA sequencing data indicated that this group is a large but heterogeneous group within the lactobacilli. The 32 *Lactobacillus* species and five pediococcal species of the group described by Collins *et al.* (1991) were extended with new species. This gave rise to the creation of several subgroups, of which the position in a single '*L. casei* group' can be questioned. In general there exist few (phenotypic) arguments to consider these subgroups together in what was formerly referred to as the '*L. casei* group', although for some subgroups there exist common phenotypic properties. The subgroups indicated in Fig. 2.2 are to some extend arbitrarily and may vary with the number of sequences included.

Besides a subgroup composed of the 'controversial' species listed above, we can discriminate the *Lactobacillus sakei* subgroup. The species *L. sakei* (*L.* '*sake*' before) comprises today also the former species *Lactobacillus* '*bavaricus*' (Kagermeier-Callaway and Lauer 1995) which lacks lactic acid racemase (Stetter and Stetter, 1980). Dellaglio *et al.* (1975) found the type strain of *Lactobacillus homohiochii*, to be genomically related to *L. sakei*. In Fig. 2.2, however, the sequence from the type strain LMG 9478 culture collection falls with *Lactobacillus fructivorans*. More taxonomic work seems necessary here before a formal transfer of *Lactobacillus homohiochii* to *Lactobacillus sakei* can be proposed. Torriani *et al.* (1996) described *Lactobacillus sakei* subsp. *carnosus*, creating automatically *Lactobacillus sakei* subsp. *sakei*.

Related to *Lactobacillus sakei* is *Lactobacillus curvatus*. Torriani *et al.* (1996) described *Lactobacillus curvatus* subsp. *melibiosus* creating automatically *L. curvatus* subsp. *curvatus*. This reclassification was questioned by Koort *et al.* (2004) who assigned *L. curvatus* subsp. *melibiosus* to *L. sake* subsp. *carnosus*, based on a polyphasic study. The sequence analysis in Fig. 2.2 seems to support this position. In the same *L. sakei* subgroup we also find *Lactobacillus fuchuensis*, isolated from vacuum-packaged refrigerated beef, and *Lactobacillus graminis*, a facultatively heterofermentative *Lactobacillus* surviving at low pH in grass silage.

Another subgroup of the *L. casei* group is composed of *Lactobacillus camelliae*, *Lactobacillus manihotivorans*, *Lactobacillus pantheris*, *Lactobacillus sharpeae* and *Lactobacillus thailandensis*. *Lactobacillus manihotivorans* was described for a starch-hydrolysing lactic acid bacterium isolated from cassava, *L. camelliae*, *L. pantheris* and *L. thailandensis* were isolated from fermented tea leaves (miang) produced in the northern part of Thailand and *Lactobacillus sharpeae* was described for strains isolated from municipal sewage and from spoiled retail samples of vacuum-packaged, cooked delicatessen meats (Holley *et al.*, 1996).

A last subgroup is the *Lactobacillus coryneformis* cluster (Pot, 2007) which also contains *Lactobacillus bifermentans*, and *Lactobacillus rennini*. *Lactobacillus coryneformis* comprises two subspecies, *L. coryneformis* subsp. *coryneformis* and *L. coryneformis* subsp. *torquens*, already described by Abo-Elnaga and Kandler in 1965. *Lactobacillus bifermentans* was described as an organism that caused small cracks by gas formation in Edam and Gouda cheeses. These authors confirmed the exotic ability of *Lactobacillus* '*bifermentans*' to carry out (at high glucose concentrations) a homolactic fermentation, as well as its ability to ferment lactate to acetic acid, ethanol, traces of propionic acid, CO_2 and H_2 at a pH above 4.0. The atypical formation of H_2 in lactobacilli was the main reason to revive the original name *Lactobacillus bifermentans* (Kandler *et al.*, 1983b). *Lactobacillus rennini* was more recently isolated from rennin, an enzyme used in cheese making that coagulates milk. The species was associated with cheese spoilage.

The *Lactobacillus plantarum* group

Again we see two possible subgroups (Fig. 2.2). The '*Lactobacillus plantarum* group' *sensu stricto* encompasses *L. plantarum* subsp. *plantarum*, *L. plantarum* subsp. *argentoratensis*, *Lactobacillus paraplantarum*, *Lactobacillus pentosus* and *Lactobacillus arizonensis*. *Lactobacillus plantarum* contains many atypical characteristics for lactobacilli, such as motility, pseudocatalase activity or reduction of nitrate. Given the considerable variation in the species, some strains may actually be related to *L. pentosus*, in earlier days considered homologous with *L. plantarum*, and thus not included on the Approved Lists of Bacterial Names (Skerman *et al.*, 1980). The name, however, was revived by Zanoni *et al.* (1987) and it is currently still considered a valid species. Identification is difficult by 16S DNA sequencing (Fig. 2.2). In order to deal with this observed heterogeneity, subspecies description may be useful: *L. plantarum* subsp. *argentoratensis* was described for 14 strains isolated from vegetables in Strasbourg (= Argentoratus, the Roman name of this city in France) and which can be discriminated by limited phenotypic properties from *L. plantarum* subsp. *plantarum*. In general the sequencing of the protein-encoding genes recA and cpn60 as well as AFLP profiling are useful for identification and classification of strains belonging to the *L. plantarum* group (Torriani *et al.* 2001; Bringel *et al.*, 2005). The taxonomic status of the species *L. arizonensis*, isolated from fermented jojoba, is still a matter of discussion. Kostinek *et al.* (2005) re-examined the taxonomic position by a polyphasic approach and concluded that the *L. arizonensis* type strain could not be distinguished from the type strain of *Lactobacillus plantarum* or from various other *L. plantarum* reference strains. Relying on the dendrogram in Fig. 2.2, this reclassification is quite unlikely, but sequencing errors or problems with the reference strains used cannot be excluded. Another taxonomic study comparing strains from both investigator groups may be necessary to make final conclusions. *Lactobacillus paraplantarum* was described for four facultatively heterofermentative lactobacilli isolated from beer and human faeces, having physiological characteristics similar to those of *L. plantarum*, although they do not catabolize alpha-methyl-D-mannoside.

A second subgroup encompasses lactobacilli from a variety of food origins exclusively: *Lactobacillus versmoldensis*, isolated from raw fermented sausages, *Lactobacillus kimchii*, a single isolate derived from the Korean fermented-vegetable food kimchi, *Lactobacillus alimentarius* and *Lactobacillus farciminis* from marinated food products, *Lactobacillus nantensis*, *Lactobacillus mindensis* and *Lactobacillus paralimentarius* from sourdough and *Lactobacillus tucceti*, isolated from sausage by Reuter in 1970. This species has not been formally validated.

Lactobacillus paralimentarius from sourdough shows similar phenotypic characteristics to *Lactobacillus alimentarius* from marinated foods but representative strains share less than 20.5% DNA homology.

Besides the concentration of sourdough species in the *L. reuteri* cluster, this subgroup contains again three additional species from sourdough. A polyphasic analysis of isolates from sourdough is given by De Vuyst *et al.* (2002). Settanni *et al.* (2005) describe an *in situ* detection method for 16 sourdough *Lactobacillus* species by multiplex PCR.

A bit apart from these two subgroups we find three species form fermented drinks: *Lactobacillus malefermentans*, *L. collinoides* and *L. paracollinoides*. *L. malefermentans* was isolated from beer, *L. collinoides* was described for a group of bacteria commonly found in fermenting apple juice and strains of *L. paracollinoides* were originally isolated from different brewery environments in Japan. This subcluster is phylogenetically a somewhat volatile group of organisms that shares 16S rRNA sequence relatedness with the large *Lactobacillus buchneri* group, described below.

The *Lactobacillus buchneri* group

A large group of *Lactobacillus* species, mostly linked to food fermentations. *L. buchneri* is one of the oldest species (Henneberg, 1903) and was formerly also named '*Bacillus buchneri*', '*Bacterium buchneri*', '*Ulvina buchneri*' and '*Lactobacterium buchneri*'. The species is phenotypically very similar to *Lactobacillus brevis*, but *L. buchneri* ferments melezitose. *L. buchneri* strains have been isolated form a variety of biological materials but the type strain was isolated from tomato pulp. *Lactobacillus lindneri*, also described in the early days of LAB taxonomy (Henneberg, 1901) is well known as beer spoiler. Back *et al.* revived the species in 1996. Spicher and Schröder (1978) showed that sourdough isolates named '*L. brevis* var. lindneri' were identical to *L. sanfranciscensis* described in 1971 for isolates from San Francisco sourdough. More sourdough-related species in the *L. buchneri* cluster are *Lactobacillus hammesii* (French sour dough), *Lactobacillus spicheri* (from rice sourdough), *Lactobacillus namurensis* (from Belgian sourdough), *Lactobacillus acidifarinae* and *Lactobacillus zymae* (from Greek and Belgian artisanal wheat sourdoughs). The last four species form a nice subcluster within the *L. buchneri* cluster (Fig. 2.2).

Another well-known species in this group is *L. brevis*. The species is quite heterogeneous, as shown by several methods, and is difficult to identify by phenotypic approaches. Farrow and Collins (1988) investigated the genetic interrelationships of some strains from the human oral cavity and saliva which were assigned to *Lactobacillus brevis* by Hayward and Davis (Hayward and Davis, 1956; Hayward, 1957). Based on nucleic acid and biochemical data, the strains could be assigned to *L. brevis*, *L. fermentum*, *L. oris* and other heterofermentative lactobacilli. More recently, a polyphasic study tackled this heterogeneity by investigating representative strains of *L. brevis* and related taxa using partial sequence analysis of the *pheS* housekeeping gene. Species-specific clusters were delineated for all taxa studied except for two *L. brevis* strains isolated from cheese and wheat respectively. Using 16S rRNA gene sequence analysis, fAFLP, DNA–DNA hybridization and G + C content, both strains, sharing 99.9% 16S rRNA gene sequence similarity, were shown to belong to a new species, *Lactobacillus parabrevis* (Vancanneyt *et al.*, 2006b).

Lactobacillus parabuchneri was originally isolated from human saliva, but strains are also available from, for example, malt whisky fermentations. The species now contains two subspecies, *L. parabuchneri* subsp. *parabuchneri* and *L. parabuchneri* subsp. *ferintoshensis*. The latter was created after Vancanneyt *et al.* (2005a), using a polyphasic approach, showed that *Lactobacillus* '*ferintoshensis*' (Simpson *et al.* 2001) was a later heterotypic synonym of *L. parabuchneri*. Moreover, according to Rules 27(3) and 30 of the Taxonomic Code, *Lactobacillus* '*ferintoshensis*' was not validly published because, at the time of publication, the type strain was not deposited in two publicly accessible service collections, located in different countries (De Vos and Trüper, 2000). *Lactobacillus diolivorans* was isolated from maize silage inoculated with *L. buchneri* and was found to degrade 1,2-propanediol, a fermentation product of *L. buchneri*.

Lactobacillus kefiri, a heterofermentative lactobacillus, was isolated from kefir and described in 1983. Later isolates were shown to ferment only L-arabinose and gluconate and were found to be similar to the species *Lactobacillus* '*desidiosus*' (Marshall *et al.*, 1984). DNA–DNA homology studies showed 85–109% homology with *Lactobacillus* '*caucasicus*' NCDO 190. Today, both species *Lactobacillus* '*desidiosus*' and *Lactobacillus* '*caucasicus*' are regarded as *L. kefiri* (Marshall *et al.*, 1984). *Lactobacillus parakefiri*, also isolated from kefir granes, was described by Takizawa *et al.* (1994). In a single, closely related subgroup we can find *Lactobacillus farraginis* and *Lactobacillus parafarraginis*, isolated from a compost of distilled shochu residue in Japan, together with the well-known *Lactobacillus hilgardii* (Douglas and Cruess, 1936). Originally isolated from a Californian wine, *L. hilgardii* strains have now been obtained from a large variety of wines. *L. homohiochii*, already mentioned above, needs further taxonomic work. Its closest neighbour (for the sequences used in Fig. 2.2) is *Lactobacillus fructivorans* (Charlton *et al.*, 1934) from spoiled salad dressing. *L. fructivorans* includes today the former species *Lactobacillus 'trichodes'* (Fornachon *et al.*, 1949) and *Lactobacillus 'heterohiochii'* (Momose *et al.*, 1974), a result of extensive phenotypic and genotypic investigations (Vescovo *et al.*, 1979; Weiss *et al.*, 1983a). It should be noted that *Lactobacillus fructivorans, Lactobacillus homohiochii, L. sanfranciscensis* and *Lactobacillus lindneri* form a separate phylogenetic group at the basis of the *L. buchneri* cluster and could probably be considered as a separate taxonomic unit, deserving the status of a genus.

A single strain takes an intermediate position between the *L. buchneri* cluster and the *L. perolens* cluster: *Lactobacillus kunkeei*, isolated from a commercial Cabernet Sauvignon undergoing sluggish alcoholic fermentation. The strain was a heterofermentative, facultative anaerobe, with a L-Lys-D-ASP murein type (Table 2.2).

The *Lactobacillus perolens* group and *Paralactobacillus*

The species of the *L. perolens* group grow in beers with a pH higher than 4.2 and with low alcohol and hop concentrations. *L. perolens* was proposed in 1999 (Back *et al.*, 1999) for a soft drink spoilage organism. Its closest phylogenetic neighbour is *Lactobacillus harbinensis*, isolated from the traditional fermented vegetables 'Sudan cai' and named after the place 'Harbin' in north-eastern China, where these fermented vegetables are prepared (Miyamoto *et al.*, 2005). *Lactobacillus concavus* and *Pediococcus dextrinicus* form a second phylogenetic branch within the *L. perolens* group. *Lactobacillus concavus* was isolated from the walls of a distilled spirit fermenting cellar in the Hebei province in China. DNA–DNA relatedness between the type strain and *Pediococcus dextrinicus* was only 5.4% (Tong and Dong 2005). The taxonomic status of *P. dextrinicus* (Coster and White, 1964), synonym of *Pediococcus cerevisiae* subsp. *dextrinicus* (renamed by Back, 1978a,b) is questionable. Its phylogenetic position apart from the other pediococci, considered part of the *L. casei* group according to Collins *et al.* (1991), makes it unlikely that the genus name of this species can be maintained in the future. The most probable genus name might be *Paralactobacillus* since the last subcluster of the *Lactobacillus perolens* group is composed of the *Paralactobacillus selangorensis* branch. This species was described for an obligatory homofermentative, rod-shaped lactic acid bacterium, isolated from a Malaysian food ingredient called chili bo (Leisner *et al.*, 1997). Given the low similarities of the 16SrRNA gene sequences with all known lactic acid bacteria, including *Lactobacillus,* and following an additional polyphasic study including biochemical and physiological investigations, FAME analysis, determination of the DNA base composition and DNA–DNA hybridization on five strains of the new species, the new genus *Paralactobacillus* was proposed, with the species *Paralactobacillus selangorensis* (belonging to the province of Selangor, Malaysia).

Finally, it should be noted that strains assigned to *P. dextrinicus* show a considerable diversity of their 16S rRNA sequences. Sequence D87679 (sequence used in Fig. 2.2) shares at least 96.8% similarity with AF404726 and AY928532 (results not shown), while similarities of these sequences with sequences AF405370, AF405373 and AY928533 are between 36.2% and 48%. Clearly, some further taxonomic work is necessary.

The *Pediococcus* subgroup

Although this is a chapter on *Lactobacillus*, Fig. 2.2 illustrates that it is impossible to not include some pediococci in the phylogenetic scheme. Pediococci are commonly found on plant materials and in various foods, but also as spoilage agents in alcoholic beverages, such as beer and wine. Members of the genus *Pediococcus*, described by Claussen in 1903, are typically spherical-shaped and most often form characteristic tetrads. They are facultatively anaerobic, chemoorganotrophic, homofermentative or heterofermentative microorganisms that require rich culture media for growth. Phylogenetically most species occupy a separate phylogenetic branch in the former *Lactobacillus casei–Pediococcus* group as defined by Collins *et al.* (1991). The *Pediococcus* branch in Fig. 2.2 comprises the species *Pediococcus acidilactici* and *Pediococcus pentosaceus* (Back and Stackebrandt, 1978; Tanasupawat *et al.*, 1993), which share considerable phenotypic similarity, *Pediococcus ethanolidurans* (Liu et al., 2006) and *Pediococcus cellicola* (Zhang *et al.*, 2005), both isolated from a distilled-spirit-fermenting cellar and sharing close phylogenetic links (Fig. 2.2), *Pediococcus inopinatus* (Back, 1978a) and *Pediococcus damnosus* (Claussen, 1903), both from the brewery environment. The last species was responsible for up to 20% of the beer spoilage in the 1980s (Back *et al.*, 1988). Today the problem is reduced due to improved sanitary conditions in the breweries. The *Pediococcus* branch may further contain species like *Pediococcus stilesii* (Franz *et al.*, 2006) and *Pediococcus parvulus* (Gunther and White, 1961), two *Pediococcus* species isolated from plant material (maize grains and silage) and the more recent species *Pediococcus siamensis* (Tanasupawat *et al.*, 2007).

Two former species of *Pediococcus* have been transferred to other genera. *Pediococcus halophilus* (Mees, 1934) was transferred to *Tetragenococcus halophilus* by Collins *et al.* (1990) and *Pediococcus urinaeequi* (*ex* Mees 1934; Garvie, 1986) was shown to be a member of *Aerococcus* as *Aerococcus urinaeequi* (Felis *et al.*, 2005).

The *Lactobacillus salivarius* group

The *L. salivarius* group is a very heterogeneous group which will be discussed here as a single group, but which is composed of different phylogenetic branches, with various fermentation types. In a renaming strategy for the genus *Lactobacillus*, this group, without any doubt, will be split in different genera.

L. salivarius is a homofermentative *Lactobacillus* isolated from the oral cavity of man. The species contains two subspecies, *L. salivarius* subsp. *salivarius*, fermenting rhamnose, but not esculin nor the glucoside salicin, and *L. salivarius* subsp. *salicinius* which can ferment salicin and esculin, but not rhamnose. Collins *et al.* (1991) situated the species in the former *Lactobacillus casei – Pediococcus* group of the lactobacilli. According to Fig. 2.2 the species is the core of the *L. salivarius* subcluster *sensu stricto*, that also contains *L. aviarius* with its two subspecies *L. aviarius* subsp. *aviarius* and *L. aviarius* subsp. *araffinosus* and which was isolated from the intestines of chickens (Fujisawa *et al.*, 1984), *Lactobacillus hayakitensis*, most recently isolated from the intestinal tract of a thoroughbred race horse, and the two species *Lactobacillus acidipiscis* and *Lactobacillus 'cypricasei'*, which share 99.8% of 16S rRNA gene sequence similarity. *L. acidipiscis* was isolated from fermented fish in Thailand, *Lactobacillus 'cypricasei'* from Halloumi cheese. According to Naser *et al.*, 2006, *L. acidipiscis* is an earlier heterotypic synonym of *Lactobacillus 'cypricasei'* as shown by 99.8–100% homology between the *pheS*, *rpoA* and *atpA* housekeeping genes, highly similar phenotypes and 84–97.5% DNA–DNA binding. *L. 'cypricasei'* is therefore considered a later heterotypic synonym of *L. acidipiscis*.

Lactobacillus algidus was described for 40 strains isolated as part of the predominant psychrophilic flora from vacuum-packaged beef, collected from different meat shops and stored at 2°C for 3 weeks. The strains are homofermentative and have a *meso*-diaminopimelic acid type murein. The phylogenetic analysis showed a somewhat separate position (Fig. 2.2), most closely related to a subcluster around *Lactobacillus mali*. *L. mali* is also a homofermentative species, described in 1970 for lactobacilli of ciders. The species showed 70–95% DNA similarity with the type strain of *Lactobacillus 'yamanashiensis'* (originally described by Nonomura *et al.* in 1965) and was originally considered to be a subspecies of the latter. *L. mali*, however, was included on the Approved Lists of Bacterial Names (Skerman *et al.*, 1980) and not *Lactobacillus 'yamanashiensis'*, which was 'revived' by Nonomura, in 1983, with the two subspecies. According to the International Code of Nomenclature of Bacteria *L. mali* has precedence over *L. 'yamanashiensis'* (Nonomura, 1983), so Kaneuchi *et al.* (1988) consequently formalized the description of *L. mali* as an earlier heterotypic synonym of *L. 'yamanashiensis'.* Its closest neighbour is *Lactobacillus satsumensis*, referring to Satsuma, an old name for the southern part of Kyushu in Japan, where the type strain was isolated. Other strains from the species came from mashes of shochu, a traditional Japanese distilled spirit made from fermented rice, from sweet potato, barley and other starchy materials. A similar story for *Lactobacillus vini*, proposed for six facultatively anaerobic bacteria that were homofermentative for pentoses (ribose and/or L-arabinose) and isolated from fermenting grape musts in France and Spain. Genotypically *L. vini* can be distinguished from the closest relatives by DNA–DNA hybridizations. For a table detailing the phenotypic differences of *L. vini* and its closest phylogenetic neighbours *L. mali*, *Lactobacillus nagelii* and *L. satsumensis*, supplementary material is available on IJSEM Online (http://ijs.sgmjournals.org/cgi/content/full/56/3/513/DC1/1). *Lactobacillus nagelii* is another species found in wine, and was isolated from a partially fermented commercial red wine after a problematic alcoholic fermentation. This rod-shaped facultative anaerobe strain grows well in an atmosphere enriched with CO_2, and forms DL-lactate from glucose but not gas. The last species of the *L. mali* subcluster is *Lactobacillus ghanensis*, which was described for three Gram-positive strains, isolated from fermenting cocoa. They produce DL-lactic acid from glucose without gas formation and the cell walls contain peptidoglycan of the D-*meso*-diaminopimelic acid type.

Lactobacillus saerimneri is also a *meso*-diaminopimelic acid containing species isolated from pig faeces and the species was named after Saerimner, a pig occurring in Nordic mythology. *Lactobacillus ceti* was isolated from the lungs and liver of two beaked whales. *L. saerimneri* and *L. ceti* each form a single-species phylogenetic subbranch in the *L. salivarius* group (Fig. 2.2), which may give rise to two phylogenetic subclusters as more related organisms will be described.

The last subcluster of the *L. salivarius* group contains six species from animal origin. *Lactobacillus murinus* was proposed for a new species of the dominant flora of the intestinal tract of rats and mice. This species settles in the digestive tract of rodents during their first days of life and remains a part of the dominant flora throughout the whole life of the animal. The organism was thought to belong to the subgenus *Thermobacterium*, although it was able to ferment pentoses. By means of its electrophoretic mobility of soluble proteins and lactic dehydrogenase, its G + C content (43.7%), its L-Lys-D-Asp cell wall type, its immunological relationship with other *Lactobacillus* species and its amino acid requirements it was assigned to a new species, *Lactobacillus murinus*. According to Fig. 2.2, the closest phylogenetic neighbour is *Lactobacillus animalis*. *L. animalis* strains have been isolated from the dental plaque and alimentary tracts of animals. Dewhirst *et al.* (1999) showed that the 16S rRNA sequence of a strain ASF 361 (from the defined murine microbiota of the Altered Schaedler Flora) was identical to the 16S rRNA sequences of *L. murinus* and *L. animalis*. Their type strains appear to belong to a single species. Still both species are not declared synonymous, as according to Dewhirst *et al.* it is possible that while *L. animalis* strains isolated from mice may belong to *L. murinus*, strains isolated from other mammalian sources may belong to different species. Therefore, they suggest a thorough examination of *L. murinus* and *L. animalis* strains. The possibility that these species were identical was also suggested by Kandler and Weiss (1986). The problem, however, may be more complex in the mean time, since in 2006 a new species, *Lactobacillus apodemi*, was described for bacteria isolated from faeces of the Japanese large wood mouse, *Apodemus speciosus*. Comparative analysis of the 16S rRNA gene sequence showed 98.9% 16S rRNA gene sequence similarity with the type strains of *L. animalis* and *L. murinus*.

Lactobacillus agilis, originally isolated from municipal sewage (Weiss *et al.*, 1981), was shown by Baele *et al.* (2001) to be the main component of the pigeon crop flora, which has only vaguely been described as being composed mainly of lactobacilli and streptococci (Shetty *et al.*, 1990). Isolates were identified to species level using the tDNA-PCR technique, applied in combination with capillary electrophoresis (Vaneechoutte *et al.*, 1998; Baele *et al.*, 2000). Its phylogenetically closest neighbour, *Lactobacillus equi* (Fig. 2.2), was isolated from faeces of healthy horses, where, if present, it is the predominant intestinal *Lactobacillus* species (Morotomi *et al.*, 2002). Phenotypically the production of the two isomers of lactic acid, the absence of acidification of amygdaline and the absence of acidification of cellobiose allow to differentiate *L. equi* from *L. agilis* and *L. ruminis*. This brings us to the last species of this cluster of animal-related species: *L. ruminis*, isolated from bovine rumen (Sharpe and Dellaglio, 1977). According to Reuter (2001), lactobacilli were the most frequent microorganisms in the ileal samples of 24 hospitalized children (Hirtzmann and Reuter, 1963). An anaerobic lactic acid bacterium from this study was later classified as *L. ruminis*, but as a non-motile (human) variant, whereas strains from the rumen of cattle principally show motility.

The *Lactoabacillus vitulinus–cateformis* cluster and some other lactobacilli

Together with *L. ruminis*, the species *L. vitulinus* was described for strains isolated from bovine rumen. Both species were thought to belong to the subgenus *Thermobacterium* and contain *meso*-diaminopimelic acid in the wall peptidoglycan. As can be deduced from Fig. 2.2, both species occupy a quite separate phylogenetic position. Significant numbers of *L. vitulinus* have indeed been found in the rumen of cattle consuming either lush alfalfa or ladino clover with little change in ruminal pH (Bryant, *et al.*, 1960). *L. vitulinus* was shown to be genotypically heterogeneous (Sharpe and Dellaglio, 1977). Its closest taxonomic neighbour is *Lactobacillus catenaformis*. Together they form a separate branch, which is phylogenetically too fare away to be considered a part of the genus *Lactobacillus* (Fig. 2.2). *L. catenaformis*, with the former lactobacilli '*Lactobacillus minutus*' and '*Lactobacillus rogosae*', were shown already some time before not to belong to the lactobacilli (Stackebrandt and Teuber, 1988; Kandler and Weiss, 1986). The species *L. catenaformis* and *Lactobacillus vitulinus* are not related to the genus *Lactobacillus* (Pot *et al.*, 1994b; Hammes and Vogel, 1995; Hammes and Hertel, 2003) and their closest relatives seem to be *Catenibacterium* (Kageyama and Benno, 2000a) and *Coprobacillus* (Kageyama and Benno, 2000b), more closely related to *Clostridium innocuum*, *Clostridium ramosum*, *Erysipelothrix rhusiopathiae* and, peripherally, the mycoplasms (Stackebrandt and Woese, 1979). '*L. minutus*' was transferred to the genus *Atopobium* as *Atopobium minutum*.

For '*Lactobacillus rogosae*' there is no type strain available in any established culture collection. Since it is therefore impossible to include this species in any scientific study, Felis *et al.* (2004) requested the rejection of the species name, at least until strains under accession process in the ATCC are available and are shown to be in agreement with the phenotype of the species as described by Holdeman and Moore (1974).

Lactobacilli as probiotics

Following the report of an Expert Consultation at a meeting convened by the FAO/WHO in October, 2001 (http://www.mesanders.com/docs/probio_report.pdf), probiotics are "live microorganisms which when administered in adequate amounts confer a health benefit on the host".

Lactic acid bacteria and lactobacilli in particular have been used in food preparations for thousands of years. Where Pasteur (1857) was one of the first microbiologists to describe the effects of the 'lactic bacteria', Ellie Metchnikov suggested for the first time a possible link between fermented milk and longevity (Metchnikov, 1907).

In the meantime it is generally accepted that lactic acid bacteria can have a positive impact on the hosts' health. Effects are either microbiological (production of bacteriocins, short chain fatty acids, competition with pathogens for epithelial receptors, nutrients, etc...), physiological (production of enzymes, vitamins, impact on bile salt metabolism, etc...) or immunological (anti-inflammatory, immunostimulating, inducers of a regulatory profile, ...) and as such can contribute to an improved intestinal barrier and homeostasis. More recently, a lot of research has focused on the interaction of commensal bacteria and the host immune system (Mercenier *et al.*, 2003). Without any doubt this is linked to the increasing amount of probiotic preparations on the market. In order to make a health claim for these products, proper scientific documentation must be provided for each strain in each particular application. Since most immune effects were shown to be strain specific (Foligné *et al.*, 2007), it is important to screen a large number of strains belonging to different species in order to obtain a strain with promising health applications.

Still the identification of a strain on the species level remains important. In the USA the 'Generally Recognized As Safe' (GRAS) status as wel as the Qualified Presumption of Safety (QPS) status in Europe are both based on the species name of the strain.

Probiotic strains have been isolated from (but are not restricted to) quite a large number of species of the genus *Lactobacillus*. Most species occur in the *L. acidophilus* group: *L. acidophilus, L. amylovorus, L. crispatus, L. delbrueckii* (subsp. *bulgaricus*) *L. gallinarum, L. gasseri, L. helveticus, L. johnsonii. L.* (*para*)*casei, L. plantarum, L. reuteri*, and *L. salivarius*, as representatives of the respective phylogenetic groups, all have known probiotic strains. The *Lactobacillus perolens* and *Lactobacillus vitulinus-catenaformis* groups do not contain any know probiotics. In the *Lactobacillus buchneri* group we find species like *Lactobacillus kefiri* and *Lactobacillus parakefiri*, who were isolated from kefir products with known health benefits. Few studies, however, have looked at the probiotic potential of these species alone. *L. fermentum*, as well as many species from e.g. sourdough fermentation may have potential health benefits too. In a sourdough application, however, the end product is heated (baked), a process during which the bacteria are killed. By definition these strains cannot be considered 'probiotic' in the strict sense.

L. rhamnosus (*Lactobacillus casei* group) is also widely used as probiotic organism, although this organism can also be linked to human clinical infections (Vancanneyt *et al.*, 2006a).

The proper identification and characterization of probiotic bacteria is therefore quite important. Many studies have tried to define criteria and propose methods for a quick and reliable identification, either based on culture dependent or culture independent methods (Holzapfel *et al.*, 1996; Temmerman *et al.*, 2003a, 2003b, 2003c, 2004)

In a recent EU research programme PROSAFE (http://ec.europa.eu/research/quality-of-life/ka1/volume1/qlk1-2001-01273.htm), lactic acid bacteria were collected and compared form different origins, including human clinical specimens and probiotic products. In the recommendations (Vankerckhoven *et al.*, 2008) of the workgroup on taxonomy, it is recognized that it is not obvious for companies lacking taxonomic expertise to select and apply the method(s) best suited for obtaining a correct species identity of commercial lactic acid bacteria cultures. The usefulness of (commercial) biochemical systems for their identification is limited due to high intra-specific phenotypic variability in many species and due to the fact that identification databases are often incomplete or contain species entries that are poorly documented. Although these systems may be useful to obtain a tentative identification, results should in any case be confirmed by molecular methods. DNA—DNA reassociation was confirmed as the 'gold standard' for the delineation and description of new species, but was noticed to be extremely impractical in routine identification. The use of pattern- and/or sequence-based molecular methods was therefore suggested as a valuable alternative. The availability of sequence-based databases was considered a positive argument although it was also stressed that sequences of multiple taxonomic reference strains per taxon (certainly including the type strain) should be included in identification databases to represent the expected genomic variation within each taxon.

Given these prerequisites, the 16S rRNA sequencing was currently regarded as the best possible tool for the taxonomic positioning of probiotic cultures, although, as mentioner earlier, public sequence databases such as EMBL and GenBank contain unreliable, poorly documented or incomplete sequence entries. The use of a relevant list of high quality, validated 16S rRNA gene sequences for identification purposes was explicitely stipulated.

As mentioned before, it was also noticed that 16S rDNA sequencing has a limited resolution for the discrimination of very closely related species of lactic acid bacteria, e.g. of the *L. acidophilus* and the *L. casei* group, where many probiotic species share high degrees of sequence similarity. In these cases, 16S rDNA sequencing may need to be complemented by other molecular methods. During the PROSAFE project, AFLP and rep-PCR allowed unequivocal differentiation of *Lactobacillus* and *Bifidobacterium* species (Huys *et al.*, 2006).

As for future developments, it was stated that the use of housekeeping genes as taxonomic markers (Naser *et al.*, 2005) holds great promise in the development of standardised methods for the identification of lactic acid bacteria used in the food industry. Already now the whole-genome of a considerable number of commercial probiotic strains has been sequenced, yielding the perfect information source for identification, not only of the species name, but more importantly of safety and functionality aspects. Genome sequencing becomes more and more common and the cost is going down. As more genomes will become publically available, it will become more efficient to evaluate these sequences for safety and functionality aspects. Still some wetwork will be necessary as the presence of e.g. an *in silico* resistance gene does not imply automatically a functional antibiotic resistance.

On a shorter term actually, several recommendations were made for immediate research in the area of lactic acid bacteria identification:

- validation of sequence-based identification schemes
- exploration of whole-cell analysis (e.g. based on MALDI-TOF MS) for rapid and cost-effective species identification of commercial probiotic cultures
- development and validation of new molecular strategies to characterise and monitor probiotic candidate strains during human intervention studies
- validation of the PROSAFE database as a probiotic reference system for academia and companies.

Finally, it was concluded that if probiotic producers do not have the internal facilities to perform the sequencing under optimal conditions, they should seek help at accredited external labs or labs that have an established internal quality control system. International public culture collections such as BCCM™/LMG Bacteria Collection, Belgium (http:// bccm.belspo.be/db/lmg_search_form.php), Culture Collection University of Goteborg, Sweden (http://www.ccug.se), German Collection of Microorganisms and Cell Cultures, Germany (http://www.dsmz.de), and ATCC®, USA (http://www.atcc.org) offer such identification services. These institutions will also take care of a possible safe deposit or a patent deposit.

References

Abo-Elnaga, I.G., and Kandler, O. (1965a). Zur Taxonomie der Gattung *Lactobacillus* Beijerinck. I Das Subgenus *Streptobacterium* Orla-Jensen. Zentralbl. Bakteriol. Parasitenkd. Infektionskr. Hyg. II Abt. *119*, 1–36.

Abo-Elnaga, I.G., and Kandler, O. (1965b). Zur Taxonomie der Gattung *Lactobacillus* Beijerinck. II Das Subgenus *Betabacterium* Orla-Jensen. Zentralbl. Bakteriol. Parasitenkd. Infektionskr. Hyg. II Abt. *119*, 117–129.

Aguirre, M., and Collins, M.D. (1993). Lactic acid bacteria and human clinical infection. J. Appl. Bacteriol. 75, 95–107.

Andrighetto, C., De Dea, P., Lombardi, A., Neviani, E., Rossetti, L., and Giraffa, G. (1998). Molecular identification and cluster analysis of homofermentative thermophilic lactobacilli isolated from dairy products. Res. Microbiol. *149*, 631–643.

Antunes, A., Rainey, F.A., Nobre, M.F., Schumann, P., Ferreira, A.M., Ramos, A., Santos, H., and Da Costa, M.S. (2002). *Leuconostoc ficulneum* sp. nov., a novel lactic acid bacterium isolated from a ripe fig, and reclassification of *Lactobacillus fructosus* as *Leuconostoc fructosum* comb. nov. Int. J. Syst. Evol. Microbiol. 52, 647–655.

Aslam, Z., Im, W.T., Ten, L.N., Lee, M.J., Kim, K.H., and Lee, S.T. (2006). *Lactobacillus siliginis* sp. nov., isolated from wheat sourdough in South Korea. Int. J. Syst. Evol. Microbiol. 56, 2209–2213.

Axelsson, L. (1998). Lactic acid bacteria: classification and physiology. In Lactic Acid Bacteria. Microbiology and Functional Aspects, 2nd edn, S. Salminen and A. von Wright eds. (Marcel Dekker Inc, New York), pp. 1–72.

Axelsson, L. (2004). Lactic acid bacteria: Classification and physiology. In Lactic Acid Bacteria. Microbiological and Functional Aspects, pp. 1–66. Edited by S. Salminen, A. von Wright and A. Ouwehand. New York: Marcel Dekker, Inc.

Back, W. (1978a). Zur Taxonomie der Gattung *Pediococcus*. Phänotypische und genotypische Abgrenzung der bisher bekannten Arten sowie Beschreibung einer neuen bierschädlichen Art: *Pediococcus inopinatus*. Brauwissenschaft *31*, 237–250, 312–316, 336–343.

Back, W. (1978b). Elevation of *Pediococcus cerevisiae* subsp. *dextrinicus* Coster and White to species status [*Pediococcus dextrinicus* (Coster and White) comb. nov.]. Int. J. Syst. Bacteriol. *28*, 523–527.

Back, W., and Stackebrandt, E. (1978). DNS/DNS-Homologiestudien innerhalb der Gattung *Pediococcus*. Arch. Microbiol. *118*, 79–85.

Back, W., Breu, S., and Weigand, C. (1988). Infektionsursachen im Jahre 1987. Brauwelt *31/32*, 1358–1362.

Back, W., Bohak, I., Ehrmann, M., Ludwig, W., and Schleifer, K.H. (1996). Revival of the species *Lactobacillus lindneri* and the design of a species specific oligonucleotide probe. Syst. Appl. Microbiol. *19*, 322–325.

Back, W., Bohak, I., Ehrmann, M., Ludwig, W., and Schleifer, K.H. (1997). Validation List N° 61. Int. J. Syst. Bacteriol. *47*, 601–602.

Back, W., Bohak, I., Ehrmann, M., Ludwig, W., Pot, B., Kersters, K., and Schleifer, K.H. (1999). *Lactobacillus perolens* sp. nov., a soft drink spoilage bacterium. Syst. Appl. Microbiol. *22*, 354–359.

Back, W., Bohak, I., Ehrmann, M., Ludwig, W., Pot, B., Kersters, K., and Schleifer, K.H. (2000). Validation List N° 72. Int. J. Syst. Evol. Microbiol. *50*, 3–4.

Baele, M., Baele, P., Vaneechoutte, M., Storms, V., Butaye, P., Devriese, L.A., Verschraegen, G., Gillis, M., and Haesebrouck F. (2000) Application of tDNA–PCR for the identification of enterococci. J. Clin. Microbiol. *38*, 4201–4207.

Baele, M., Devriese, L.A., and Haesebrouck, F. (2001). *Lactobacillus agilis* is an important component of the pigeon crop flora. J. Appl. Microbiol. *91*, 488–491.

Baele, M., Vancanneyt, M., Devriese, L.A., Lefebvre, K., Swings, J., and Haesebrouck, F. (2003). *Lactobacillus ingluviei* sp. nov., isolated from the intestinal tract of pigeons. Int. J. Syst. Evol. Microbiol. 53, 133–136.

Beck, R., Weiss, N., and Winter, J. (1988). *Lactobacillus graminis* sp. nov., a new species of facultatively hetero-fermentative lactobacilli surviving at low pH in grass silage. Syst. Appl. Microbiol. *10*, 279–283.

Beck, R., Weiss, N., and Winter, J. (1989). Validation List N° 28. Int. J. Syst. Bacteriol. 39, 93–94.

Beijerinck, M.W. (1901). Anhäufungsversuche mit Ureumbakterien. Ureumspaltung durch Urease und durch Katabolismus. Zentralbl. Bakteriol. Parasitenkd. Infektionskr. Hyg. I.I. Abt. *7*, 33–61.

Bentley, R.W., Leigh, J.A., and Collins, M.D. (1991). Intrageneric structure of *Streptococcus* based on comparative analysis of small-subunit rRNA sequences. Int. J. Syst. Bacteriol. *41*, 487–494.

Bergey, D.H., Breed, R.S., Hammer, B.W., Huntoon, F.M., Murray, E.G.D., and Harrison, F.C. (eds) (1934). Bergey's Manual of Determinative Bacteriology, 4th edn, Williams and Co., Baltimore, pp. 1–664.

Bergey, D.H., Harrison, F.C., Breed, R.S., Hammer, B.W., and Huntoon, F.M. (eds) (1923). Bergey's Manual of Determinative Bacteriology, 1st ed., The Williams and Wilkins Co., Baltimore, pp. 1–442.

Bergey, D.H., Harrison, F.C., Breed, R.S., Hammer, B.W., and Huntoon, F.M. (eds) (1925). Bergey's Manual of Determinative Bacteriology, 2nd edn, The Williams and Wilkins Co., Baltimore, pp. 1–462.

Bhowmik, T., and Marth, E.H. (1990). β-Galactosidase of *Pediococcus* species: induction, purification and partial characterization. Appl. Microbiol. Biotechnol. *33*, 317–323.

Biavati, B. (2001). International Committee on Systematic Bacteriology. Subcommittee on the taxonomy of the *Bifidobacterium*, *Lactobacillus* and related organisms. Minutes of the meetings, 26 and 29 August 1996, Budapest, Hungary. Int. J. Syst. Evol. Microbiol. *51*, 257–258.

Bisset, D.L., and Anderson, R.L. (1974). Lactose and D-galactose metabolism in group N streptococci: presence of enzymes for both the D-galactose-1-phosphate and D-tagatose-6-phosphate pathways. J. Bacteriol. *117*, 318–320.

Björkroth, J., and Holzapfel, W. (2003). Genera *Leuconostoc*, *Oenococcus* and *Weissella*. *In* The Prokaryotes: An Evolving Electronic Resource for the Microbiological Community. Edited by M. Dworkin. New York, http://link.springer-ny.com/link/service/books/10125/: Springer-Verlag. Epub March 28.

Björkroth, J., and Korkeala, H.J. (1997). Use of rRNA gene restriction patterns to evaluate lactic acid bacterium contamination of vacuum-packaged sliced cooked whole-meat product in a meat processing plant. Appl. Environ. Microbiol. *63*, 448–453.

Björkroth, K.J., Geisen, R., Schillinger, U., Weiss, N., De Vos, P., Holzapfel, W.H., Korkeala, H.J., and Vandamme, P. (2000). Characterization of *Leuconostoc gasicomitatum* sp. nov., associated with spoiled raw tomato-marinated broiler meat strips packaged under modified-atmosphere conditions. Appl. Environ. Microbiol. 66, 3764–3772.

Björkroth, K.J., Schillinger, U., Geisen, R., Weiss, N., Hoste, B., Holzapfel, W.H., Korkeala, H.J., and Vandamme, P. (2002). Taxonomic study of *Weissella confusa* and description of *Weissella cibaria* sp. nov., detected in food and clinical samples. Int. J. Syst. Evol. Microbiol. 52, 141–148.

Blaiotta, G., Moschetti, G., Simeoli, E., Andolfi, R., Villani, F., and Coppola, S. (2001). Monitoring lactic acid bacteria strains during 'Cacioricotta' cheese production by restriction endonuclease analysis and pulsed-field gel electrophoresis. J. Dairy Res. *68*, 139–144.

Bohak, I., Back, W., Richter, L., Ehrmann, M., Ludwig, W., and Schleifer, K.H. (1998). *Lactobacillus amylolyticus* sp. nov., isolated from beer malt and beer wort. Syst. Appl. Microbiol. *21*, 360–364.

Bohak, I., Back, W., Richter, L., Ehrmann, M., Ludwig, W., and Schleifer, K.H. (1999). Validation List N° 68. Int. J. Syst. Bacteriol. *49*, 1–3.

Bohak, I., Thelen, K., and Beimfohr, C. (2006). Description of *Lactobacillus backi* sp. nov., an obligate beer spoiling bacterium. *Monatsschrift Brauwiss.*, March/April, 78–82.

Bringel, F., Castion, I. A., Olukoya, D.K., Felis, G.E., Torriani, S., and Dellaglio, F. (2005). *Lactobacillus plantarum* subsp. *argentoratensis* subsp. nov., isolated from vegetable matrices. Int. J. Syst. Evol. Microbiol. *55*, 1629–1634.

Bryant, M.P., Barrentine, B.F., Sykes, J.F., Robinson, I.M., Shawver, C.V., and Williams, L.W. (1960).

Predominant bacteria in the rumen of cattle on bloat-provoking ladino clover pasture. J. Dairy Sci. *43*, 1435–1444.

Brygoo, E.R., and Aladame, N. (1953). Etude d'une espèce nouvelle anaerobic stricte du genre *Eubacterium*: *E. crispatum* n. sp. Ann. Inst. Pasteur (Paris) *84*, 640–651.

Cachat, E., and Priest, F.G. (2005). *Lactobacillus suntoryeus* sp. nov., isolated from malt whisky distilleries. Int. J. Syst. Evol. Microbiol. 55, 31–34.

Cai, Y., Okada, H., Mori, H., Benno, Y., and Nakase, T. (1999). *Lactobacillus paralimentarius* sp. nov., isolated from sourdough. Int. J. Syst. Bacteriol. *49*, 1451–1455.

Canchaya, C., Claesson, M.J., Fitzgerald, G.F., van Sinderen, D., and O'Toole, P.W. (2006). Diversity of the genus *Lactobacillus* revealed by comparative genomics of five species. Microbiology *152*, 3185–3196.

Cantino, P.D., Bryant, H.N., de Queiroz, K., Donoghue, M.J., Eriksson, T., Hillis, D.M., and Lee, M.S.Y. (1999). Species names in phylogenetic nomenclature. Syst. Biol. *48*, 790–807.

Carr, J.G., and Davies, P.A. (1970). Homofermentative lactobacilli of ciders including *Lactobacillus mali* nov. spec. J. Appl. Bacteriol. 33, 768–774.

Carr, J.G., Davies, P.A. (1972). The ecology and classification of strains of *Lactobacillus collinoides* nov. spec.: A bacterium commonly found in fermenting apple juice. J. Appl. Bacteriol. 35, 463–471.

Castillo, I., Requena, T., Fernandez de Palencia, P., Fontecha, J., and Gobbetti, M. (1999). Isolation and characterization of an intracellular esterase from *Lactobacillus casei* subsp. *casei*. IFPL731. J. Appl. Microbiol. *86*, 653–659.

Cato, E.P., Moore, W.E.C, and Johnson, J.L. (1983). Synonymy of strains of '*Lactobacillus acidophilus*' group A2 (Johnson *et al.*, 1980) with the type strain of *Lactobacillus crispatus* (Brygoo and Aladame 1953) Moore and Holdeman 1970. Int. J. Syst. Bacteriol. 33, 426–428.

Charlton, D.B., Nelson, M.E., and Werkman, C.H. (1934). Physiology of *Lactobacillus fructivorans* sp. nov. isolated from spoiled salad dressing. Iowa State J. Sci. 9, 1–11.

Chassy, B.M., and Alpert, C.A. (1989). Molecular characterization of the plasmid-encoded lactose-PTS of *Lactobacillus casei*. FEMS Microbiol. Rev. *63*, 157–166.

Chenoll, E., Macián, M.C., and Aznar, R. (2006). *Lactobacillus rennini* sp. nov., isolated from rennin and associated with cheese spoilage. Int. J. Syst. Evol. Microbiol. 56, 449–452.

Chenoll, E., Macián, M.C., and Aznar, R. (2006). *Lactobacillus tucceti* sp. nov., a new lactic acid bacterium isolated from sausage. Syst. Appl. Microbiol. 29, 389–395.

Chich, J.F., Marchesseau, K., and Gripon, J.C. (1997). Intracellular esterase from *Lactococcus lactis* subsp. *lactis* NCDO 763: Purification and characterization. Int. Dairy J. 7, 169–174.

Cho, J.C., and Tiedje, J.M. (2001). Bacterial Species Determination from DNA–DNA hybridization by using genome fragments and DNA microarrays. Appl. Environ. Bacteriol. *67*, 3677–3682.

Christensen, J.E., Dudley, E.G., Pederson, J.A., and Steele, J. (1999). Peptidases and amino acid catabolism in lactic acid bacteria. Anton. Leeuwen. *76*, 217–246.

Christensen, J.E., Lin, D., Plava, A., and Steele, J.L. (1995). Sequence analysis, distribution and expression of an aminopeptidase N-encoding gene from *Lactobacillus helveticus* CNRZ32. Gene *155*, 89–93.

Claussen, N.H. (1903). Etudes sur les bactéries dites sarcines et sur les maladies qu'elles provoquent dans la bière. *C.R.* Trav. Lab. Carlsberg, 6, 64–83.

Collins, M.D., and Jones, D. (1981). The distribution of isoprenoid quinone structural types in bacteria and their taxonomic implications. Microbiol. Rev. *45*, 316–354.

Collins, M.D., and Wallbanks, S. (1992). Comparative sequence analysis of the 16S rRNA genes of *Lactobacillus minutus*, *Lactobacillus rimae*, and *Streptococcus parvulus*: proposal for the creation of a new genus *Atopobium*. FEMS Microbiol. Lett. *95*, 235–240.

Collins, M.D., and Wallbanks, S. (1993). Validation List N° 44. Int. J. Syst. Bacteriol. *43*, 188–189.

Collins, M.D., Farrow, J.A.E, Phillips, B.A., Ferusu, S., and Jones, D. (1987). Classification of *Lactobacillus divergens*, *Lactobacillus piscicola*, and some catalase-negative, asporogenous, rod-shaped bacteria from poultry in a new genus, *Carnobacterium*. Int. J. Syst. Bacteriol. *37*, 310–316.

Collins, M.D., Ash, C., Farrow, J.A.E, Wallbanks, S., and Williams, A.M. (1989a). 16S Ribosomal ribonucleic acid sequence analyses of lactococci and related taxa. Description of *Vagococcus fluvialis* gen. nov., sp. nov. J. Appl. Bacteriol. *67*, 453–460.

Collins, M., Phillips, B.A., and Zanoni, P. (1989b). Deoxyribonucleic acid homology studies of *Lactobacillus casei*, *Lactobacillus paracasei* sp. nov., subsp. *paracasei* and subsp. *tolerans*, and *Lactobacillus rhamnosus* sp. nov., comb. nov. Int. J. Syst. Bacteriol. 39, 105–108.

Collins, M.D., Williams, A.M., and Wallbanks, S. (1990). The phylogeny of *Aerococcus* and *Pediococcus* as determined by 16S rRNA sequence analysis: description of *Tetragenococcus* gen. nov. FEMS Microbiol. Lett. *70*, 255–262.

Collins, M.D., Rodrigues, U.M., Ash, C., Aguirre, M., Farrow, J.A.E, Martinez-Murcia, A., Phillips, B.A., Williams, A.M., and Wallbanks, S. (1991). Phylogenetic analysis of the genus *Lactobacillus* and related lactic acid bacteria as determined by reverse transcriptase sequencing of 16S rRNA. FEMS Microbiol. Lett. *77*, 5–12.

Collins, M.D., Samelis, J., Metaxopoulos, J., and Wallbanks, S. (1993). Taxonomic studies on some *Leuconostoc*-like organisms from fermented sausages: description of a new genus *Weissella* for the *Leuconostoc paramesenteroides* group of species. J. Appl. Bacteriol. 75, 595–603.

Collins, M.D., Samelis, J., Metaxopoulos, J., and Wallbanks, S. (1994). Validation List N° 49. Int. J. Syst. Bacteriol. *44*, 370–371.

Coppola, R., Succi, M., Sorrentino, E., Iorizzo, M., and Grazia, L. (2003). Survey of lactic acid bacteria during the ripening of Caciocavallo cheese produced in Molise. Lait *83*, 211–222.

Corsetti, A., Settanni, L., Van Sinderen, D., Felis, G.E., Dellaglio, F., and Gobbetti, M. (2005). *Lactobacillus rossii* sp. nov., isolated from wheat sourdough. Int. J. Syst. Evol. Microbiol. *55*, 35–40.

Coster, E., and White, H.R. (1964). Further studies of the genus *Pediococcus*. J. Gen. Microbiol. *37*, 15–31.

Curk, M.C., Hubert, J.C., and Bringel, F. (1996). *Lactobacillus paraplantarum* sp. nov., a new species related to *Lactobacillus plantarum*. Int. J. Syst. Bacteriol. *46*, 595–598.

Czerny, M., and Schieberle, P. (2002). Important aroma compounds in freshly ground wholemeal and white wheat flour-identification and quantitative changes during sourdough fermentation. J. Agric. Food Chem. *50*, 6835–6840.

Dahllof, I., Baillie, H., and Kjelleberg, S. (2000). rpoB-based microbial community analysis avoids limitations inherent in 16S rRNA gene intraspecies heterogeneity. Appl. Environ. Microbiol. *66*, 3376–3380.

Dal Bello, F., Walter, J., Roos, S., Jonsson, H., and Christian, H. (2005). Inducible gene expression in *Lactobacillus reuteri* LTH5531 during type I.I. sourdough fermentation. Appl. Environ. Microbiol. *71*, 5873–5878.

De Angelis, M., Curtin, A.C., McSweeney, P.L., Faccia, M., and Gobbetti, M. (2002). *Lactobacillus reuteri* D.S.M 20016: purification and characterization of a cystathionine gamma-lyase and use as adjunct starter in cheesemaking. J. Dairy Res. *69*, 255–267.

De Figueroa, R.M., Cerutti de Guglielmone, G., Betino de Cárdenas, I.L., and Oliver, G. (1998). Flavour compound production and citrate metabolism in *Lactobacillus rhamnosus* ATCC. 7469. Milchwissenschaft 53, 617–619.

De Parasis, J., and Roth, D.A. (1990). Nucleic acid probes for identification of phytobacteria: identification of genus-specific 16S rRNA sequences. Phytopathology *80*, 618–621.

De Quieroz, K., and Gauthier, J. (1992). Phylogenetic taxonomy. Annu. Rev. Ecol. Syst. 23, 449–480.

De Urraza, P.J., Gomez-Zavaglia, A., Lozano, M.E., Romanowski, V., and De Antoni, G.L. (2000). DNA fingerprinting of thermophilic lactic acid bacteria using repetitive sequence-based polymerase chain reaction. J. Dairy Res. *67*, 381–392.

De Vos, P., and Trüper, H.G. (2000) Judicial Commission of the International Committee on Systematic Bacteriology. I.Xth International (IUMS) Congress of Bacteriology and Applied Microbiology. Minutes of the meetings, 14, 15 and 18 August 1999, Sydney, Australia. Int. J. Syst. Evol. Microbiol. 50, 2239–2244.

De Vuyst, L., Schrijvers, V., Paramithiotis, S., Hoste, B., Vancanneyt, M., Swings, J., Kalantzopoulos, G., Tsakalidou, E., and Messens, W. (2002). The biodiversity of lactic acid bacteria in greek traditional wheat sourdoughs is reflected in both composition and metabolite formation. Appl. Environ. Microbiol. 68, 6059–6069.

De Vuyst, L., Vancanneyt, M. (2007). Epub 2006 Sep 11. Biodiversity and identification of sourdough lactic acid bacteria. Food Microbiol. *24*, 120–127.

De Vuyst, L., Schrijversi, V., Paramithiotis, S., Hoste, V., Vancanneyt, M., Swings, J., Kalantzopoulos, G., Tsakalidou, E., and Messens, W. (2003). The biodiversity of lactic acid bacteria in Greek traditional wheat sourdoughs is reflected in both composition and metabolite formation. Appl. Environ. Microbiol. *68*, 6059–6069.

Dellaglio, F., Bottazzi, V., and Trovatelli, L.D. (1973). Deoxyribonucleic acid homology and base composition in some thermophilic lactobacilli. J. Gen. Microbiol. *74*, 289–297.

Dellaglio, F., Bottazzi, V., and Vescovo, M. (1975). Deoxyribonucleic acid homology among *Lactobacillus* species of the subgenus Streptobacterium Orla-Jensen. Int. J. Syst. Bacteriol. *25*, 160–172.

Dellaglio, F., Dicks, L.M.T, Du Toit, M., and Torriani, S. (1991). Designation of ATCC 334 in place of ATCC 393 (NCDO 161) as the neotype strain of *Lactobacillus casei* subsp. *casei* and rejection of the name *Lactobacillus paracasei* (Collins *et al.*, 1989) Request for an opinion. Int. J. Syst. Bacteriol. *41*, 340–342.

Dellaglio, F., Felis, G.E., Castioni, A., Torriani, S., and Germond, J. (2005). *Lactobacillus delbrueckii* subsp. *indicus* subsp. nov., isolated from Indian dairy products. Int. J. Syst. Evol. Microbiol. 55, 401–404.

Dellaglio, F., Torriani, S., and Felis, G.E., (2004a). Reclassification of *Lactobacillus cellobiosus* Rogosa *et al.*, 1953 as a later synonym of *Lactobacillus fermentum* Beijerinck 1901. Int. J. Syst. Evol. Microbiol. 54, 809–812.

Dellaglio, F., Felis, G.E., and Germond, J.E. (2004b). Should names reflect the evolution of bacterial species? Int. J. Syst. Evol. Microbiol. 54, 279–281.

Dellaglio, F., and Felis, G.E. (2005).The taxonomy of *Lactobacillus* and *Bifidobacterium*. *In*: Probiotics and prebiotics: scientific aspects. Gerald, W. Tannock (ed). Caister Academic Press.

Dellaglio, F., Felis, G.E., and Torriani, S. (2002). The status of the species *Lactobacillus casei* (Orla-Jensen 1916) Hansen and Lessel 1971 and *Lactobacillus paracasei* Collins *et al.*, 1989. Request for an opinion. Int. J. Syst. Evol. Microbiol. 52, 285–287.

Dellaglio, F., and Torriani, S. (1986). DNA:DNA homology, physiological characteristics and distribution of lactic acid bacteria isolated from maize silage. J. Appl. Bacteriol. *60*, 83–92.

Dellaglio, F., Trovatelli, L.D., and Sara, P.G. (1981). Deoxyribonucleic acid homology among representative strains of the genus *Pediococcus*. Zentralbl. Bakteriol. Parasitenkd. Infektionskr. Hyg. I Abt. Orig. C 2, 140–150.

Dellaglio, F., Vancanneyt, M., Endo, A., Vandamme, P., Felis, G.E., Castioni, A., Fujimoto, J., Watanabe, K., Okada, S. (2006). *Lactobacillus durianis* Leisner *et al.*, 2002 is a later heterotypic synonym of *Lactobacillus vaccinostercus* Kozaki and Okada 1983. Int. J. Syst. Evol. Microbiol. 56, 1721–1724.

Dent, V.E., Williams, R.A.D. (1982). *Lactobacillus animalis* sp. nov., a new species of *Lactobacillus* from the alimentary canal of animals. Zentralbl. Bakteriol. Parasitenkd. Infektionskr. Hyg. I Abt. Orig., C 3, 377–387.

Dent, V.E., Williams, R.A.D. (1983). Validation List N° 10. Int. J. Syst. Bacteriol. 33, 438–440.

Dewhirst, F.E., Chien, C.C., Paster, B.J., Ericson, R.L., Orcutt, R.P., Schauer, D.B., and Fox, J.G. (1999). Phylogeny of the Defined Murine Microbiota: Altered Schaedler Flora. Appl. Environ. Microbiol. 65, 3287–3292.

Dewhirst, F.E., Paster, B.J., Tzellas, N., Coleman, B., Downes, J., Spratt, D.A., and Wade, W.G. (2001) Characterization of novel human oral isolates and cloned 16S rDNA sequences that fall in the family *Coriobacteriaceae*: description of *Olsenella* gen. nov., reclassification of *Lactobacillus uli* as *Olsenella uli* comb. nov. and description of *Olsenella profusa* sp. nov. Int. J. Syst. Evol. Microbiol. *51*, 1797–1804.

Di Cagno, R., De Angelis, M., Lavermicocca, P., De Vincenzi, M., Giovannini, C., Faccia, M., and Gobbetti, M. (2002). Proteolysis by sourdough lactic acid bacteria: effects on wheat flour protein fractions and gliadin peptides involved in human cereal intolerance. Appl. Environ. Microbiol. *68*, 623–633.

Diaz-Muniz, I., Banavara, D.S., Budinich, M.F., Rankin, S.A., Dudley, E.G., and Steele, J.L. (2006). *Lactobacillus casei* metabolic potential to utilize citrate as an energy source in ripening cheese: a bioinformatics approach. J. Appl. Microbiol. *101*, 872–882.

Dicks, L.M.T. (1995). Relatedness of *Leuconostoc* species of the *Leuconostoc* sensu stricto line of descent, *Leuconotoc oenos* and *Weissella paramesenteroides* revealed by numerical analysis of total soluble cell protein pattern. System Appl. Microbiol. 18, 99–102.

Dicks, L.M.T, Silvester, M., Lawson, P.A., and Collins, M.D. (2000). *Lactobacillus fornicalis* sp. nov., isolated from the posterior fornix of the human vagina. Int. J. Syst. Evol. Microbiol. *50*, 1253–1258.

Dicks, L.M.T, Van Vuuren, H.J.J. (1987). Relatedness of heterofermentative *Lactobacillus* species revealed by numerical analysis of total soluble cell protein patterns. Int. J. Syst. Bacteriol. *37*, 437–440.

Dicks, L.M.T., Du Plessis, E.M., Dellaglio, F., and Lauer, E. (1996). Reclassification of *Lactobacillus casei* subsp. *casei* ATCC 393 and *Lactobacillus rhamnosus* ATCC 15820 as *Lactobacillus zeae* nom. rev., designation of ATCC 334 as the neotype of *Lactobacillus casei* subsp. *casei*, and rejection of the name *Lactobacillus paracasei*. Int. J. Syst. Bacteriol. *46*, 337–340.

Douglas, H.C., and Cruess, W.V. (1936). A *Lactobacillus* from California wine: *Lactobacillus hilgardii*. Food Res. *1*, 113–119.

Edwards, C.G., Collins, M.D., Lawson, P.A., and Rodriguez, A.V. (2000). *Lactobacillus nagelii* sp. nov., an organism isolated from a partially fermented wine. Int. J. Syst. Evol. Microbiol. *50*, 699–702.

Edwards, C.G., Haag, K.M., Collins, M.D., Hutson, R.A., Huang, Y.C. (1998a). *Lactobacillus kunkeei* sp. nov.: a spoilage organism associated with grape juice fermentations. J. Appl. Microbiol. *84*, 698–702.

Edwards, C.G., Haag, K.M., Collins, M.D., Hutson, R.A., Huang, Y.C. (1998b). Validation List N° 67. Int. J. Syst. Bacteriol. *48*, 1083–1084.

Eggerth, A.H. (1935). The Gram-positive non-spore-bearing anaerobic bacilli of human feces. J. Bact., 30, 277–290.

Ehrmann, M.A., Müller, M.R.A, and Vogel, R.F. (2003). Molecular analysis of sourdough reveals *Lactobacillus mindensis* sp. nov. Int. J. Syst. Evol. Microbiol. 53, 7–13.

Ehrmann, M.A., and Vogel, R.F. (2005) Taxonomic note ‚*Lactobacillus pastorianus*‘ (Van Laer, 1892) a former synonym for *Lactobacillus paracollinoides*. Syst. Appl. Microbiol. *28*(1), 54–56.

Ehrmann, M.A., Brandt, M., Stolz, P., Vogel, R.F. and Korakli, M. (*2007*). *Lactobacillus secaliphilus* sp. nov., isolated from type I.I. sourdough fermentation Int J. Syst. Evol. Microbiol. 57 (2007), 745–750; D.O.I 10.1099/ijs.0.64700 0.

El Soda, M., Law, J., Tsakalidou, E., and Kalantzopoulos, G. (1995). Lipolytic activity of cheese related microorganisms and its impact on cheese flavour. In Food Flavors: Generation, Analysis and Process Influence. G. Charalambous ed, (Elsevier Science B.V.), pp 1823–1847.

El-Soda, M., El-Wahab, H.A., Ezzat, N., Desmazeaud, M.J., and Ismail, A. (1986). The esterolyticand lipolyticac tivities of the lactobacilli. I.I. Detection of esterase system of *Lactobacillus helveticus*, *Lactobacillus bulgaricus*, *Lactobacillus lactis* and *Lactobacillus acidophilus*. Lait 66, 431–443.

Embley, T.M., Faquir, N., Bossart, W., and Collins, M.D. (1989). *Lactobacillus vaginalis* sp. nov. from the human vagina. Int. J. Syst. Bacteriol. 39, 368–370.

Endo, A., and Okada, S. (2005). *Lactobacillus satsumensis* sp. nov., isolated from mashes of shochu, a traditional Japanese distilled spirit made from fermented rice and other starchy materials. Int. J. Syst. Evol. Microbiol. 55, 83–85.

Endo, A., and Okada, S. (2007a). *Lactobacillus composti* sp. nov., a lactic acid bacterium isolated from a compost of distilled shochu residue. Int. J. Syst. Evol. Microbiol. 57, 870–872.

Endo, A., and Okada, S. (2007b). *Lactobacillus farraginis* sp. nov. and *Lactobacillus parafarraginis* sp. nov., heterofermentative lactobacilli isolated from a compost of distilled shochu residue. Int. J. Syst. Evol. Microbiol. 57, 708–712.

Enright, M.C., and Spratt, B.G. (1998). A multilocus sequence typing scheme for *Streptococcus pneumoniae*: identification of clones associated with serious invasive disease. Microbiol. *144*, 3049–3060.

Enright, M.C., Spratt, B.G., Kalia, A., Cross, J.H., and Bessen, D.E. (2001). Multilocus sequence typing of *Streptococcus pyogenes* and the relationships between emm type and clone. Infect. Immun. *69*, 2416–2427.

Entani, E., Masai, H., Suzuki, K-I. (1986). *Lactobacillus acetotolerans*, a new species from fermented vineagar broth. Int. J. Syst. Bacteriol. 36, 544–549.

Falsen, E., Pascual, C., Sjödén, B., Ohlén, M., and Collins, M.D. (1999). Phenotypic and phylogenetic characterization of a novel *Lactobacillus* species from human sources: description of *Lactobacillus iners* sp. nov. Int. J. Syst. Bacteriol. *49*, 217–221.

Farrow, J.A.E, and Collins, M.D. (1988). *Lactobacillus oris* sp. nov. from the human oral cavity. Int. J. Syst. Bacteriol. *38*, 116–118.

Farrow, J.A.E, Phillips, B.A., and Collins, M.D. (1988). Nucleic acid studies on some heterofermentative lactobacilli: description of *Lactobacillus malefermentans* sp. nov. and *Lactobacillus parabuchneri* sp. nov. FEMS Microbiol. Lett. 55, 163–168.

Farrow, J.A.E, Phillips, B.A., and Collins, M.D. (1989). Validation List N° 30. Int. J. Syst. Bacteriol. *39*, 371.

Felis, G.E., Torriani, S., Dellaglio, F. (2004). The status of the species *Lactobacillus rogosae* Holdeman and Moore 1974. Request for an Opinion. Int. J. Syst. Evol. Microbiol. *54*, 1903–1904.

Felis, G., and Dellaglio, F. (2007). Taxonomy of Lactobacilli and Bifidobacteria. Curr. Issues Intestinal Microbiol. *8*, 44–61.

Felis, G.E., Torriani, S., Dellaglio, F. (2005). Reclassification of *Pediococcus urinaeequi* (*ex* Mees 1934) Garvie 1988 as *Aerococcus urinaeequi* comb. nov. Int. J. Syst. Evol. Microbiol. 55, 1325–1327.

Felis, G.E., Vancanneyt, M., Snauwaert, C., Swings, J., Torriani, S., Castioni, A., Dellaglio, F. (2006). Reclassification of *Lactobacillus thermotolerans* Niamsup *et al.*, 2003 as a later synonym of *Lactobacillus ingluviei* Baele *et al.*, 2003. Int. J. Syst. Evol. Microbiol. *56*, 793–795.

Felis, G.E., Dellaglio, F., Mizzi, L., and Torriani, S. (2001). Comparative sequence analysis of a *recA* gene fragment brings new evidence for a change in the taxonomy of the *Lactobacillus casei*-group. Int. J. Syst. Evol. Microbiol. *51*, 2113–2117.

Foligne, B., Nutten, S., Grangette, C., Dennin, V., Goudercourt, D., Poiret, S., Dewulf, J., Brassard, D., Mercenier, A., and Pot, B. (2007). Correlation between *in vitro* and *in vivo* immunomodulatory properties of lactic acid bacteria. World J. Gastroenterol. *13*, 236–43.

Fornachon, J.C., Douglas, H.C., and Vaughn, R.H. (1949). *Lactobacillus trichodes* nov. spec., a bacterium causing spoilage in appetizer and dessert wines. Hilgardia, *19*, 119–132.

Foucaud, C., Kunji, E.R.S., Hahting, A., Richard, J., Konings, W.N., Desmazeaud, M., and Poolman, B. (1995). Specificity of peptide transport system in *Lactococcus lactis*: evidence for a third system which transports hydrophobic di- and tripeptides. J. Bacteriol. *177*, 4652–4657.

Fox, G.E., Wisotzkey, J.D., Jurtshuk, P. Jr. (1992). How close is close: 16S rRNA sequence identity may not be sufficient to guarantee species identity. Int. J. Syst. Bacteriol. *42*, 166–170.

Fox, P.F., and Wallace, J.M. (1997). Formation of flavour compounds in cheese. Adv. Appl. Microbiol. *45*, 17–85.

Fox, P.F., Law, J., McSweeney, P.L.H., and Wallace, J. (1993). Biochemistry of cheese ripening. In Cheese: Chemistry, physics and microbiology (2nd ed.), Vol. 1, Fox, P.F. ed., (Chapman & Hall, London), pp. 389–438.

Franz, C.M.A.P., Vancanneyt, M., Vandemeulebroecke, K., De Wachter, M., Cleenwerck, I., Hoste, B., Schillinger, U., Holzapfel, W.H., and Swings, J. (2006). *Pediococcus stilesii* sp. nov., isolated from maize grains. Int. J. Syst. Evol. Microbiol. *56*, 329–333.

Fujisawa, T., Adachi, S., Toba, T., Arihara, K., and Mitsuoka, T. (1988). *Lactobacillus kefiranofaciens* sp. nov. isolated from kefir grains. Int. J. Syst. Bacteriol. *38*, 12–14.

Fujisawa, T., Benno, Y., Yaeshima, T., and Mitsuoka, T. (1992). Taxonomic study of the *Lactobacillus acidophilus* group, with recognition of *Lactobacillus gallinarum* sp. nov. and *Lactobacillus johnsonii* sp. nov. and synonymy of *Lactobacillus acidophilus* group A3 (Johnson *et al.*, 1980) with the type strain of *Lactobacillus amylovorus* (Nakamura 1981). Int. J. Syst. Bacteriol. *42*, 487–491.

Fujisawa, T., Shirasaka, S., Watabe, J., and Mitsuoka, T. (1984). *Lactobacillus aviarius* sp. nov.: a new species isolated from the intestines of chickens. Syst. Appl. Microbiol. 5, 414–420.

Fujisawa, T., Shirasaka, S., Watabe, J., and Mitsuoka, T. (1985). Validation List N° 17. Int. J. Syst. Bacteriol. *35*, 223–225.

Fujisawa, T., Shirasaka, S., Watabe, J., and Mitsuoka, T. (1986). Validation List N° 20. Int. J. Syst. Bacteriol. *36*, 354–356.

Fujisawa, T., Itoh, K., Benno, Y., and Mitsuoka, T. (1990). *Lactobacillus intestinalis* (ex Hemme 1974) sp. nov., nom. rev., isolated from the intestines of mice and rats. Int. J. Syst. Bacteriol. *40*, 302–304.

Gänzle, M.G., Vermeulen, N., and Vogel, R.F. (2007). Carbohydrate, peptide and lipid metabolism of lactic acid bacteria in sourdough. Food Microbiol. *24*, 128–138.

Gallo, G., De Angelis, M., McSweeney, P.L.H., Corbo, M.R., and Gobbetti, M. (2005). Partial purification and characterization of an X-prolyl dipeptidyl aminopeptidase from *Lactobacillus sanfranciscensis* C.B.1. Food Chem. 9, 535–544.

Gancheva, A., Pot, B., Vanhonacker, K., Hoste, B., and Kersters, K. (1999). A polyphasic approach towards the identification of strains belonging to *Lactobacillus acidophilus* and related species. Syst. Appl. Microbiol. *22*, 573–585.

Garrity, G.M., Bell, J.A., and Lilburn, T.G. (2004). (D.O.I:http://dx.doi.org/10.1007/bergeysoutline200405) Taxonomic Outline of the Prokaryotes. Bergey's Manual of Systematic Bacteriology, 2nd edn, Release 5.0, Springer-Verlag, New York.

Garvie, E.I. (1986). Genus *Pediococcus* Clausen 1903, 68A.L. *In* Bergey's Manual of Systematic Bacteriology, Vol. 2. ed. P.H.A. Sneath, N.S. Mair, M.E. Sharpe, and J.G. Holt. The Williams and Wilkins Co., Baltimore. pp. 1075–1079.

Garvie, E.I. (1988). Validation List N° 25. Int. J. Syst. Bacteriol. *38*, 220–222.

Gasser F. (1970). Electrophoretic characterization of lactic dehydrogenases in the genus *Lactobacillus*. J. Gen. Microbiol. 62, 223–239.

Gasser, F., and Janvier, M. (1980). Deoxyribonucleic acid homologies of *Lactobacillus jensenii*, *Lactobacillus leichmannii*, and *Lactobacillus acidophilus*. Int. J. Syst. Bacteriol. *30*, 28–30.

Gasser, F., Mandel, M., and Rogosa, M. (1970). *Lactobacillus jensenii* sp. nov., a new representative of the subgenus *Thermobacterium*. J. Gen. Microbiol. 62, 219–222.

Gevers, D., Huys, G., and Swings, J. (2001). Applicability of rep-P.C.R fingerprinting for identification of *Lactobacillus species*. FEMS Microbiol. Lett. 205(1), 31–36.

Gevers, D., Cohan, F.M., Lawrence, J.G., Spratt, B.G., Coenye, T., Feil, E.J., Stackebrandt, E., Van de Peer, Y., Vandamme, P., Thompson, F.L., and Swings, J. (2005). Opinion: Re-evaluating prokaryotic species. Nat. Rev. Microbiol. 3, 733–739.

Gilbert, C., Atlan, D., Blanc, B., Portailer, R., Germond, J.E., Lapierre, L., and Mollet, B. (1996). A new cell surface proteinase: sequencing and analysis of the prtB gene from *Lactobacillus delbruekii* subsp. bulgaricus. J. Bacteriol. *178*, 3059–3065.

Gilbert, C., Blanc, B., Frot-Coutez, J., Portalier, R., and Atlan, D. (1997). Comparison of cell surface proteinase activities within the *Lactobacillus* genus. J. Dairy Res. *64*, 561–571.

Giraffa, G., De Vecchi, P., and Rossetti, L. (1998). Note: identification of *Lactobacillus delbrueckii* subspecies *bulgaricus* and subspecies *lactis* dairy isolates by amplified rDNA restriction analysis. J. Appl. Microbiol. 85, 918–924.

Gobbetti, M. (1998). The sourdough microflora: interactions of lactic acid bacteria and yeasts. Trends Food. Sci. Technol. *9*, 267–274.

Gobbetti, M., and Corsetti, A. (1996). Co-metabolism of citrate and maltose by *Lactobacillus brevis* subsp. *linderi* C.B.1 citrate-negative strain: effect on growth, end-products and sourdough fermentation. Z. Lebensm. Unters. Forsch. *203*, 82–87.

Gobbetti, M., Corsetti, A., and Rossi, J., (1994). The sourdough microflora. Interactions between lactic acid bacteria and yeasts: metabolism of carbohydrates. Appl. Microbiol. Biotechnol. *41*, 456–460.

Gobbetti, M., Corsetti, A., and Rossi, J. (1995). Maltose-fructose co-fermentation by *Lactobacillus brevis* subsp. *lindneri* C.B.1 fructose-negative strain. Appl. Microbiol. Biotechnol. *42*, 939–944.

Gobbetti, M., Fox, P.G., Smacchi, E., Stepaniak, L., and Damiani, P. (1996b). Purification and characterization of a lipase from *Lactobacillus plantarum* 2739. J. Food Biochem. *20*, 227–246.

Gobbetti, M., Lavermicocca, P., Minervini, F., De Angelis, M., and Corsetti, A. (2000). Arabinose fermentation by *Lactobacillus plantarum* in sourdough with added pentosans and a-L-arabinofuranosidase: a tool to increase the production of actetic acid. J. Appl. Microbiol. *88*, 317–324.

Gobbetti, M., Smacchi, E., and Corsetti, A. (1996a). The proteolytic system of *Lactobacillus sanfrancisco* C.B.1: purification and characterization of a proteinase, a dipeptidase, and an aminopeptidase. Appl. Environ. Microbiol. *62*, 3220–3226.

Gobbetti, M., Smacchi, E., and Corsetti, A. (1997). Purification and characterization of a cell surface-associated esterase from *Lactobacillus fermentum* D.T.41. Int. Dairy J. *7*, 13–21.

Groot, M.N.N., and de Bont, J.A.M. (1998). Conversion of phenylalanine to benzaldehyde initiated by an aminotransferase in *Lactobacillus plantarum*. Appl. Environ. Microbiol. *64*, 3009–3013.

Gummalla, S., and Broadbent, J.R. (1996). Indole production by *Lactobacillus* spp. in cheese: A possible role for tryptophanase. J. Dairy Sci. *79*, 101–109.

Gummalla, S., and Broadbent, J.R. (1999). Tryptophan catabolism by *Lactobacillus casei* and *Lactobacillus helveticus* cheese flavour adjuncts. J. Dairy Sci. *82*, 2070–2077.

Gunther, H.L., and White, H.R. (1961). The cultural and physiological characters of the pediococci. J. Gen. Microbiol. *26*, 185–197.

Guopeng Z' and Holley, R.A. (1999). Development and PFGE monitoring of dominance among spoilage lactic acid bacteria from cured meats. Food Microbiol. *16*, 633–644.

Guschin, D.Y., Mobarry, B.K., Proudnikov, D., Stahl, D.A., Rittmann, B.E., and Mirzabekov, A.D. (1997). Oligonucleotide microchips as genosensors for determinative and environmental studies in microbiology. Appl. Environ. Microbiol. *63*, 2397–2402.

Hammes, W.P., and Hertel, C. (2003). The Genera *Lactobacillus* and *Carnobacterium*. *In* The Prokaryotes: An Evolving Electronic Resource for the Microbiological Community. Edited by M. Dworkin. New York. http://link.springer-ny.com/link/service/books/10125/: Springer-Verlag. Epub December 15th, 2003.

Hammes, W.P. and Vogel, R.F. (1995). The genus *Lactobacillus*. In: The Genera of Lactic Acid Bacteria, vol. 2, pp. 19–54. Edited by B.J.B. Wood and W.H. Holzapfel. Blackie Academic and Professional (U.K.).

Hammes, W.P., Weiss, N., and Holzapfel, W.P. (1991). The genera *Lactobacillus* and *Carnobacterium*. *In* The Prokaryotes. A Handbook on the Biology of Bacteria: Ecophysiology, Isolation Identification, Applications. ed. A. Balows, H.G. Trüper, M. Dworkin, W. Harder and K.H. Schleifer. Springer, New York. pp. 1535–1594.

Hammes, W.P., and Gänzle, M. (1998). Sourdough breads and related products. In Microbiology of Fermented Foods. Vol 1, B.J.B. Woods ed., (Blackie Academic and Professional, London, United Kingdom), pp. 199–216.

Hammes, W.P., Korakli, M., Schwarz, E., and Wolf, G. (1999). Production of mannitol by *Lactobacillus sanfranciscensis*. 17th I.C.C Conference 'Cereals across the Continents' Valencia, Spain.

Hammes, W.P., Stolz, P., and Gänzle, M. (1996). Metabolism of lactobacilli in traditional sourdoughs. Adv. Food. Sci. *18*, 176–184.

Hansen, P.A. (1968). Type strains of *Lactobacillus* species. A report by the taxonomic subcommittee on lactobacilli and closely related organisms. American Type Culture Collection, Rockville, Maryland, USA, p. 76.

Hansen, P.A., and Lessel, E.F. (1971). *Lactobacillus casei* (Orla-Jensen) comb. nov. Int. J. Syst. Bacteriol. *21*, 69–71.

Hansen, P.A., and Mocquot, G. (1970). *Lactobacillus acidophilus* (Moro) comb. nov. Int. J. Syst. Bacteriol. *20*, 325–327.

Hansen, A, and Schieberle, P. (2005). Generation of aroma compounds during sourdough fermentation: applied and fundamental aspects. Trends Food Sci. Technol. *16*, 85–94.

Hauduroy, P., Ehringer, G., Urbain, G., Guillot, G., and Magrou, J. (eds) (1937). Dictionnaire des bactéries pathogènes, Masson et Cie, Paris, pp. 1–597.

Hayward, A.C. 1957. A comparison of *Lactobacillus* species from human saliva with those from other natural sources. Br. Dent. J. *102*, 450–451.

Hayward, A.C., and G.H.G. Davis. 1956. The isolation and classification of *Lactobacillus* strains from Italian saliva samples. Br. Dent. J. *101*, 43–46.

Hemme, D. (1974). Taxonomie des lactobacilles homofermentaires du tube digestif du rat. Description de deux nouvelles espèces *Lactobacillus intestinalis*, *Lactobacillus murini*. Thèse, Univ. Paris VII.

Hemme, D., Raibaud, P., Ducluzeau, R., Galpin, J.V., Sicard, P., Van Heijenoort, J. (1980). *Lactobacillus murinus* n. sp., une nouvelle espèce de la flore dominante autochtone du tube digestif du rat et de la souris. Ann. Microbiol. 131A, 297–308.

Hemme, D., Raibaud, P., Ducluzeau, R., Galpin, J.V., Sicard, P., Van Heijenoort, J. (1982). Validation List N° 9. Int. J. Syst. Bacteriol. *32*, 384–385.

Henneberg, W. (1901). Zur Kenntnis der Milchsäurebakterien der Brennereimaische der Milch, des Bieres, (*Bacillus Delbrücki* (Leichman), *Bacillus Delbrücki* var. a n. sp., *Pediococcus lactis acidi* (Lindner), *Bacillus lactis acidi* (Leichmann), *Bacterium lactis acidi* (Leichmann), *Saccharobacillus pastorianus* (v. Laer), *Saccharobacillus pastorianus* var. a n. sp., *Saccharobacillus pastorianus* var. berolinensis n. sp., *Bacillus lindneri* n. sp. Wschr. Brau. No. 30 Jahrg., 18, 381–384.

Henneberg, W. (1903). Zur Kenntnis der Milchsäurebakterien der Brennereimaische der Milch, des Bieres, der Presshefe, der Melasse, des Sauerkohls, der sauren Gurken und des Sauerteiges, sowie einige Bemerkungen über die Milchsäurebakterien des menschlichen Magens Z. Spiritusind. 26: Nr. 22–31; see: Zentralbl. Bakteriol. Parasitenkd. Infektionskr. Hyg. I.I. Abt., 11, 163.

Hickey, M.W., Hillier, A.J., and Jago, G.R. (1986). Trransport and metabolism of lactose, glucose and galactose in heterofermentative lactobacilli. Appl. Environ. Microbiol. *51*, 825–831.

Hirtzmann, M., and Reuter, G. (1963). Klinische Erfahrungen mit einer neuen, automatisch gesteuerten Kapsel zur Gewinnung von Darminhalt und bakteriologische Untersuchung des Inhaltes höherer Darmabschnitte. Med. Klinik *58*, 1408–1412.

Hiu, S.F., Holt, R.A., Sriranganathan, N., Seidler, R.J., and Fryer, J.L. (1984). *Lactobacillus piscicola*, a new species from salmonid fish. Int. J. Syst. Bacteriol. *34*, 393–400.

Holck, A., and Nes, N. (1992). Cloning, sequencing and expresssionof the gene encoding the cell-envelope associated proteinase from *Lactobacillus paracasei* subsp. *paracasei* NCDO151. J. Gen. Microbiol. *138*, 1353–1364.

Holdeman, L.V., Moore, W.E.C. (1974). New genus, *Coprococcus*, twelve new species, and emended descriptions of four previously described species of bacteria from human feces. Int. J. Syst. Bacteriol. *24*, 260–277.

Holley, R.A., Lamoureux, M., and Dussault, F. (1996). Identification of lactic spoilage bacteria from vacuum-packed cooked luncheon meat and induction of repairable injury by mild thermal stress. Lebensmittel-Wissenschaft und-Technologie 29 (1–2), 114–122.

Holzapfel, H.W., Schillinger, U., Du Toit, M., Dicks, L.M. (1996). Systematics of probiotic lactic acid bacteria with reference to modern phenotypic and genomic methods. Deutsche Veterinärmedizische Gesellschaft. Symposium. Proceedings of the Probiotics in Man and Animal. p. 11.

Holzapfel, W.H., and Gerber, E.S. (1983). *Lactobacillus divergens* sp. nov., a new heterofermentative *Lactobacillus* species producing L(+)-lactate. Syst. Appl. Microbiol. 4, 522–534.

Holzapfel, W.H., and Gerber, E.S. (1984). Validation List N° 14. Int. J. Syst. Bacteriol. *34*, 270–271.

Holzapfel, W.H., Kandler, O. (1969). Zur Taxonomie der Gattung *Lactobacillus* Beijerinck. V.I. *Lactobacillus coprophilus* subsp. *confusus* nov. subsp., eine neue Unterart der Untergattung *Betabacterium*. Zentralbl. Bakteriol. Parasitenkd. Infektionskr. Hyg. I.I. Abt., 123, 657–666.

Holzapfel, W.H., Van Wyk, E.P. (1982). *Lactobacillus kandleri* sp. nov., a new species of the subgenus *Betabacterium*, with glycine in the peptidoglycan. Zentralbl. Bakteriol. Parasitenkd. Infektionskr. Hyg. Abt. 1 Orig., C3, 495–502.

Holzapfel, W.H., Van Wyk, E.P. (1983). Validation List N° 10. Int. J. Syst. Bacteriol. *33*, 438–440.

Homan, W.L., Tribe, D., Poznanski, S., Li, M., Hogg, G., Spalburg, E., van Embden, J.D.A., Willems, R.J.L (2002). Multilocus sequence typing scheme for *Enterococcus faecium*. J. Clin. Microbiol. *40*, 1963–1971.

Hugenholtz, J. (1993). Citrate metabolism in lactic acid bacteria. FEMS. Microbiol. Rev. *12*, 165–178.

Hugenholtz, J., Perdon, L., and Abee, T. (1993). Growth and energy generation by *Lactococcus lactis* subsp. *lactis* biovar *diacetylactis* during citrate metabolism. Appl. Environ. Microbiol. 59, 4216–422.

Hutkins, R.W., and Morris, H.A. (1987). Carbohydrate metabolism in *Streptococcus thermophilus*: a review. J. Food Prot. *50*, 876–884.

Huys, G., Vancanneyt, M., D'Haene, K., Vankerckhoven, V., Goossens, H., Swings, J. (2006). Accuracy of species identity of commercial bacterial cultures intended for probiotic or nutritional use. Res. Microbiol. *157*, 803–810.

Jessen, B. (1995). Starter cultures for meat fermentions. In Fermented Meats. G. Campbell-Platt, P.E. Cook, eds. (Blackie Academic and Professional, London), pp. 130–159.

Jin-Woo, B., Sung-Keun, R., Ja Ryeong, P., Won-Hyong, C., Young-Do, N., Insun, L., Hongik, K., and Yong-Ha, P. (2005). Development and evaluation of genome-probing microarrays for monitoring lactic acid bacteria. Appl. Environ. Microbiol. *71*, 8825–8835.

Johnson, J.L., Phelps, C.F., Cummins, C.S., London, J., and Gasser, F. (1980). Taxonomy of the *Lactobacillus acidophilus* group. Int. J. Syst. Bacteriol. *30*, 53–68.

Jones, N., Bohnsack, J.F., Takahashi, S., Oliver, K.A., Chan, M.S., Kunst, F., Glaser, P., Rusniok, C., Crook, D.W., Harding, R.M., Bisharat, N., Spratt, B.G. other authors (2003). Multilocus Sequence Typing System for Group B *Streptococcus*. J. Clin. Microbiol. *41*, 2530–2536.

Kagermeier-Callaway, A.S., and Lauer, E. (1995). *Lactobacillus sake* Katagiri, Kitahara, and Fukami 1934 is the senior synonym for *Lactobacillus bavaricus* Stetter and Stetter 1980. Int. J. Syst. Bacteriol. *45*, 398–399.

Kageyama, A. and Benno, Y. (2000a). *Catenibacterium mitsuokai* gen. nov., sp. nov., a gram-positive anaerobic bacterium isolated from human faeces. Int. J. Syst. Evol. Microbiol. *50*, 1595–1599.

Kageyama, A. and Benno, Y. (2000b). *Coprobacillus catenaformis* gen. nov., sp. nov., a new genus and species isolated from human feces. Microbiol. Immunol. *44*, 23–28.

Kandler, O. (1983). Carbohydrate metabolism in lactic acid bacteria. Anton. Leeuwen. *49*, 209–224.

Kandler, O., and Kunath, P. (1983a). *Lactobacillus kefir* sp. nov., a component of the microflora of kefir. Syst. Appl. Microbiol. 4, 286–294.

Kandler, O., and Kunath, P. (1983b). Validation List N° 11. Int. J. Syst. Bacteriol. 33, 672–674.

Kandler, O., Schillinger, U., and Weiss, N. (1983a). *Lactobacillus halotolerans* sp. nov., nom. rev. and *Lactobacillus minor* sp. nov., nom. rev. Syst. Appl. Microbiol. *4*, 280–285.

Kandler, O., Schillinger, U., and Weiss, N. (1983b). *Lactobacillus bifermentans* sp. nov., nom. rev., an organism forming C.O.2 and H2 from lactic acid. Syst. Appl. Microbiol. *4*, 408–412.

Kandler, O., Schillinger, U., and Weiss, N. (1983c). Validation List N° 12. Int. J. Syst. Bacteriol. 33, 896–897.

Kandler, O., Stetter, K.O., and Köhl, R. (1980). *Lactobacillus reuteri* sp. nov., a new species of heterofermentative lactobacilli. Zentralbl. Bakteriol. *1*, 264–269.

Kandler, O., Stetter, K.O., and Köhl, R. (1982). Validation List N° 8. Int. J. Syst. Bacteriol. 32, 266–268.

Kandler, O., and Weiss, N. (1986). Genus *Lactobacillus* Beijerinck 1901, 212A.L., p. 1209–1234. *In* P.H. A. Sneath, N.S. Mair, N.E. Sharpe, and J.G. Holt (ed, Bergey's manual of systematic bacteriology, vol. 2. Williams and Wilkins, Baltimore, Md.

Kaneuchi, C., Seki, M., and Komagata, K. (1988). Taxonomic study of *Lactobacillus mali* Carr and Davis 1970 and related strains: validation of *Lactobacillus mali* Carr and Davis 1970 over *Lactobacillus yamanashiensis* Nonomura 1983. Int. J. Syst. Bacteriol. 38, 269–272.

Katagiri, H., Kitahara, K., and Fukami, K. (1934). The characteristics of the lactic acid bacteria isolated from moto, yeast mashes for sake manufacture. I.V. Classification of the lactic acid bacteria. Bull. Agri. Chem. Soc. Jap. *10*, 156–157.

Kato, Y., Sakala, R.M., Hayashidani, H., Kiuchi, A., Kaneuchi, C., and Ogawa, M. (2000). *Lactobacillus algidus* sp. nov., a psychrophilic lactic acid bacterium isolated from vacuum-packaged refrigerated beef. Int. J. Syst. Evol. Microbiol. *50*, 1143–1149.

Kawamura, Y., Itoh, Y., Mishima, N., Ohkusu, K., Kasai, H., and Ezaki, T. (2005). High genetic similarity of *Streptococcus agalactiae* and *Streptococcus difficilis*: *S. difficilis* Eldar *et al.*, 1995 is a later synonym of *S. agalactiae* Lehmann and Neumann 1896 (Approved Lists 1980). Int. J. Syst. Evol. Microbiol. 55, 961–965.

Khalid, N.M., and Marth, E. H. (1990). Lactobacillitheir enzymes and role in ripening and spoilage of cheese: A review. J. Dairy Sci. *73*, 2669–2684.

Kim, B., Lee, J., Jang, J., Kim, J., and Han, H. (2003). *Leuconostoc inhae* sp. nov., a lactic acid bacterium isolated from kimchi. Int. J. Syst. Evol. Microbiol. 53, 1123–1126.

Kimoto, H., Nomura, M., and Suzuki, I. (1999). Growth energetics of *Lactococcus lactis* subsp. *lactis* biovar *diacetylactis* in cometabolism of citrate and glucose. Int. Dairy J. *9*, 857–863.

King, S.J., Leigh, J.A., Heath, P.J., Luque, I., Tarradas, C., Dowson, C.G., Whatmore, A.M. (2002). Development of a multilocus sequence typing scheme for the pig pathogen *Streptococcus suis*: identification of virulent clones and potential capsular serotype exchange. J. Clin. Microbiol. *40*, 3671–3680.

Kitahara, K. (1938). Studies in lactic acid-forming bacteria in milk and milk products. J. Agric. Chem. Soc. Japan, *14*, 1449–1465.

Kitahara, K., Kaneko, T., and Goto, O. (1957a). Taxonomic studies on the hiochi-bacteria, specific saprophytes of saké. I. Isolation and grouping of bacterial strains. J. Gen. Appl. Microbiol. 3, 102–110.

Kitahara, K., Kaneko, T., and Goto, O. (1957b). Taxonomic studies on the hiochi-bacteria, specific saprophytes of sake. I.I. Identification and classification of hiochi-bacteria. J. Gen. Appl. Microbiol. 3, 111–120.

Klein, G. (2001). International Committee on Systematic Bacteriology. Subcommittee on the taxonomy of *Bifidobacterium*, *Lactobacillus* and related organisms. Minutes of the meetings, 22 and 23 September 1999, Veldhoven, The Netherlands. Int. J. Syst. Evol. Microbiol. *51*, 259–261.

Klein, G., Dicks, L.M.T, Pack, A., Hack, B., Zimmermann, K., Dellaglio, F., and Reuter, G. (1996). Emended descriptions of *Lactobacillus sake* (Katagiri, Kitahara, and Fukami) and *Lactobacillus curvatus* (Abo-Elnaga and Kandler): numerical classification revealed by protein fingerprinting and identification based on biochemical patterns and DNA–DNA hybridizations. Int. J. Syst. Bacteriol. *46*, 367–376.

Klein, G., Pack, A., Bonaparte, C., and Reuter, G. (1998). Taxonomy and physiology of probiotic lactic acid bacteria. Int. J. Food Microbiol. *41*, 103–125.

Kleynmans, U., Heinzl, H., Hammes, W.P. (1989a). *Lactobacillus suebicus* sp. nov., an obligately heterofermentative *Lactobacillus* species isolated from fruit mashes. Syst. Appl. Microbiol. *11*, 267–271.

Kline, L., and Sugihara, T.F. (1971). Microorganisms of the San Francisco sour dough bread process. I.I.

Isolation and characterization of undescribed bacterial species responsible for the souring activity. Appl. Microbiol. *21*, 459–465.

Kodama R. (1956). Studies on the nutrition of lactic acid bacteria. Part I. *Lactobacillus fructosus* nov. sp., a new species of lactic acid bacteria. J. Agr. Chem. Soc. Jpn., *30*, 705–708.

Konings, W.N., Poolman, P., and Driessen, A.J.M. (1989). Bioenergetics and solute transport in lactococci. Crit. Rev. Microbiol. *16*, 419–476.

Konstantinov, S.R., Poznanski, E., Fuentes, S., Akkermans, A.D.L, Smidt, H., De Vos, W.M. (2006). *Lactobacillus sobrius* sp. nov., abundant in the intestine of weaning piglets. Int. J. Sys. Evol. Microbiol. 5629–5632.

Koort, J., Vandamme, P., Schillinger, U., Holzapfel, W., and Björkroth, J. (2004). *Lactobacillus curvatus* subsp. *melibiosus* is a later synonym of *Lactobacillus sakei* subsp. *carnosus*. Int. J. Syst. Evol. Microbiol. *54*, 1621–1626.

Koort, J., Murros, A., Coenye, T., Eerola, S., Vandamme, P., Sukura, A., and Björkroth, J. (2005a). *Lactobacillus oligofermentans* sp. nov., associated with spoilage of modified-atmosphere-packaged poultry products. Appl. Environ. Microbiol. *71*, 4400–4406.

Koort, J., Murros, A., Coenye, T., Eerola, S., Vandamme, P., Sukura, A., and Björkroth, J. (2005b). Validation List N° 106. Int. J. Syst. Evol. Microbiol. *55*, 2235–2238.

Kostinek, M., Pukall, R., Rooney, A.P., Schillinger, U., Hertel, C., Holzapfel, W.H., and Franz, C.M. (2005). *Lactobacillus arizonensis* is a later heterotypic synonym of *Lactobacillus plantarum*. Int. J. Syst. Evol. Microbiol. *55*, 2485–2489.

Kozaki, M., and Okada, S. (1983). Validation List N° 10. Int. J. Syst. Bacteriol. 1983, *33*, 438–440.

Kröckel, L., Schillinger, U., Franz, C.M.A.P., Bantleon, A., and Ludwig, W. (2003). *Lactobacillus versmoldensis* sp. nov., isolated from raw fermented sausage. Int. J. Syst. Evol. Microbiol. *53*, 513–517.

Kröckel, L. (1995). Bacterial fermentation of meats. In Fermented Meats. G. Campbell-Platt, and P.E. Cook, eds. (Blackie Academic and Professional, London), pp. 69–109.

Krooneman, J., Faber, F., Alderkamp, A.C., Oude Elferink, S.J.H.W., Driehuis, F., Cleenwerck, I., Swings, J., Gottschal, J.C., and Vancanneyt, M. (2002). *Lactobacillus diolivorans* sp. nov., a 1,2-propanediol-degrading bacterium isolated from aerobically stable maize silage. Int. J. Syst. Evol. Microbiol. *52*, 639–646.

Kunji, E.R.S., Mierau, I., Hagting, A., Poolman, P., and Konings, W. (1996). The proteolytic system of lactic acid bacteria. Anton. Leeuwen. *70*, 187–221.

Kuznetsov, V.D. (1959). A new species of lactic acid bacteria. *Mikrobiologiya*, 28, 248–351.

Laignelet, B., and Dumas, C. (1984). Oxydation des lipides et distribution des lipides oxydés au cours du pétrissage de la pate de farine de blétendre. Lebensm. Wiss. Technol. *17*, 226–230.

Larrouture, C., Ardaillon, V., Pépin, M., and Montel, M.C. (2000). Ability of meat starter cultures to catabolize leucine and evaluation of the degradation products by using an HPLC method. Food Microbiol. *17*, 563–570.

Lauer, E., Helming, C., and Kandler, O. (1980). Heterogeneity of the species *Lactobacillus acidophilus* (Moro) Hansen and Moquot as revealed by biochemical characteristics and DNA/DNA hybridization. Zentralbl. Bakteriol. Parasitenkd. Infektionskr. Hyg. I Abt. Orig., *Reihe* C., 1, 150–168.

Lauer, E., and Kandler, O. (1980). *Lactobacillus gasseri* sp. nov., a new species of the subgenus *Thermobacterium*. Zentralbl. Bakteriol. Parasitenkd. Infektionskr. Hyg. I Abt. Orig., C 1, 75–78.

Law, B.A.M. and Kolstadt, J. (1983). Proteolytic systems in lactic acid bacteria. Anton. Leeuwen. *49*, 225–245.

Lawson, P.A., Falsen, E., Cotta, M.A. and Whitehead, T.R. (2007). *Vagococcus elongatus* sp. nov., isolated from a swine-manure storage pit. Int. J. Syst. Evol. Microbiol. *57*, 751–754.

Lawson, P.A., Falsen, E., Truberg-Jensen, K., *Aerococcus sanguicola* sp. nov., isolated from a human clinical source. Int. J. Syst. Evol. Microbiol. *51*, 475–479.

Lee, S.Y., and Lee, B.H. (1990). Esterolyticand lipolyticac tivities of *Lactobacillus casei* subsp. *casei* LLG. J. Food Sci. 55, 119–126.

Leichmann, G. (1896). Über die freiwillige Säurung der Milch. Zentralbl. Bakteriol. Parasitenkd. Infektionskr. Hyg. I.I. Abt., *2*, 777–780.

Leisner, J.J., Rusul, G., Wee, B.W., Boo, H.C., and Muhammad, K. (1997). Microbiology of chili bo, a popular Malaysian food ingredient. J. Food Protein *60*, 1235–1240.

Leisner, J.J., Vancanneyt, M., Goris, J., Christensen, H., and Rusul, G. (2000). Description of *Paralactobacillus selangorensis* gen. nov., sp. nov., a new lactic acid bacterium isolated from chili bo, a Malaysian food ingredient. Int. J. Syst. Evol. Microbiol. *50*, 19–24.

Leisner, J.J., Vancanneyt, M., Lefebvre, K., Vandemeulebroecke, K., Hoste, B., Vilalta, N.E., Rusul, G., and Swings, J. (2002). *Lactobacillus durianis* sp. nov., isolated from an acid-fermented condiment (tempoyak) in Malaysia. Int. J. Syst. Evol. Microbiol. *52*, 927–931.

Li, Y., Raftis, E., Canchaya, C., Fitzgerald, G.F., van Sinderen, D., O'toole, P.W. (2006). Polyphasic analysis indicates that *Lactobacillus salivarius* subsp. *salivarius* and *Lactobacillus salivarius* subsp. *salicinius* do not merit separate subspecies status. Int. J. Syst. Evol. Microbiol. *56*, 2397–403.

Lick, S., Brockmann, E., and Heller, K.J. (2000). Identification of *Lactobacillus delbrueckii* and subspecies by hybridization probes and P.C.R. Syst. Appl. Microbiol. *23*, 251–259.

Liu, B., and Dong, X. (2002). *Lactobacillus pantheris* sp. nov., isolated from faeces of a jaguar. Int. J. Syst. Evol. Microbiol. *52*1745–1748.

Liu, L., Zhang, B., Tong, H., and Dong, X. (2006). *Pediococcus ethanolidurans* sp. nov., isolated from the walls of a distilled-spirit-fermenting cellar. Int. J. Syst. Evol. Microbiol. *56*, 2405–2408.

London, J., Chase, N.M., and Kline, K. (1975). Aldolase of lactic acid bacteria: immunological relationships among aldolases of streptococci and gram-positive nonsporeforming anaerobes. Int. J. Syst. Bacteriol. *25*, 114–123.

London, J., Chase, N.M. (1976). Aldolases of the lactic acid bacteria. Demonstration of immunological relationship among eight genera of Gram positive bacteria using anti-pediococcal aldolase serum. Arch. Microbiol. 110, 121–128.

London, J., Chase, N.M. (1983). Relationship among lactic acid bacteria demonstrated with glyceraldehyde-3-phophate dehydrogenase as an evolutionry probe. Int. J. Syst. Bacteriol. *33*, 723–737.

London, J., and Kline, K. (1973). Aldolase of lactic acid bacteria: a case history in the use of an enzyme as an evolutionary marker. Bacteriol. Rev., 37, 453–478.

London, J., Meyer, E.Y., Kulczyk, S.R. (1971a). Comparative biochemical and immunological study of malic enzyme from two species of lactic acid bacteria: Evolutionary implications. J. Bacteriol. *106*, 126–137.

London, J., Meyer, E.Y., Kulczyk, S.R. (1971b). Detection of relationships between *Streptococcus faecalis* and *Lactobacillus casei* by immunological studies with two forms of malic enzymes. J. Bact. *108*, 196–201.

Loponen, J., Mikola, M., Katina, K., Sontag-Strohm, F., and Salovaara, H. (2004). Degradation of H.M.W glutenins during wheat sourdough fermentations. Cereal Chem. *81*, 87–93.

Makarova, K., Slesarev, A., Wolf, Y., Sorokin, A., Mirkin, B., Koonin, E., Pavlov, A., Pavlova, N., Karamychev, V., Polouchine, N., Shakhova, V., Grigoriev, I., Lou, Y., Rohksar, D., Lucas, S., Huang, K., Goodstein, D.M., Hawkins, T., Plengvidhya, V., Welker, D., Hughes, J., Goh, Y., Benson, A., Baldwin, K., Lee, J.H., Diaz-Muniz, I., Dosti, B., Smeianov, V., Wechter, W., Barabote, R., Lorca, G., Altermann, E., Barrangou, R., Ganesan B., Xie, Y., Rawsthorne, H., Tamir, D., Parker, C., Breidt, F., Broadbent, J., Hutkins, R., O'Sullivan, D., Steele, J., Unlu, G., Saier, M., Klaenhammer, T., Richardson, P., Kozyavkin, S., Weimer, B., and Mills, D. (2006). Comparative genomics of the lactic acid bacteria. Proc. Natl. Acad. Sci. USA *103*, 15611–15616.

Makarova, K.S., and Koonin, E.V. (2007). Evolutionary genomics of lactic acid bacteria. J. Bacteriol. *189*, 1199–1208.

Marshall, V.M., Cole, W.M., Farrow, J.A., (1984). A note on the heterofermentative *Lactobacillus* isolated from kefir grains. J. Appl. Bacteriol. 56(3), 503–505.

Martinez-Anaya, M.A., Llin, M.L., Macias, M.L., and Collar, C. (1994). Regulation of acetic acid production by homo- and heterofermentative lactobacilli in whole-wheat sourdoughs. Z. Lebensm. Unters. Forsch. *199*, 186–190.

Martinez-Murcia, A.J., Benlloch, S., and Collins, M.D. (1992). Phylogenetic interrelationships of members of the genera *Aeromonas* and *Plesiomonas* as determined by 16S ribosomal DNA sequencing: lack of congruence with results of DNA-DNA hybridizations. Int. J. Syst. Bacteriol. 42, 412–421.

Martinez-Murcia, A.J., and Collins, M.D. (1990). A phylogenetic analysis of the genus *Leuconostoc* based on reverse transcriptase sequencing of 16 S rRNA. FEMS Microbiol. Lett. 58, 73–83.

Marty-Teysset, C., Posthuma, C., Lolkema, J.S., Schmitt, P., Divies, C., and Konings, W.N. (1996). Proton motive force generation by citro-lactic fermentation in *Leuconostoc mesenteroides*. J. Bacteriol. *178*, 2178–2185.

Masco, L., Huys, G., Gevers, D., Verbrugghen, L., and Swings, J. (2003). Identification of *Bifidobacterium* species using rep-P.C.R fingerprinting. Syst Appl Microbiol. *26*(4), 557–63.

McKay, L., Miller, I.I.I, A., Sandine, W.E., and Elliker, P.R. (1970). Mechanisms of lactose utilization by lactic streptococci: enzymatic and genetic analyses. J. Bacteriol. *102*, 804–809.

McSweeney, P.L.H., and Sousa, M.J. (2000). Biochemical pathways for the production of flavour compounds in cheeses during ripening: A review. Lait *80*, 293–324.

Mercenier, A., Pavan, S., Pot, B., (2003). Probiotics as biotherapeutic agents: present knowledge and future prospects. Curr. Pharm. Des. *9*, 175–191.

Mees, R.H. (1934). Onderzoekingen over biersarcina. Thesis Technical University Delft, Holland, pp. 1–110.

Meroth, C.B., Hammes, W.P., and Hertel, C. (2004a). Characterisation of the microbiota of rice sourdoughs and description of *Lactobacillus spicheri* sp. nov. Syst. Appl. Microbiol. *27*, 151–159.

Meroth, C.B., Hammes, W.P. and Hertel, C. (2004b). Validation List N° 97. Int. J. Syst. Evol. Microbiol. *54*, 631–632.

Metchinov, E. (1907). The Prolongation of Life. Heinemann, London.

Mierau, I., Kunji, E.R.S., Venema, G., and Kok, J. (1997). Casein and peptide degradation in lactic acid bacteria. Biotechnol. Genet. Eng. Rev. *14*, 279–301.

Miller, I.I.I. A., Morgan, M.E., and Libbey, L.M. (1974). *Lactobacillus maltaromicus*, a new species producing a malty aroma. Int. J. Syst. Bacteriol. *24*, 346–354.

Mitsuoka, T., and Fujisawa, T. (1987). *Lactobacillus hamsteri*, a new species from the intestine of hamsters. Proc. Jpn. Acad. Ser. B Phys. Biol. Sci. *63*, 269–272.

Mitsuoka, T., and Fujisawa, T. (1988). Validation List N° 25. Int. J. Syst. Bacteriol. *38*, 220–222.

Miyamoto, M., Seto, Y., Hao, D.H., Teshima, T., Sun, Y.B., Kabuki, T., Yao, L.B., and Nakajima, H. (2005). *Lactobacillus harbinensis* sp. nov., consisted of strains isolated from traditional fermented vegetables 'Suan cai' in Harbin, Northeastern China and *Lactobacillus perolens* D.S.M 12745. Syst. Appl. Microbiol. *28*, 688–694.

Miyamoto, M., Seto, Y., Hao, D.H., Teshima, T., Sun, Y.B., Kabuki, T., Yao, L.B., and Nakajima, H. (2006). Validation List N° 107. Int. J. Syst. Evol. Microbiol. *56*, 1–6.

Momose, H., Yamanaka, E., Akiyama, H., and Nosiro, K. (1974). Taxonomic study on hiochi bacteria, with special reference to deoxyribonucleic acid base composition and chemical composition of bacterial cell wall. J. Gen. Appl. Microbiol. 20, 179–185.

Moore, W.E.C, and Holdeman, L.V. (1970). *Propionibacterium, Arachnia, Actinomyces, Lactobacillus* and *Bifidobacterium*. *In*: E.P. Cato, C.S. Cummins, L.V. Holdeman, J.L. Johnson, W.E.C. Moore, R.M. Smibert and L.D.S. Smith (eds): Outline of Clinical

Methods in Anaerobic Bacteriology, 2nd revision, Virginia Polytechnic Institute, Anaerobe Laboratory, Blacksburg, Virginia, pp. 15–22.

Moore, W.E.C, and Holdeman, L.V. (1972). *Lactobacillus*. *In*: L.V. Holdeman and W.E.C. Moore (eds): Anaerobe Laboratory Manual, Virginia Polytechnic Institute, Anaerobe Laboratory, Blacksburg, Virginia, pp. 61–66.

Mora, D., Scarpellini, M., Franzetti, L., Colombo, S. and Galli, A. (2003). Reclassification of *Lactobacillus maltaromicus* (Miller *et al.*, 1974) D.S.M 20342^{T} and D.S.M 20344 and *Carnobacterium piscicola* (Collins *et al.*, 1987) D.S.M 20730^{T} and D.S.M 20722 as *Carnobacterium maltaromaticum* comb. nov. Int. J. Syst. Evol. Microbiol. 53, 675–678.

Mori, K., Yamazaki, K., Ishiyama, T., Katsumata, M., Kobayashi, K., Kawai, Y., Inoue, N., and Shinano, H. (1997). Comparative sequence analyses of the genes coding for 16S rRNA of *Lactobacillus casei*-related taxa. Int. J. Syst. Bacteriol. *47*, 54–57.

Morita, H., Shiratori, C., Murakami, M., Takami, H., Kato, Y., Endo, A., Nakajima, F., Takagi, M., Akita, H., Okada, S., Masaoka, T. (2007). *Lactobacillus namurensis* sp. nov., isolated from intestines of healthy Thoroughbreds. Int. J. Syst. Evol. Microbiol. 57, 2836–2839.

Morlon-Guyot, J., Guyot, J.P., Pot, B., Jacobe De Haut, I., and Raimbault, M. (1998). *Lactobacillus manihotivorans* sp. nov., a new starch-hydrolysing lactic acid bacterium isolated during cassava sour starch fermentation. Int. J. Syst. Bacteriol. *48*, 1101–1109.

Moro, E. (1900). Über den *Bacillus acidophilus* n. sp. *Jahrb. Kinderheilk.*, 52, 38–55.

Morotomi, M., Yuki, N., Kado, Y., Kushiro, A., Shimazaki, T., Watanabe, K., and Yuyama, T. (2002). *Lactobacillus equi* sp. nov., a predominant intestinal *Lactobacillus* species of the horse isolated from faeces of healthy horses. Int. J. Syst. Evol. Microbiol. 52, 211–214.

Moschetti, G., G. Blaiotta, M. Aponte, G. Mauriello, F. Villani, and S. Coppola. 1997. Genotyping of *Lactobacillus delbrueckii* subsp. *bulgaricus* and determination of the number and forms of rrn operons in *Lactobacillus delbrueckii* and its subspecies. Res. Microbiol. 148, 501–510.

Müller, M.R.A, Ehrmann, M.A., and Vogel, R.F. (2000). *Lactobacillus frumenti* sp. nov., a new lactic acid bacterium isolated from rye-bran fermentations with a long fermentation period. Int. J. Syst. Evol. Microbiol. *50*, 2127–2133.

Mukai, T., Arihara, K., Ikeda, A., Nomura, K., Suzuki, F., and Ohori, H. (2003). *Lactobacillus kitasatonis* sp. nov., from chicken intestine. Int. J. Syst. Evol. Microbiol. *53*, 2055–2059.

Nakajima, H., Kunji, E.R.S., Poolman, B., and Konings, W.N. (1998). Amino acid transport in *Lactobacillus helveticus*. FEMS Microbiol. Lett. *158*, 249–253.

Nakamura, L.K. (1981). *Lactobacillus amylovorus*, a new starch-hydrolyzing species from cattle waste-corn fermentations. Int. J. Syst. Bacteriol. *31*, 56–63.

Nakamura, L.K., and Crowell, C.D. (1979). *Lactobacillus amylophilus*, a new starch-hydrolyzing species from swine waste-corn fermentation. Dev. Ind. Microbiol. 20, 532–540.

Nakamura, L.K., and Crowell, C.D. (1981). Validation List N° 6. Int. J. Syst. Bacteriol. *31*, 215–218.

Naser, S.M., Thompson, F., Hoste, B., Gevers, D., Dawyndt, P., Vancanneyt, M., and Swings, J. (2005). Application of multilocus sequence analysis (MLSA) for rapid identification of *Enterococcus* species based on rpoA and pheS genes. Microbiol. 151, 2141–2150.

Naser, S.M., Hagen, K.E., Vancanneyt, M., Cleenwerck, I., Swings, J., Tompkins, T.A. (2006a). *Lactobacillus suntoryeus* Cachat and Priest 2005 is a later synonym of *Lactobacillus helveticus* (Orla-Jensen 1919) Bergey *et al.*, 1925 (Approved Lists 1980). Int. J. Syst. Evol. Microbiol. *56*, 355–360.

Naser, S.M., Vancanneyt, M., Hoste, B., Snauwaert, C., and Swings, J. (2006b) *Lactobacillus cypricasei* Lawson *et al.*, 2001 is a later heterotypic synonym of *Lactobacillus acidipiscis* Tanasupawat *et al.*, 2000. Int. J. Syst. Evol. Microbiol. *56*, 1681–1683.

Naser, S.M., Vancanneyt, M., Hoste, B., Snauwaert, C., Vandemeulebroecke K., and Swings, J. (2006c). Reclassification of *Enterococcus flavescens* Pompei *et al.*, 1992 as a later synonym of *Enterococcus casseliflavus* (*ex* Vaughan *et al.*, 1979) Collins *et al.*, 1984 and *Enterococcus saccharominimus* Vancanneyt *et al.*, 2004 as a later synonym of *Enterococcus italicus* Fortina *et al.*, 2004. Int. J. Syst. Evol. Microbiol. *56*, 413–416.

Naser, S.M., Dawyndt, P., Hoste, B., Gevers, D., Vandemeulebroecke, K., Cleenwerck, I., Vancanneyt, M., and Swings, J. (2007). Identification of lactobacilli by pheS and rpoA gene sequence analyses. Int. J. Syst. Evol. Microbiol. 57, 2777–2789.

Neubauer, H., Glaasker, E., Hammes, W.P., Poolman, B., and Konings, W.N. (1994). Mechanism of maltose uptake and glucose excretion in *Lactobacillus sanfrancisco*. J. Bacteriol. *176*, 3007–3012.

Niamsup, P., Sujaya, I.N., Tanaka, M., Sone, T., Hanada, S., Kamagata, Y., Lumyong, S., Assavanig, A., Asano, K., Tomita, F., and Yokota, A. (2003). *Lactobacillus thermotolerans* sp. nov., a novel thermotolerant species isolated from chicken faeces. Int. J. Syst. Evol. Microbiol. 53, 263–268.

Nielsen, D.S., Schillinger, U., Franz, C.M.A.P., Bresciani, J., Amoa-Awua, W., Holzapfel, W.H., Jakobsen, M. (2007). *Lactobacillus ghanensis* sp. nov., a motile lactic acid bacterium isolated from Ghanain cocoa fermentations. Int. J. Syst. Evol. Microbio. 57, 1468–1472.

Nikolaitchouk, N., Wacher, C., Falsen, E., Andersch, B., Collins, M.D. and Lawson, P.A. (2001). *Lactobacillus coleohominis* sp. nov., isolated from human sources. Int. J. Syst. Evol. Microbiol. *51*, 2081–2085.

Niven, C.F. Jr., and Evans, J.B. (1957). *Lactobacillus viridescens* nov. spec. a heterofermentative species that produces a green discoloration of cured meat pigments. J. Bacteriol. *73*, 758–759.

Nokuthula, F.K., Geornaras, I., von Holy, A., and Hastings, J.W. (2000). Characterization and Determination of origin of lactic acid bacteria from a sorghum-based fermented weaning food by analysis of soluble proteins and amplified fragment length polymorphism fingerprinting. Appl Environ. Microbiol. *66*, 1084–1092.

Nonomura, H. (1983). *Lactobacillus yamanashiensis* subsp. *yamanashiensis* and *Lactobacillus yamanashiensis* subsp. *mali* sp. and subsp. nov., nom. rev. Int. J. Syst. Bacteriol. 33, 406–407.

Nonomura, H., and Ohara, Y. (1965). Die Klassifikation der Äpfelsäure-Milchsäure-Bakterien. *Mitt. Klosterneuburg Ser. A Rebe Wein*, 17A, 449–465.

Okada, S., Suzuki, Y., Kozaki, M., (1979). A new heterofermentative *Lactobacillus* species with *meso*-diaminopimelic acid in peptidoglycan. *Lactobacillus vaccinostercus* Kozaki and Okada sp. nov., J. Gen. Appl. Microbiol. 25, 215–221.

Olsen, I., Johnson, J.L., Moore, L.V.H, Moore, W.E.C. (1991). *Lactobacillus uli* sp. nov. and *Lactobacillus rimae* sp. nov. from the human gingival crevice and emended descriptions of *Lactobacillus minutus* and *Streptococcus parvulus*. Int. J. Syst. Bacteriol. *41*, 261–266.

Orla-Jensen, S. (1916). Maelkeri-Bakteriologi. Schønberske Forlag, Copenhagen.

Orla-Jensen, S. (1919). The Lactic acid bacteria. Copenhagen: Andr. Fred Høst and Son. Mém. Acad. Roy. Sci., Danemark, Sect. Sci., *8* Sér., 5, 81–197.

Orla-Jensen, S., Otte, N.C., Snog-Kjaer (1936). The vitamin and nitrogen requirements of the lactic acid bacteria. Mém. Acad. Roy. Sci., Danemark, Sect. Sci., *9 Sér.*, 6, 1–52.

Orla-Jensen, S. (1942). The lactic acid bacteria. Ed. E. Munksgaard. 2nd ed. Copenhagen, pp. 106–107.

Orla-Jensen, S. (1943). The lactic acid bacteria. Copenhagen: Munksgaard. Ergaenzungsband. Mem. Acad. Roy. Sci. Danemark, Sect. Sci. Biol. 2(3), pp. 1–145.

Orrhage' K., Sjöstedt, S., and Nord, C.E. (2000). Effect of supplements with lactic acid bacteria and oligofructose on the intestinal microflora during administration of cefpodoxime proxetil. J. Antimicrob. Chemoth. *46*, 603–612.

Osawa, R., Fujisawa, T., and Pukall, R. (2006). *Lactobacillus apodemi* sp. nov., a tannase-producing species isolated from wild mouse faeces. Int. J. Syst. Evol. Microbiol. *56*, 1693–1696.

Pasteur, L. (1857). Mémoire sur la fermentation appelée lactique. *In* Memoires de 1a Société des Science, de l'Agriculture et des arts de Lille, séance du 3 août 1857, 2e Série, V., pp. 13–26.

Pedersen, C., and Roos, S. (2004). *Lactobacillus saerimneri* sp. nov., isolated from pig faeces. Int. J. Syst. Evol. Microbiol. *54*, 1365–1368.

Pederson, J.A., Mileski, G.J., Weimer, B.C., and Steele, J.L. (1999). Genetic Characterization of a Cell Envelope-Associated Proteinase from *Lactobacillus helveticus* CNRZ32. J. Bacteriol. *181*, 4592–4597.

Pepe, O., Villani, F., Oliviero, D., Greco, T., and Coppola, S. (2003). Effect of proteolytic starter cultures as leavening agents of pizza dough. Int. J. Food Microbiol. *84*, 319–326.

Pette, J.W., Van Beynum J. (1943). Boekelscheurbacterien. Rijkslandbouwproef-station te hoorn. Versl. Landbouwkd. Onderz. *490*, 315–346.

Posthuma, C.C., Bader, R., Engelamnn, R., Postma, P.W., Hengstenberg, W., and Powels, P.H. (2002). Expression of the xylulose 5-phosphate phosphoketolase gene, *xpk*A, from *Lactobacillus pentosus* M.D.363 is induced by sugars that are fermented via the phosphoketolase pathway and is repressed by glucose mediated by CcpA and the mature phosphoenolpuryvate phopshotransferase system. Appl. Environ. Microbiol. *68*, 831–837.

Pot, B., Devriese, L.A. Ursi, D., Vandamme, P., Haesebrouck, F., Kersters, K. 1996. Phenotypic identification and differentiation of *Lactococcus* strains isolated from animals. Syst. Appl. Microbiol. *19*, 213–222.

Pot, B., Devriese, L.A., Hommez, J., Miry, C., Vandemeulebroecke, K., Kersters, K., and Haesebrouck, F. (1994a). Characterisation and identification of *Vagococcus fluvialis* strains isolated from domestic animals. J. Appl. Bacteriol. *77*, 362–369.

Pot, B., Ludwig, W., Kersters, K., Schleifer K-H. (1994b). Taxonomy of lactic acid bacteria. (Chapter 2, pp. 13–90). *In* Bacteriocins of lactic acid bacteria. Microbiology, Genetics and Applications. L. De Vuyst, and E.J. Vandamme (Eds). Blackie Academic and Professional (An imprint of Chapman and Hall), London.

Pot, B., Hertel, C., Ludwig, W., Descheemaeker, P., Kersters, K., and Schleifer, K.H. (1993). Identification and classification of *Lactobacillus acidophilus*, *Lactobacillus gasseri*, and *Lactobacillus johnsonii* strains by S.D.S-PAGE and rRNA-targeted oligonucleotide probe hybridization. J. Gen. Microbiol. *139*, 513–517.

Pot, B., Vandamme, P., and Kersters, K. (1992). Analysis of electrophoretic whole-organism protein fingerprints. *In* Chemical Methods in Bacterial Systematics. Ed. M. Goodfellow, and A.G. O'Donnell. J. Wiley and sons, Chichester.

Pot, B. (2007). The taxonomy of lactic acid bacteria. p. 1–152; *In*: Bactéries lactiques. De la génétique aux ferments. Corrieu, G., Luquet, F.M. (eds). Collections Sciences et Techniques Agroalimentaires. Lavoisier, France.

Reuter, G. (1970). Laktobazillen und eng verwandte Mikroorganismen in Fleisch und Fleischerzeugnissen. 2. Mitteilung: Die Charakterisierung der isolierten Laktobazillenstämme. Fleischwirtschaft, 50, 954–962.

Reuter, G. (1983a). *Lactobacillus alimentarius* sp. nov., nom. rev. and *Lactobacillus farciminis* sp. nov., nom. rev. Syst. Appl. Microbiol. 4, 277–279.

Reuter, G. (1983b). Validation List N° 11. Int. J. Syst. Bacteriol. 33, 672–674.

Reuter, G. (2001) The *Lactobacillus* and *Bifidobacterium* Microflora of the Human Intestine: Composition and Succession. Curr. Issue Intestin. Microbiol. 2(2), 43–53.

Röcken, W., and Voisey, P.A. (1995). Sourdough fermentation in breadmaking. J. Appl. Bacteriol. Symp. Suppl. *79*, 38S-48S

Rodas, A.M., Chenoll, E., Macián, M.C., Ferrer, S., Pardo I., and Aznar, R. (2006). *Lactobacillus vini* sp. nov., a wine lactic acid bacterium homofermentative for pentoses. Int. J. Syst. Evol. Microbiol. *56*, 513–517.

Rogosa, M., and Hansen, P.A. (1971). Nomenclatural considerations of certain species of *Lactobacillus* Beijerinck. Int. J. Syst. Bacteriol. *21*, 177–186.

Rogosa, M., Wiseman, R.F., Mitchell, J.A., and Disraely, M.N. (1953). Species differentiation of oral lactobacilli from man including descriptions of *Lactobacillus salivarius* nov. spec. and *Lactobacillus cellobiosus* nov. spec. J. of Bacteriol., *65*, 681–699.

Roos, S., Engstrand, L., and Jonsson, H. (2005). *Lactobacillus gastricus* sp. nov., *Lactobacillus antri* sp. nov., *Lactobacillus kalixensis* sp. nov. and *Lactobacillus ultunensis* sp. nov., isolated from human stomach mucosa. Int. J. Syst. Evol. Microbiol. *55*, 77–82.

Roos, S., Karner, F., Axelsson, L., and Jonsson, H. (2000). *Lactobacillus mucosae* sp. nov., a new species with *in vitro* mucus-binding activity isolated from pig intestine. Int. J. Syst. Evol. Microbiol. *50*, 251–258.

Rosselló-Mora, R. (2003). Opinion: the species problem, can we achieve a universal concept? System. Appl. Microbiol. *26*, 323–326.

Rosselló-Mora, R., and Amann, R. (2001). The species concept for prokaryotes. FEMS Microbiol. Rev. 25, 39–67.

Russel, C., and Walker, T.K. (1953). *Lactobacillus malefermentans* n. sp. isolated from beer. J. Gen. Microbiol. 8, 160–162.

Sakala, R.M., Kato, Y., Hayashidani, H., Murakami, M., Kaneuchi, C., and Ogawa, M. (2002). *Lactobacillus fuchuensis* sp. nov., isolated from vacuum-packaged refrigerated beef. Int. J. Syst. Evol. Microbiol. 52, 1151–1154.

Sanders, M.E. (1994) Lactic acid bacteria as promoters of human health. *In:* Functional Foods: Designer Foods, Pharmafoods and Nutraceuticals, (Goldberg, I., ed.), pp. 294–322. Chapman and Hall, London, U.K.

Sarantinopoulos, P., Kalantzopoulos, G., and Tsakalidou, E. (2001). Citrate Metabolism by *Enterococcus faecalis* FAIR-E 229. Appl Environ Microbiol. 67, 5482–5487.

Sarra, P.G., Magri, M., Bottazzi, V., Dellaglio, F. (1980). Genetic heterogeneity among *Lactobacillus acidophilus* strains. Ant. v. Leeuwenhoek J. Microbiol. Serol., 46, 169–176.

Scheirlinck, I., Van der Meulen, R., Van Schoor, A., Cleenwerck, I., Huys, G., Vandamme, P., De Buyst, L., Vancanneyt, M. (2007). *Lactobacillus namurensis* sp. nov., isolated from a traditional Belgian sourdough. Int. J. Syst. Evol. Microbiol. 57, 223–227.

Schleifer, K., Kraus, J., Dvorak, C., Kilpper-Bälz, R., Collins, M.D., and Fischer, W. (1985). Transfer of *Streptococcus lactis* and related streptococci to the genus *Lactococcus* gen. nov. Syst. Appl. Microbiol. 6, 183–195.

Schleifer, K.H., and Kandler, O. (1972). Peptidoglycan types of bacterial cell walls and their taxonomic implications. Bacteriol. Rev., *36*, 407–477.

Schleifer, K.H., and Ludwig, W. (1995). Phylogeny of the genus *Lactobacillus* and related genera. Syst. Appl. Microbiol. *14*, 461–467.

Schleifer, K.H., and Stackebrandt, E. (1983). Molecular systematics of prokaryotes. *Ann.* Rev. Microbiol. 37, 143–187.

Schleifer, K.H., Kraus, J., Dvorak, C., Kilpper-Bälz, R., Collins, M.D., and Fischer, W. (1986). Validation List N° 20. Int. J. Syst. Bacteriol. *36*, 354–356.

Schmitt, P., Mathot, A., and Divies, C. (1989). Fatty acid composition of the *genus Leuconostoc*. Milchwissenschaft, *44*, 556–559.

Settanni, L., van Sinderen, D., Rossi, J., and Corsetti, A. (2005). Rapid Differentiation and In Situ Detection of 16 Sourdough *Lactobacillus* Species by Multiplex PCR Appl. Environ. Microbiol. *71*, 3049–3059.

Sharpe, M.E. (1955). A serological classification of lactobacilli. J. Gen. Microbiol. *12*, 107–122.

Sharpe, M.E. (1970). Cell wall and cell membrane antigens used in the classification of lactobacilli. Int. J. Syst. Bacteriol. *20*, 509–518.

Sharpe, M.E. (1979). Identification of the lactic acid bacteria. In Identification methods for bacteriologists. Society for applied Bacteriology Technical Series No. 14, 2nd ed. Ed. Skinner and Lovelock. Academic Press, London. pp. 233–259.

Sharpe, M.E. (1981). The genus *Lactobacillus*. *In* The Prokaryotes. A Handbook on Habitats, Isolation and Identification of Bacteria. Ed. M.P. Starr, H. Stolp, H.G. Trüper, A. Balows and H.G. Schlegel. Springer, New York. pp. 1653–1659.

Sharpe, M.E., Dellaglio, F. (1977). Desoxyribonucleic acid homology in anaerobic lactobacilli and in possibly related species. Int. J. System. Bact., *27*, 19–21.

Sharpe, M.E., Garvie, E.I., and Tilbury, R. (1972). Some slime-forming heterofermentative species of the genus *Lactobacillus*. *Appl. Microbiol.* 23, 389–397.

Sharpe, M.E., Latham, M.J., Garvie, E.I., Zirngibl, J., and Kandler, O. (1973). Two new species of *Lactobacillus* isolated from the bovine rumen, *Lactobacillus ruminis* sp. nov. and *Lactobacillus vitulinus* sp. nov. J. Gen. Microbiol. 1973, *77*, 37–49.

Shaw, B.G., and Harding, C.D. (1985). Atypical lactobacilli from vacuum-packaged meats: composition by DNA hybridization, cell composition and biochemical tests with a description of *Lactobacillus carnis* sp. nov. Syst. Appl. Microbiol. 6, 291–297.

Shaw, B.G., and Harding, C.D. (1986). Validation List N° 20. Int. J. Syst. Bacteriol. *36*, 354–356.

Shaw, B.G., and Harding, C.D. (1989). *Leuconostoc gelidum* sp. nov. and *Leuconostoc carnosum* sp. nov. from chill-stored meats. Int. J. Syst. Bacteriol. 39, 217–223.

Shetty, S., Sridar, K.R., Shenoy, K.B., and Hedge, H.N. (1990) Observations on bacteria associated with pigeon crop. Folia Microbiologica 35, 240–244.

Simonds, J., Hansen, P.A., and Lakshmanan, S. (1971). Deoxyribonucleic acid hybridization among strains of lactobacilli. J. Bact., 107, 382–384.

Simpson, K.L., Pettersson, B., and Priest, F.G. (2001). Characterization of lactobacilli from Scotch malt whisky distilleries and description of *Lactobacillus ferintoshensis* sp. nov., a new species isolated from malt whisky fermentations. Microbiology, *147*, 1007–1016.

Simpson, K.L., Pettersson, B., and Priest, F.G. (2002). Validation List N° 86. Int. J. Syst. Evol. Microbiol. 52, 1075–1076.

Skerman, V.B.D, McGowan, V., Sneath, P.H.A (ed.) (1980). Approved lists of bacterial names. Int. J. Syst. Bacteriol. *30*, 225–420.

Smit, G., Verheul, A., van Kranenburg, R., Ayad, E., Siezen, R., and Engels, W. (2000). Cheese flavour development by enzymatic conversion of peptides and amino acids. J. Food Res. Int. *33*, 153–160.

Somers, E.B., Johnson, M.E., Wong, A.C.L (2001). Biofilm Formation and Contamination of Cheese by Nonstarter Lactic Acid Bacteria in The Dairy Environment. J. Dairy Sci. *84*, 1926–1936.

Spicher, G., and Schröder, R. (1978). Die Mikroflora des Sauerteiges, V.I. Untersuchungen über die Art der in 'Reinzuchtsauern' anzutreffenden Milchsaürebakterien (Genus *Lactobacillus* Beijerink). Z Lebensm-Unters-Forsch *167*, 342–354.

Stackebrandt, E., Frederiksen, W., Garrity, G.M., Grimont, P.A., Kampfer, P., Maiden, M.C., Nesme, X., Rossello-Mora, R., Swings, J., Truper, H.G., Vauterin, L., Ward, A.C., and Whitman, W.B. (2002). Report of the Ad Hoc Committee for the Re-Evaluation of the species definition in bacteriology. Int. J. Syst. Evol. Microbiol. *52*, 1043–1047.

Stackebrandt, E., and Goebel, B.M. (1994). Taxonomic note: a place for DNA-DNA reassociation and 16S rRNA sequence analysis in the present species definition in bacteriology. Int. J. Syst. Bacteriol. *44*, 846–849.

Stackebrandt, E., and Teuber, M. (1988). Molecular taxonomy and phylogenetic position of lactic acid bacteria. Biochimie *70*, 317–324.

Stackebrandt, E., and Woese, C.R. (1979). A phylogenetic dissection of the family Micrococcaceae. Curr. Microbiol. 2, 317–322.

Stetter, H., and Stetter, K.O. (1980). *Lactobacillus bavaricus* sp. nov., a new species of the subgenus *Streptobacterium*. Zentralbl. Bakteriol. Parasitenkd. Infektionskr. Hyg. Abt. 1 Orig. Reihe C. 1, 70–74.

Stiles, M.E., and Holzapfel, W.H. (1997). Lactic acid bacteria of foods and their current taxonomy. Int. J. Food Microbiol. *36*, 1–29.

Stolz, P., Böcker, G., Hammes, W.P., and Vogel, R.F. (1995a). Utilization of electron acceptors by lactobacilli isolated from sourdough. I. *Lactobacillus sanfranciscensis*. Z. Lebensm. Unters. Forsch. *201*, 91–96.

Stolz, P., Vogel, R.F., and Hammes, W.P. (1995b). Utilization of electron acceptors by lactobacilli isolated from sourdough I. *Lactobacillus pontis*, *Lactobacillus reuteri*, *Lactobacillus amylovorus* and *Lactobacillus fermentum*. Z. Lebensm. Unters. Forsch. *201*, 402–410.

Stolz, P., Böcker, G., Vogel, R.F., and Hammes, W.P. (1993). Utilisation of maltose and glucose by lactobacilli isolated from sourdough. FEMS Microbiol. Lett. *109*, 237–242.

Stolz, P., Hammes, W.P., and Vogel, R.F. (1996). Maltose-phosphorylase and hexokinase activity in lactobacilli from traditionally prepared sourdoughs. Adv. Food Sci. *18*, 1–6.

Stucky, K., Hagting, A., Klein, J.R., Marten, H., Henrich, B., Konings, W.N., and Plapp, R. (1995). Cloning and characterization of *brnQ*: a gene encoding a low affinity branched amino acid carrier of *Lactobacillus delbrueckii* subsp. *lactis*. Mol. Gen. Genet. *249*, 682–690.

Sugihara, F., and Kline, L. (1975). Further studies on a growth medium for *Lactobacillus sanfrancisco*. J. Milk Food Technol. *38*, 667–672.

Suzuki, K., Funahashi, W., Koyanagi, M., and Yamashita, H. (2004). *Lactobacillus paracollinoides* sp. nov., isolated from brewery environments. Int. J. Syst. Evol. Microbiol. *54*, 115–117.

Švec, P., Devriese, L.A., Sedlacek, I., Baele, M., Vancanneyt, M., Haesebrouck, F., Swings, J., and Doskar, J. (2001). *Enterococcus haemoperoxidus* sp. nov. and *Enterococcus moraviensis* sp. nov., isolated from water. Int. J. Syst. Evol. Microbiol. *51*, 1567–1574.

Švec, P., Vancanneyt, M., Devriese, L.A., Naser, S.M., Snauwaert, C., Lefebvre, K., Hoste, B., and Swings, J. (2005a). *Enterococcus aquimarinus* sp. nov., isolated from sea water. Int. J. Syst. Evol. Microbiol. *55*, 2183- 2187.

Švec, P., Vancanneyt, M., Koort, J., Naser, S., Hoste, B., Vihavainen, E., Vandamme, P., Swings, J., and Björkroth, J. (2005b). *Enterococcus devriesei* sp. nov., associated with animal sources. Int. J. System. Evol. Microbiol. 55, 2479–2484.

Swezey, J.L., Nakamura, L.K., Abbott, T.P., and Peterson, R.E. (2000) *Lactobacillus arizonensis* sp. nov., isolated from jojoba meal. Int. J. Syst. Evol. Microbiol. *50*, 1803–1809.

Takizawa, S., Kojima, S., Tamura, S., Fujinaga, S., Benno, Y., and Nakase, T. (1994). *Lactobacillus kefirgranum* sp. nov. and *Lactobacillus parakefir* sp. nov., two new species from kefir grains. Int. J. Syst. Bacteriol. *44*, 435–439.

Tammam, J.D., Williams, A.G., Noble, J., and Lloyd, D. (2000). Amino acid fermentation in non-starter *Lactobacillus* spp. Isolated from Cheddar cheese. Lett. Appl. Microbiol. *30*, 370–374.

Tanasupawat, S., Okada, S., Kozaki, M., and Komagata, K. (1993) Characterization of *Pediococcus pentosaceus* and *Pediococcus acidilactici* strains and replacement of the type strain of *P. acidilactici* with the proposed neotype D.S.M 20284. Request for an opinion. Int. J. Syst. Bacteriol. *43*, 860–863.

Tanasupawat, S., Pakdeeto, A., Thawai, C., Yukphan, P., and Okada, S. (2007). Identification of lactic acid bacteria from fermented tea leaves (miang) in Thailand and proposals of *Lactobacillus thailandensis* sp. nov., *Lactobacillus camelliae* sp. nov., and *Pediococcus siamensis* sp. nov. J. Gen. Appl. Microbiol. *53*, 7–15.

Tanasupawat, S., Shida, O., Okada, S., and Komagata, K. (2000). *Lactobacillus acidipiscis* sp. nov. and *Weissella thailandensis* sp. nov., isolated from fermented fish in Thailand. Int. J. Syst. Evol. Microbiol. *50*, 1479–1485.

Temmerman, R., Masco, L., Vanhoutte, T., Huys, G., and Swings, J. (2003b). Development and validation of a nested-PCR-denaturing gradient gelelectrophoresis method for taxonomic characterization of bifidobacterial communities. Appl. Environ. Microbiol. *69*, 6380–6385.

Temmerman, R., Huys, G., and Swings, J. (2004). Identification of lactic acid bacteria: culture-dependent and culture independent methods. Trends Food Sci. Technol. *15*, 348–359.

Temmerman, R., Scheirlinck, I., Huys, G., Swings, J. (2003a). Culture-independent analysis of probiotic products by denaturing gradient gel electrophoresis. Appl. Environ. Microbiol. *69*, 220–226.

Temmerman, R., Pot, B., Huys, G., Swings, J. (2003c). Identification and antibiotic susceptibility of bacterial isolates from probiotic products. Int. J. Food Microbiol. *81*, 1–10.

Tenover, F.C., Arbeit, R.D., and Goering, R.V. (1997). The Molecular Typing Working Group of the Society for Healthcare Epidemiology of America. How to select and interpret molecular typing methods for epidemiological studies of bacterial infections: a review for healthcare epidemiologists. Infect. Control Hosp. Epidemiol., *18*, 426–39.

Thiele, C., Grassl, S., and Gaenzle, M.G. (2004). Gluten hydrolysis and depolymerization during sourdough fermentation. J. Agric. Food Chem. 52, 1307–1314.

Thomas, T.D., and Pritchard, G.G. (1987). Proteolytic enzymes of dairy starter cultures. FEMS Microbiol. Rev. *46*, 245–268.

Tong, H., and Dong, X. (2005). *Lactobacillus concavus* sp. nov., isolated from the walls of a distilled spirit fermenting cellar in China. Int. J. Syst. Evol. Microbiol. 55, 2199–2202.

Torriani, S., Felis, G.E., and Dellaglio, F. (2001). Differentiation of *Lactobacillus plantarum*, *Lactobacillus pentosus*, and *Lactobacillus paraplantarum* by *recA* Gene Sequence Analysis and Multiplex P.C.R Assay with *recA* Gene-Derived Primers. Appl. Environ. Microbiol. *67*, 3450–3454.

Torriani, S., Van Reenen, C.A., Klein, G., Reuter, G., Dellaglio, F., and Dicks, L.M.T. (1996). *Lactobacillus curvatus* subsp. *curvatus* subsp. nov. and *Lactobacillus curvatus* subsp. *melibiosus* subsp. nov. and *Lactobacillus sake* subsp. *sake* subsp. nov. and *Lactobacillus sake* subsp. *carnosus* subsp. nov., new subspecies of *Lactobacillus curvatus* Abo-Elnaga and Kandler 1965 and *Lactobacillus sake* Katagiri, Kitahara, and Fukami 1934 (Klein *et al.*, 1996, emended descriptions), respectively. Int. J. Syst. Bacteriol. *46*, 1158–1163.

Torriani, S., Zapparoli, G., and Dellaglio, F. (1999). Use of P.C.R-based methods for rapid differentiation of *Lactobacillus delbrueckii* subsp. bulgaricus and *Lactobacillus delbrueckii* subsp. lactis. Appl. Environ. Microbiol. *65*, 4351–4356.

Troili-Petersson, G. (1903). Studien über die Mikroorganismen des schwedischen Güterkäses. Zentralbl. Bakteriol. Parasitenkd. Infektionskr. Hyg. I.I. Abt., 11, 120–143.

Turner, M.S., Woodberry, T., Hafner, L.M., and Giffard, P.M. (1999). The bspA locus of *Lactobacillus fermentum* B.R.11 encodes a L-cystine uptake system. J. Bacteriol. *181*, 2191–2198.

Tynkkynen, S., Buits, G., Kunji, E., Kok, J., Poolman, B., Venema, G., and Haandrikman, A. (1993). Genetic and biochemical characterisation of the oligopeptide transport system of *Lactococcus lactis*. J. Bacteriol. *175*, 7523–7532.

Valcheva, R., Ferchichi, M.F., Korakli, M., Ivanova, I., Gänzle, M.G., Vogel, R.F., Prévost, H., Onno, B., and Dousset, X. (2006). *Lactobacillus nantensis* sp. nov., isolated from French wheat sourdough. Int. J. Syst. Evol. Microbiol. 56, 587–591.

Valcheva, R., Korakli, M., Onno, B., Prévost, H., Ivanova, I., Ehrmann, M.A., Dousset, X., Gänzle, M., and Vogel, R. (2005). *Lactobacillus hammesii* sp. nov., isolated from French sourdough. Int. J. Syst. Evol. Microbiol. 55, 763–767.

van der Hamer, E. (1960). The carbohydrate metabolism of the lactic acid bacteria. Thesis, University of Utrecht, The Netherlands.

van Kranenburg, R., Kleerebezem, M., van Hylckama Vlieg, J., Ursing, B.M., Bockhost, J., Smit, B.A., Ayad, E.H.E., Smit, G., and Siezen, R.J. (2002). Flavour formation from amino acids by lactic acid bacteria: predictions from genome sequence analysis. Int. Dairy J. *12*, 111–121.

Van Laer, H. (1892). Contributions à l'historie des ferments des hydrates de carbone. Acad. Roy. Belg. Cl. Sci. Collect. Octavo Mem., 47, 1–37.

Vancanneyt, M., Engelbeen, K., De Wachter, M., Vandemeulebroecke, K., Cleenwerck, I. and Swings, J. (2005a). Reclassification of *Lactobacillus ferintoshensis* as a later heterotypic synonym of *Lactobacillus parabuchneri*. Int. J. Syst. Evol., 55, 2195–2198.

Vancanneyt, M., Huys, G., Lefebvre, K., Vankerckhoven, V., Goossens, H., and Swings, J. (2006a). Intraspecific Genotypic Characterization of *Lactobacillus rhamnosus* Strains Intended for Probiotic Use and Isolates of Human Origin. Appl. Environ. Microbiol. 72, 5376–83.

Vancanneyt, M., Mengaud, J., Cleenwerck, I., Vanhonacker, K., Hoste, B., Dawyndt, P., Degivry, M.C., Ringuet, D., Janssens, D., and Swings, J. (2004a). Reclassification of *Lactobacillus kefirgranum* Takizawa *et al.*, 1994 as *Lactobacillus kefiranofaciens* subsp. *kefirgranum* subsp. nov. and emended description of *L. kefiranofaciens* Fujisawa *et al.*, 1988. Int. J. Syst. Evol. Microbiol. *54*, 551–556.

Vancanneyt, M., Naser, S.M., Engelbeen, K., De Wachter, M., Van Der Meulen, R., Cleenwerck, I., Hoste, B., De Vuyst, L., and Swings, J. (2006b). Reclassification of *Lactobacillus brevis* strains L.M.G 11494 and L.M.G 11984 as *Lactobacillus parabrevis* sp. nov. Int. J. Syst. Evol. Microbiol. 56, 1553–1557.

Vancanneyt, M., Neysens, P., De Wachter, M., Engelbeen, K., Snauwaert, C., Cleenwerck, I., Van der Meulen, R., Hoste, B., Tsakalidou, E., De Vuyst, L., and Swings, J. (2005b). *Lactobacillus acidifarinae* sp. nov. and *Lactobacillus zymae* sp. nov., from wheat sourdoughs. Int. J. Syst. Evol. Microbiol. 55, 615–620.

Vandamme, P., Pot, B., Gillis, M., de Vos, P., Kersters, K., and Swings, J. (1996a). Polyphasic taxonomy, a consensus approach to bacterial systematics. Microbiol. Rev. *60*, 407–438.

Vandamme, P., Pot, B., Falsen, E., Kersters, K., and Devriese, L.A. (1996b). Taxonomic study of Lancefield streptococcal groups C., G., and L (*Streptococcus dysgalactiae*) and proposal of *S. dysgalactiae* subsp. *equisimilis* subsp. nov. Int. J. Syst. Bacteriol. *46*, 774–781.

Vaneechoutte, M., Boerlin, P., Tichy, H.V., Bannerman, E., Jaeger, B., and Bille, J. (1998). Comparison of

P.C.R-based DNA fingerprinting techniques for the identification of *Listeria* species and their use for atypical *Listeria* isolates. Int. J. Syst. Bact., *48*, 127–139.

Vankerckhoven, V., Huys, G., Vancanneyt, M., Vael, C., Klare, I., Romond, M.B., Entenza, J.M., Moreillon, P., Wind R, D., Knol, J., Wiertz, E., Pot, B., Vaughan, E.E., Kahlmeter, G., and Goossens, H. (2008). Biosafety assessment of probiotics used for human consumption: recommendations from the E.U.-PROSAFE project. Trends Food. Sci. Technol. *19*, 102–114.

Vaughn, R.H., Douglas, H.C., and Fornachon, J.C. (1949). The taxonomy *of Lactobacillus hilgardii*, and related heterofermentative Lactobacilli. Hilgardia, *19*, 133-139.

Vela, A.I., Fernández, E., Lawson, P.A., Latre, M.V., Falsen, E., Domínguez, L., Collins, M.D., Fernández-Garayzábal, J.F. (2008). *Lactobacillus ceti* sp. nov., isolated from beaked whales (*Ziphius cavirostris*). Int. J. Syst. Evol. Microbiol. *58*, 891–894.

Ventura, M., Elli, M., Reniero, R., and Zink, R. (2001) Molecular microbial analysis of *Bifidobacterium* isolates from different environments by the species-specific amplified ribosomal DNA restriction analysis (ARDRA). FEMS Microbiol. Ecol., *36*, 113.

Vermeulen, N., Pavlovic, M., Ehrmann, M.A., Gaenzle, M.G., and Vogel, R.F. (2005). Functional characterization of the proteolytic system of *Lactobacillus sanfranciscensis* D.S.M20451T. Appl. Environ. Microbiol. *71*, 6260–6266.

Vermeulen, N., Thiele, C., Gaenzle, M.G., and Vogel, R.F. (2003). Cysteine metabolism by cereal-associated lactobacilli. In Sourdough, from Fundamentals to Applications. de Vuyst, L. ed., [Vrije Universiteit Brussel (V.U.B), IMDO, Brussels], pp. 58–59.

Vescovo, M., Dellaglio, F., Botazzi, V., and Sarra, P.G. (1979). Deoxyribonucleic acid homology among *Lactobacillus* species of the subgenus *Betabacterium* Orla-Jensen. Microbiologica, 2, 317–330.

Viljanen, M.J., Murros, A., Palva, A., Björkroth, K.J. (2008). *Lactobacillus sobrius* Konstantinov *et al.* 2006 is a later synonym of *Lactobacillus amylovorus* Nakamura 1981. Inst. J. Syst. Evol. Microbiol. *58*, 910–913.

Vogel, R.F., Böcker, G., Stolz, P., Ehrmann, M., Fanta, D., Ludwig, W., Pot, B., Kersters, K., Schleifer, K.H, and Hammes, W.P. (1994). Identification of lactobacilli from sourdough and description of *Lactobacillus pontis* sp. nov. Int. J. Syst. Bacteriol. *44*, 223–229.

Vogel, R.F., Knorr, R., Müller, M.R.A., Steudel, U., Gänzle, M.G., and Ehrmann, M. (1999). Non-dairy lactic fermentations: the cereal world. Anton. Leeuwen. *76*, 403–411.

Wallbanks, S., Martinez-Murcia, A.J., Fryer, J.L., Phillips, B.A., M.D. Collins. (1990). 16S rRNA sequence determination for members of the genus *Carnobacterium* and related lactic acid bacteria and description of *Vagococcus salmoninarum* sp. nov. Int. J. Syst. Bacteriol. *40*, 224–230.

Wayne, L.G. (1994). Actions of the Judicial Commission of the International Committee on Systematic Bacteriology on Requests for Opinions published between January 1985 and July 1993 Int. J. Syst. Bacteriol. *44*, 177–178.

Wayne, L.G., Brenner, D.J., Colwell, R.R., Grimont, P.A.D, Kandler, O., Krichevsky, M.I., and Trüper, H.G. (1987). Report of the ad hoc committee on reconciliation of approaches to bacterial systematics. Int. J. Syst. Bacteriol. *37*, 463–464.

Weimer, B., Seefeldt, K., and Dias, B. (1999) Sulfur metabolism in bacteria associated with cheese. Anton. Leeuwen. *76*, 247–261.

Weisburg, W.G., Barns, S.M., Pelletier, D.A., and Lane, D.J. (1991). 16S ribosomal DNA amplification for phylogenetic study. J. Bacteriol. 173, 697–703.

Weiss, A., Lettner, H.P., Krame, W., Mayer, H.K., and Kneifel, W. (2005) Molecular Methods Used for the Identification of Potentially Probiotic *Lactobacillus reuteri* Strains. Food Technol. Biotechnol., *43*, 295–300.

Weiss, N., and Schillinger, U. (1984a). *Lactobacillus sanfrancisco* sp. nov., nom. rev. Syst. Appl. Microbiol. 5, 230–232.

Weiss, N., and Schillinger, U. (1984b). Validation List N° 16. Int. J. Syst. Bacteriol. *34*, 503–504.

Weiss, N., Schillinger, U., and Kandler, O. (1983a). *Lactobacillus trichodes*, and *Lactobacillus heterohiochii*, subjective synonyms of *Lactobacillus fructivorans*. Syst. Appl. Microbiol. 4, 507–511.

Weiss, N., Schillinger, U., and Kandler, O. (1983b). *Lactobacillus lactis*, *Lactobacillus leichmannii* and *Lactobacillus bulgaricus*, subjective synonyms of *Lactobacillus delbrueckii* and description of *Lactobacillus delbrueckii* subsp. *lactis* comb. nov. and *Lactobacillus delbrueckii* subsp. *bulgaricus* comb. nov. Syst. Appl. Microbiol. 4, 552–557.

Weiss, N., Schillinger, U., and Kandler, O. (1984). Validation List N° 14. Int. J. Syst. Bacteriol. *34*, 270–271.

Weiss, N., Schillinger, U., Laternser, M., and Kandler, O. (1981). *Lactobacillus sharpeae* sp. nov. and *Lactobacillus agilis* sp. nov., two new species of homofermentative, *meso*-diaminopimelic acid-containing lactobacilli. Zentralbl. Mikrobiol. Parasitenkd. Infektionskr. Hyg. Abt. 1 Orig., C2, 242–253.

Weiss, N., Schillinger, U., Laternser, M., and Kandler, O. (1982). Validation List N° 8. Int. J. Syst. Bacteriol. *32*, 266–268.

Wibowo, D., Eschenbruch, R., Davis, C.R., Fleet, G.H., and Lee, T.H. (1985). Occurrence and growth of lactic acid bacteria in wine: a review. Am. J. Enol. Vitic. 36, 302–313.

Wiese, B.J., Strohmar, W., Rainey, F.A., and Diekmann, H. (1996). *Lactobacillus panis* sp. nov., from sourdough with a long fermentation period. Int. J. Syst. Bacteriol. *46*, 449–453.

Williams, A.M., and Collins, M.D. (1990). Molecular taxonomic studies on *Streptococcus uberis* types I an I.I. Description of *Streptococcus parauberis* sp. nov. J. Appl. Bacteriol. *68*, 485–490.

Williams, A.M., Fryer, J.L., and Collins, M.D. (1990). *Lactococcus piscium* sp. nov. a new *Lactococcus* species from salmonid fish. FEMS Microbiol. Lett. 68, 109–114.

Williams, A.M., Rodrigues, U.M., and Collins, M.D. (1991). Intragenic relationships of enterococci as determined by reverse transcriptase sequencing of small-subunit rRNA. Res. Microbiol. 142, 67–74.

Woese, C.R. (1998). Default taxonomy: Ernst Mayr's view of the microbial world. Proc. Natl. Acad. Sci. 95, 11043–11046.

Wolfrum, G., and Vogel, R.F. (1999). Growth and metabolism of *Lactobacillus pontis* T.M.W 1.109 isolated from cereal fermentations described by differential equations. 17th I.C.C Conference 'Cereals across the Continents' Valencia, Spain.

Yang, D., and Woese, C.R. (1989). Phylogenetic structure of the '*Leuconostocs*': an interesting case of a rapidly evolving organism. Syst. Appl. Microbiol. *12*, 145–149.

Yavuz, E., Gunes, H., Bulut, C., Harsa, S. and A.F. Yenidunya (2004). RFLP of 16S-I.T.SrDNA region to differentiate lactobacilli at species level. World J. Microbiol. Biotechnol. *20*, 535–537.

Ye, R.W., Wang, T., Bedzyk, L., and Croker, K.M. (2001). Applications of DNA microarrays in microbial systems. J. Microbiol. Methods *47*, 257–272.

Yoon, J.H., Kang, S.S., Mheen, T.I., Ahn, J.S., Lee, H.J., Kim, T.K., Park, C.S., Kho, Y.H., Kang, K.H., and Park, Y.H. (2000). *Lactobacillus kimchii* sp. nov., a new species from kimchi. Int. J. Syst. Evol. Microbiol. *50*, 1789–1795.

Yvon, M., Berthelot, S., and Gripon, J.C. (1998). Adding a-ketoglutarate to semi-hard cheese curd highly enhances the conversion of amino acids to aroma compounds. Int. Dairy J. *8*, 889–898.

Zanoni, P., Farrow, J.A.E, Phillips, B.A., and Collins, M.D. (1987). *Lactobacillus pentosus* (Fred, Peterson, and Anderson) sp. nov., nom. rev. Int. J. Syst. Bacteriol. *37*, 339–341.

Zhang, B., Tong, H., and Dong, X. (2005). *Pediococcus cellicola* sp. nov., a novel lactic acid coccus isolated from a distilled-spirit-fermenting cellar. Int. J. Syst. Evol. Microbiol. 55, 2167–2170.

Zotta, T., Piraino, P., Ricciardi, A., McSweeney, P.L.H., and Parente, E. (2006). Proteolysis in model sourdough fermentations. J. Agric. Food Chem. 54, 2567–2574.

Comparative and Functional Genomics of the Genus *Lactobacillus*

3

Jan-Peter Van Pijkeren and Paul W. O'Toole

Abstract

Lactobacilli are members of the lactic acid bacteria group and constitute an ecologically and phylogenetically very diverse group. Some strains are of industrial importance since they are applied in a range of fermentation processes, whereas other strains are exploited for their probiotic properties. To date, ten *Lactobacillus* genomes encompassing nine species have been sequenced, and their genome content broadly reflect the diversity of this genus. With the exception of members of the '*acidophilus*- complex', there is no long range synteny based on whole-genome alignments. The species are diverse in their metabolic capacity, and some species appear to be in an ongoing phase of specialization, largely determined by preferred ecological niches. Each of these species produces proteins which enable them to compete or survive within their preferred habitat. A repertoire of diverse adhesins has been functionally characterized in several gastrointestine-associated lactobacilli. The comparative genomics of different *Lactobacillus* strains has revealed novel insights in the complexity of this diverse genus.

Introduction and overview of *Lactobacillus* genomics

Lactic acid bacteria (LAB) are a group of Gram-positive organisms which produce lactic acid by fermentation (Kandler and Weiss, 1986). Genera of lactic acid bacteria include *Lactococcus, Enterococcus, Oenococcus, Pediococcus, Streptococcus, Leuconostoc* and *Lactobacillus*, all members of the order *Lactobacillales* (Kandler and Weiss, 1986). With over 100 species, the genus *Lactobacillus* represents the largest group within the family *Lactobacillaceae* (Benson *et al.*, 2000; Wheeler *et al.*, 2000). Lactobacilli can be found in a variety of ecological niches, such as plants or plant derived materials, animals (including humans) and in raw milk (reviewed in Hammes and Vogel, 1995). In addition, lactobacilli have been detected in insects (Reeson *et al.*, 2003; Mohr and Tebbe, 2006). The ability to flourish in such a variety of habitats is a direct reflection of the versatility of this genus. Therefore, it is not surprising that lactobacilli have been exploited for many decades in industry for fermentation of meat, dairy products and beverages. Also, it has been suggested by Nakamura *et al.* that some lactobacilli might have potential to be applied in industry for the production of propanediol, a product generated by some species when grown anaerobically with glycerol as sole carbon source (Nakamura and Whited, 2003).

The Nobel Laureate Elias Metchnikoff (1845–1916) stated that the consumption of yoghurt could increase the lifespan. He believed that the lactic acid bacteria present in fermented yoghurt would prevent decay of the intestine (Metchnikoff, 1907). Seventy-five years later Underdahl reported that diarrhoea could be prevented or reduced when pigs were fed with *Streptococcus* prior to induction of diarrhoea by *E. coli* (Underdahl *et al.*, 1982, 1983). Bacteria which are beneficial for the host upon consumption are currently defined as 'probiotic': 'live organisms which, when administered in adequate amounts, confer a health benefit on the host' (Reid *et al.*, 2003). Most probiotic products for human consumption contain bacteria which are commensal organisms, such as strains of the genus *Lactobacillus* and *Bifidobacterium*. Beneficial effects of probiotic bacteria include reduction of nosocomial- and chemotherapy-associated diarrhoea (Szajewska *et al.*, 2001, 2006) and some strains display immunomodulatory effects (Mohamadzadeh *et al.*, 2005; O'Hara *et al.*, 2006). The latter will be discussed in more detail elsewhere in this volume, in the chapter 'Interactions of *Lactobacillus* spp. with the immune system'.

Research on probiotic bacteria, and in particular on lactobacilli, has grown exponentially in the last decade. Illustrative of this is the observation that 180 research articles dating from 1980 to 2000 were found using the search string 'probiotic *Lactobacillus*' in Pubmed, in contrast to nearly 1200 research articles in the period 2000–2007. The combination of the industrial importance of lactobacilli, and the lack of understanding of the mechanisms of action of the probiotic functions of some strains, has motivated sequencing the chromosomes of several *Lactobacillus* species. These first *Lactobacillus* to be sequenced was *Lactobacillus plantarum* WCFS1 (Kleerebezem *et al.*, 2003), a single colony isolate of *L. plantarum* NCIMB 8826. This strain originates from human saliva and the main genome characteristic is its potentially large metabolic capacity. The second genome to be published was *Lactobacillus johnsonii* NCC 533, previously known as *Lactobacillus acidophilus* strain La1, which was isolated from human faeces (Pridmore *et al.*, 2004). Analysis of this genome highlighted the lack of genes encoding metabolic pathways, suggesting a dependence on the host or other intestinal microbes, and a diverse set of surface proteins potentially contributing to the interaction of this bacterium with its environment. By sequencing the genome of *Lactobacillus acidophilus* NCFM (Altermann *et al.*, 2005) several surface proteins potentially responsible for host interaction were identified. Also, some strain-specific genes and gene clusters were annotated which were predicted to encode unique features exhibited by this strain (Altermann *et al.*, 2005). The genome of the meat-isolate *Lactobacillus sakei* 23K was sequenced (Chaillou *et al.*, 2005) and revealed a range of metabolic capacities and genes encoding proteins which potentially enhances the ability to survive on meat surfaces, a reflection of the organism's adaptation to a specific environment. *Lactobacillus salivarius* UCC118 is a human isolate and was the first sequenced strain harbouring a mega-plasmid (Claesson *et al.*, 2006). Based on gene content, the mega-plasmid does not seem to be required for survival but has several genes which can be related to enhanced fitness or survival in the host. *Lactobacillus bulgaricus* is commonly used in yoghurt fermentation. Its genome sequence (van de Guchte *et al.*, 2006) showed specific features indicating that this strain is in an ongoing process of adapting to a lactose-rich milk environment. Makarova and co-workers sequenced nine bacterial genomes (Makarova *et al.*, 2006), all members of the group of lactic acid bacteria, which included four *Lactobacillus* strains. These *Lactobacillus* strains are of importance to the industry because of their application in various fermentation processes. This study has led to a better understanding of the evolutionary progression within the group *Lactobacillales*.

In the era of genomics, an incredible amount of information has become available. An organism's genome content can reveal many secrets such as its catabolic and metabolic capacities or the potential ability to withstand certain environmental conditions, such as high salt concentrations or oxygen exposure. The application of comparative genome analyses provides insight into the adaptive ability of an organism for a particular environment, often reflected by a relative small genome and limited metabolic capacities. To date, ten fully sequenced *Lactobacillus* genomes are publicly available. The aim of this review is to discuss each of these genomes with regard to their genome content and to discuss the functionally characterized proteins encoded by these sequenced genomes.

Genome features

The nine *Lactobacillus* species with their general genome characteristics are listed in Table 3.1. The genome size of *L plantarum* WCFS1 is 3.3 Mb, the largest chromosome of the sequenced lactobacilli to date. The G+C% of the *L. plantarum* genome is the highest (44.4%) in comparison with other human-derived isolates, such as *L. johnsonii*

Table 3.1 General genome characteristics of the ten sequenced *Lactobacillus* strains.

Species	Genome size (bp)	GC%	No. of genes	Unique genes[a,b]	No. of pseudogenes	No. of tRNA	No. of rRNA operons	Reference
L. plantarum WCFS1	3,308,274	44.4	3,051	1289	39	70	5	(Kleerebezem *et al.*, 2003)
L. johnsonii NCC 533	1,992,676	34.6	1,821	453	NA	79	6	(Pridmore *et al.*, 2004)
L. acidophilus NCFM	1,993,564	34.7	1,864	503	NA	61	4	(Altermann *et al.*, 2005)
L. sakei 23K	1,884,661	41.3	1,884	487	30	63	7	(Chaillou *et al.*, 2005)
L. salivarius UCC118 (pMP118)[c]	1,827,111 (242,436)	32.9 (32.1)	1,765 (242)	483 (224)	49 (20)	78 (0)	7 (0)	(Claesson *et al.*, 2006)
L. delbrueckii ssp. *bulgaricus* ATCC 11842	1,864,998	49.7	1,562	NT	270	95	9	(van de Guchte *et al.*, 2006)
L. gasseri ATCC 33323	1,894,360	35.3	1,763	NT	43	78	6	(Makarova *et al.*, 2006)
L. brevis ATCC 367	2,340,228	46.2	2,221	NT	50	65	5	(Makarova *et al.*, 2006)
L. casei ATCC 334	2,924,325	46.6	2,776	NT	82	59	5	(Makarova *et al.*, 2006)
L. delbrueckii ssp. *bulgaricus* ATCC BAA-365	1,856,951	49.7	1,725	NT	192	98	9	(Makarova *et al.*, 2006)

[a]Unique genes are defined as genes that had no hits against the non-redundant protein database with BLASTP E-value of $<1 \times 10^{-10}$.
[b]NT indicates not tested.
[c]Numbers in parenthesis correspond to pMP118.

(34.6%), *L. acidophilus* (34.7%) and *L. salivarius* (32.9%). With the exception of *L. gasseri* (G+C% of 35.3%), non-human derived strains seem to have a higher G+C% ranging from 41.3% for the meat-isolate *L. sakei* up to 49.7% for *L. delbrueckii* (Table 3.1). One of the interesting features of the *L. plantarum* genome is the presence of 25 complete phosphotransferase systems (PTS) and 30 transporter systems (Kleerebezem *et al.*, 2003). These features illustrate the ability to live in a variety of niches containing diverse nutrients. In terms of genome organisation, it was noted that many genes related to sugar transporters or metabolism are present on a 213-kb region, located close to the origin of replication (Kleerebezem *et al.*, 2003). This, combined with a skew in G+C% within this region, led to the hypothesis that many of these genes were acquired by horizontal gene transfer (HGT) and that this region represents a lifestyle adaptation by *L. plantarum* (Kleerebezem *et al.*, 2003). The genome of *L. plantarum* contains 3.051 genes, compared to an average of 1.958 genes encoded by other sequenced *Lactobacillus* genomes and is in agreement with the larger genome size of *L. plantarum*. Konstantinidis and Tiedje noted that there is a trend between genome size and gene content (Konstantinidis and Tiedje, 2004). Analyses of 115 completed prokaryotic genomes, ranging from <1500 ORFs to >6000 ORFs, showed that larger genomes preferentially have more regulatory genes, more genes related to metabolism and, to a lesser extent, more energy conservation-related genes (Konstantinidis and Tiedje, 2004). Thus the ecological flexibility of *L. plantarum* is clearly linked with its comparatively large(r) genome when compared to the other sequenced lactobacilli.

General features of the genome of *L. johnsonii* NCC 533, such as genome size, G+C% and number of genes, are proportionally comparable to *L. acidophilus* and *L. salivarius* (Table 3.1). However, *L. johnsonii* NCC 533 has several features that can be correlated with its dependence on its host and persistence in the gastrointestinal tract (GIT) (Pridmore *et al.*, 2004). In contrast with *L. plantarum*, *L. johnsonii* cannot synthesize most amino acids *de novo*, since no biosynthetic pathways are encoded by its genome. This deficiency is off-set by the presence of 16 PTS units and over 20 amino acid-permeases, although the overall percentage of proteins dedicated to transport is comparable to other sequenced lactobacilli. Also, there is a lack of enzymes that break down larger sugar-substrates required for its growth. *L. johnsonii* is thus dependent on smaller sugars for fermentation, such as mono-, di- and trisaccharides (Pridmore *et al.*, 2004). In terms of host interaction, the *L. johnsonii* genome harbours a 30-kb region which includes 9 genes potentially encoding fimbrial-like proteins, unique to lactobacilli. Moreover, *L. johnsonii* encodes an IgA protease, a surface protein present in pathogenic streptococci where it plays an indirect role in adhesion (Weiser *et al.*, 2003) – all clearly indicative of adaptation to its preferred ecological niche.

Like *L. johnsonii* NCC 533, *L. acidophilus* NCFM has a relatively small genome and displays several features which could correlate with specialization of the organism to the GIT. For example, *L. acidophilus* is likely to be auxotrophic for 14 amino acids (Altermann *et al.*, 2005) and is thus more dependent either on its host or other intestinal micro-organisms. As a relevant precedent in related bacteria, adaptation to a specific environment, at the genome level, has previously been described for *Streptococcus thermophilus* (Bolotin *et al.*, 2004). This bacterium has been subject to genome decay because of its adaptation to a milk environment which resulted in loss of many genes, including some genes linked to pathogenicity.

The adaptive ability of an organism for its environment is also clear from the genome of the meat-isolate *L. sakei* 23K. Purine nucleosides are abundant in meat, and *L. sakei* is the only lactic acid bacterium that is able to use these for growth if glucose becomes limiting (Chaillou *et al.*, 2005). Boekhorst *et al.* showed that proteins with mucus binding domains are present in many LAB (Boekhorst *et al.*, 2006a). However, none are present in *L. sakei* (Chaillou *et al.*, 2005). This is clearly not surprising since *L. sakei* will not normally encounter mucins on the surface of meat. However, *L. sakei* is predicted to display proteins on its surface which are not found in the sequenced intestinal lactobacilli and which are therefore believed to specifically play a role in the colonization of the meat surface, such as a 1,987-

amino acid biofilm formation protein (Chaillou *et al.*, 2005).

One of the features that distinguishes the human isolate *L. salivarius* from the other sequenced lactic acid bacteria, is the 242-kb megaplasmid pMP118 (Claesson *et al.*, 2006). This megaplasmid notably completes the set of genes required for encoding the pentose phosphate pathway. pMP118 also harbours genes encoding enzymes predicted to be involved in rhamnose and sialic acid catabolism though the latter pathway is possibly incomplete. Also, genes encoding sorbitol utilization proteins are present on pMP118. Thus, pMP118 can be beneficial to *L. salivarius* in terms of carbon source utilization, possibly yielding an advantage over competing organisms (Claesson *et al.*, 2006).

L. delbrueckii ssp. *bulgaricus* ATCC 11842 (*Lactobacillus bulgaricus*) is a member of the so-called *acidophilus* complex, which also includes the species *L. gasseri*, *L. acidophilus* and *L. johnsonii*. When compared to the other members of the *acidophilus* complex, it is apparent that the G+C% is much higher for *L. bulgaricus* (Table 3.1). It was hypothesized that this strain is in an ongoing process of development or adaptation (van de Guchte *et al.*, 2006). For example, 270 pseudogenes are present in the genome of *L. bulgaricus*. Moreover, the genome contains a higher number of tRNA and rRNA genes compared to similar size *Lactobacillus* genomes, suggesting that it has undergone size reduction (Table 3.1) (van de Guchte *et al.*, 2006). It is not surprising that similar genome characteristics are found in the closely related strain *L. delbrueckii* subsp. *bulgaricus* ATCC BAA-365 (Makarova *et al.*, 2006). Comparing the genomes of nine lactic acid bacteria, including four lactobacilli, revealed that the *Lactobacillales* are in a constant mode of genome development (Makarova *et al.*, 2006). The ongoing evolutionary process within *Lactobacillales* is emphasized by the high number of acquired genes related to the ability to grow in a nutrient-rich environment. Many of these genes are predicted to be acquired by horizontal gene transfer (Makarova *et al.*, 2006). Between 0% and 17% of bacterial genes in general are estimated to be acquired by horizontal gene transfer (Ochman *et al.*, 2000) while 15–20% of the genes are species specific (Nelson *et al.*, 2000). Thus, bacterial gene transfer is a factor which has had a discrete role in shaping the evolution of the lactobacilli.

Comparative genomics

The genus *Lactobacillus* is ecologically widespread and this is reflected in variability of the genome content. For example, the G+C% values range from 34.6% to 49.7% for *L. johnsonii* NCC 533 and *L. delbrueckii* ssp. *bulgaricus* ATCC BAA-365, respectively. This range is larger than that is proposed for bacteria within a genus (Schleifer and Ludwig, 1995). With several *Lactobacillus* genomes sequenced from different ecological niches, a foundation is established which contain the roots for potentially linking genome content with environmental habitats.

The first *Lactobacillus* genomes which have been compared are *L. plantarum* WCFS1 and *L. johnsonii* NCC 533 (Boekhorst *et al.*, 2004). Overall there is little DNA similarity between these genomes based on a dot-plot DNA comparison. Similarly, a dot-plot based on protein sequence did not reveal gene order conservation but conservation of gene clusters. This would be suggestive of a large phylogenetic distance as gene order would be expected to be extensively conserved in closely related species (Tamames, 2001). This was further substantiated by the fact that proteins more similar to those of *L. plantarum* WCFS1 are found in *Enterococcus faecalis*, *Listeria monocytogenes* and *Bacillus subtilis* than in *L. johnsonii*, respectively (Boekhorst *et al.*, 2004). Compared to *L. johnsonii*, *L. plantarum* has more than double the number of genes encoding proteins involved in metabolism and transport of amino acids. Also, *L. plantarum* encodes more proteins involved in sugar uptake and degradation than *L. acidophilus* (342 versus 196, respectively) (Boekhorst *et al.*, 2004). This is not surprising since, as previously mentioned, *L. plantarum* is found in a range of niches *in vivo* (Ahrne *et al.*, 1998; Stoyancheva *et al.*, 2006) and *ex vivo* (Paludan-Muller *et al.*, 1999; Ennahar *et al.*, 2003; Pepe *et al.*, 2004) and thus requires a larger metabolic repertoire than for example *L. johnsonii*, which is usually found in nutrient-rich environments. On this theme, proportionally more extracellular proteins are encoded by *L. plantarum* of which 27 (Boekhorst *et al.*, 2005) are predicted to be anchored to the cell wall by a C-terminally located LPXTG-motif. This is recognized by the membrane-associated enzyme sortase (Fig. 3.1) (Fischetti *et al.*, 1990). Sortase cleaves between the Thr- and Gly-residue within the LPXTG-motif and then covalently links the Thr-residue to the peptidoglycan by forming an amide bond with the amino group of the glycine interbridge (Fig. 3.1) (Fischetti *et al.*, 1990; Navarre and Schneewind, 1994; Ton-That *et al.*, 1999). Some of these sortase-dependent proteins are predicted to play a role in the acquisition of sugars (Siezen *et al.*, 2006) while other proteins are predicted to be involved in host-specific interactions via mucus-binding domains, which are especially abundant in host-associated lactic acid bacteria (Boekhorst *et al.*, 2006a). The latter will be further elucidated in the section Functional Genomics – *Adhesion factors* below.

Using 16S rRNA gene phylogeny, a phylogenetic tree of 111 *Lactobacillus* species was generated (Canchaya *et al.*, 2006). Although this genus is phylogenetically very diverse, species could be divided into five groups, four of which could be associated with proposed ecological niches of the organisms. For example, *Lactobacillus* species found in the GI-tract are found in group A, which encompasses species of the *acidophilus*-complex, and group E in which for example the human isolate *L. salivarius* UCC118 is located. Species associated with the environment and food, are clustered in group C and D. These groups include organisms isolated

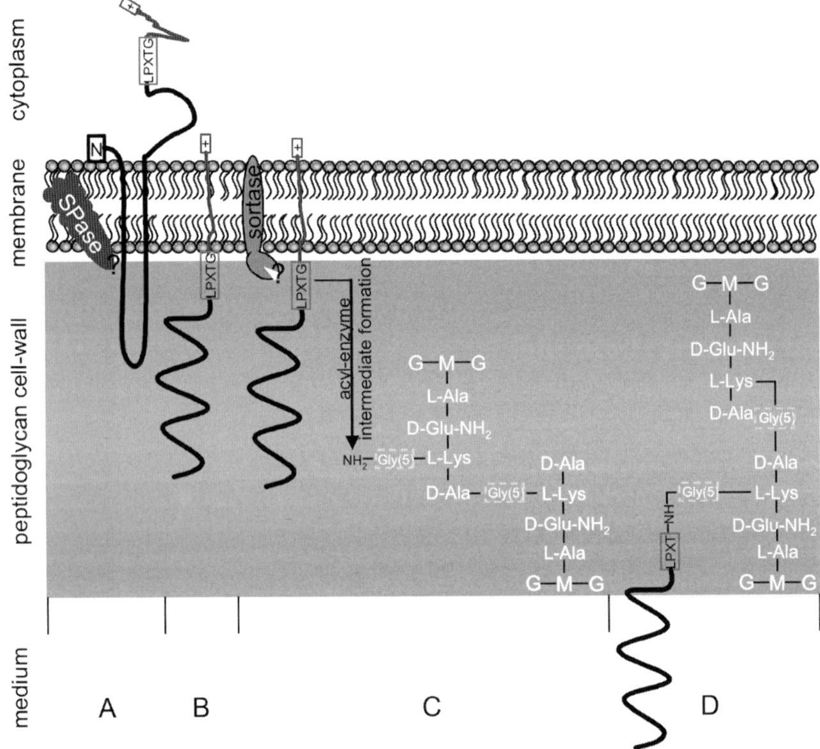

Figure 3.1 Overview of cell-wall anchoring of sortase-dependent proteins (SDPs) by the enzyme sortase. A: Translocation of the SDP to the membrane to enable cleavage of the signal peptide by the signal peptidase (SPase); B: after cleavage, the transmembrane helix (blue line) is inserted into the membrane whereby the positively charged residues at the C-terminus (boxed +) of the SDP prevent further translocation through the lipid bi-layer; C: sortase enzymes typically have a signal peptide, which could serve as a signal anchor, and could therefore stay associated with the membrane. Sortase recognizes the LPXTG-motif (X being any amino acid) and cleaves between the Thr and Gly residues after which an acyl-enzyme intermediate is formed. D: the carboxyl of the threonine is amide linked to the glycine interbridge (dashed green box) which is the final step in the sorting process. G, *N*-Acetylglucosamine; M, *N*-acetylmuramic acid. Figure is adapted from Cossart and Jonquieres and Ton-That *et al.* (Cossart and Jonquieres, 2000; Ton-That *et al.*, 1999) describing the sorting-process in *S. aureus*.

from for example sourdough, sewage and silage. No clear environmental niche could be linked with the species in group B (Canchaya *et al.*, 2006). These findings could be linked with whole-genome alignments. At DNA level, *L. johnsonii* aligns well with *L. gasseri* and *L. acidophilus* whereas at protein level *L. johnsonii* aligns also with *L. delbrueckii* (Berger *et al.*, 2007) since phylogenetic signals at protein level are more stable than at DNA level (Canchaya *et al.*, 2006). Those four species are found in Group A of the phylogenetic tree and thus display a high level of synteny whereas genomes which display a low degree of synteny are phylogenetically more distant. As expected, a whole genome alignment of protein sequences of *L. delbrueckii* ATCC 11842 and *L. delbrueckii* ATCC BAA-365 displays a high level of synteny (Fig. 3.2A). A small region corresponding to a change of genome composition is noted in the middle of the plot, which is located furthest away from the origin of replication (Fig. 3.2A). Interestingly, the genome-wide protein comparison of *L. sakei versus L. johnsonii* clearly shows an X-shaped distribution, indicative that similar sequences can be found at the same distance from the origin of replication, but not necessarily on the same side of it (Canchaya *et al.*, 2006; Berger *et al.*, 2007). Comparing the sequences of *L. delbrueckii* ATCC 11842 with *L. gasseri* ATCC 33323 also reveals an X-shaped distribution (Fig. 3.2B) which is more expected since, as previously mentioned, these species are part of the *acidophilus*-complex and thus phylogenetically more closely related than *L. sakei*

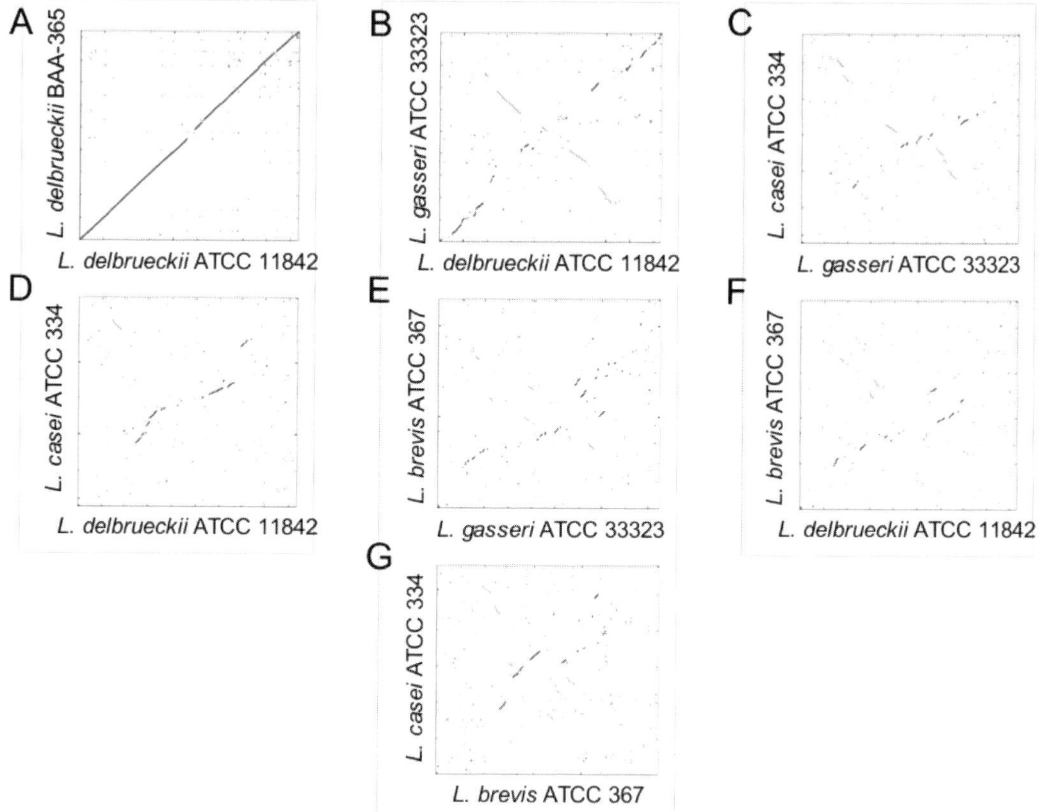

Figure 3.2 Whole-genome alignments of five *Lactobacillus* genomes encompassing four species. The between-species alignments are depicted in A to G. For each alignment, the origin of replication of the two genomes is in the lower-right corner. The red and green coloured points refer to the forward and reverse matching substring regions, respectively. The alignments are generated using the PROmer software (Delcher *et al.*, 2002) with default settings.

and *L. johnsonii*. Although the levels of synteny are not as strong as for example *L. delbrueckii* ATCC 11842 and *L. gasseri* ATCC 3323, weak X-shaped distributions in the middle of the plot can also be noted when comparing *L. gasseri* ATCC 33323 *versus L. casei* ATCC 334 (Fig. 3.2C) and *L. delbrueckii* ATCC 11842 *versus L. casei* ATCC 334 (Fig. 3.2D). These X-shaped distributions occur in the middle region of the comparison-plot, suggesting that these regions are more likely to undergo a change in genome composition than the regions closer to the origin of replication (Canchaya *et al.*, 2006). This was also revealed by microarray hybridization of several species of *L. johnsonii* where the region around the origin of replication showed greatest levels of gene conservation (Berger *et al.*, 2007). However, the X-shaped distribution is not as evident, and not in close proximity to the middle of the plot, when performing comparisons with *L. brevis* ATCC 367 (Fig. 3.2E, 2F and 2G, respectively). This suggests that there is less synteny between *L. brevis* and the other tested genomes and that the changes in DNA composition are closer to the origin of replication, respectively. Analogously, a 213-kb region in *L. plantarum* was identified close to the origin of replication with a G+C% which is 3% lower than the rest of genome, suggesting horizontal gene transfer (Kleerebezem *et al.*, 2003).

Inter-species analysis of *L. johnsonii* (Berger *et al.*, 2007) and *L. plantarum* (Molenaar *et al.*, 2005), respectively, revealed that genes encoding proteins required for growth are conserved. In both species, a high degree of variability was observed in the 'mobilome' represented by phage and transposon elements. Moreover, in different species of *L. johnsonii*, genes were present that appear to be strain specific, such as genes encoding proteins that are predicted to be involved in bacterium–host interaction, metabolism and unknown function (Berger *et al.*, 2007).

Genes which are present in different species but are derived from a common ancestor are defined as orthologues (Fitch, 1970). Based on 12 species Makarova *et al.* defined a cluster of orthologous genes specific for *Lactobacillales* (LaCOGs) (Makarova *et al.*, 2006; Makarova and Koonin, 2007) and identified a total of 3,199 LaCOGs. LaCOGs covered 86% of the genome and 11% of the identified LaCOGs appeared specific for *Lactobacillales*. Eighteen per cent of the total LaCOGs identified appeared to encode conserved proteins, of which the majority are predicted to be involved in general cellular functions, such as transcription, translation and replication, in agreement with previous findings (Molenaar *et al.*,2005; Canchaya *et al.*, 2006; Berger *et al.*, 2007). Interestingly, two genes were identified which do not appear to have orthologues outside lactobacilli and are proposed to be novel markers for the *Lactobacillales* (Makarova *et al.*, 2006; Makarova and Koonin, 2007). One of these markers contains a peptidoglycan-binding domain (LysM) and is possibly co-regulated with the housekeeping genes coding for a ribosomal protein and a cytidylate kinase. The second marker is also located within a region of conserved genes (Makarova *et al.*, 2006; Makarova and Koonin, 2007).

Members of the *Lactobacillales* appear to have primarily evolved towards a specific niche through genome decay rather than adaptation through acquiring genes for specialization. This has previously been suggested for *S. thermophilus* (Bolotin *et al.*, 2004) and has more recently been proposed by Makarova *et al.* (Makarova *et al.*, 2006; Makarova and Koonin, 2007). A model reconstructing gene gain and loss throughout the evolution of *Lactobacillales* after diverging from *Bacilli*, showed that the common ancestor of this family had approximately 2100–2200 genes. Compared to the ancestor of all *Bacilli*, it was predicted that approximately 600–1200 genes were lost whereas less than 100 genes were gained, apparently through horizontal gene transfer. Many transformations of genome content that had occurred within the family *Lactobacillales* appear to be specifically related to adaptation to a nutrient-rich environment, since in several species a number of genes related to biosynthesis were lost whereas a number of peptidases seem to be acquired. Genes encoding proteins involved in sugar metabolism and transport appear to be have been duplicated. Examples include phosphoenolpyruvate phosphotransferase systems, β-galactosidase, GpmB family sugar phosphatases, galactose mutarotase and L-lactate dehydrogenases of different classes (Makarova *et al.*, 2006).

Within the *Lactobacillales*, there seems to be different levels of evolutionary change. *S. thermophilus* and *L. delbrueckii* ssp. have relatively more tRNAs and rRNAs and a high number of pseudogenes (Makarova *et al.*, 2006; van de Guchte *et al.*, 2006). For example, 12% of the total number of genes in the *L. delbrueckii* genome are pseudogenes, possibly indicative of ongoing specialization. Of the 81 pseudogenes for which a function could be assigned, 51 did not have paralogues, thus meaning that the respective function was completely lost. Among these are genes coding for proteins involved in carbohydrate metabolism and amino acid biosynthesis (van de Guchte *et al.*, 2006). This is in contrast to the genome of *L. johnsonii*, for example, which does not have any pseudogenes, indicating that this bacterium is not in a state of ongoing specialization. This is further substantiated by the finding that a low number of genes are predicted to be obtained throughout evolution (Makarova *et al.*, 2006; Makarova and Koonin, 2007). The availability of several *Lactobacillus* genome sequences, combined with the application of comparative genome analyses, has revealed valuable information of this complex taxonomy which is probably merely indicative of the usefulness of this approach in such a diverged group of species.

Functional genomics

Since publication of the genome sequences of the 10 *Lactobacillus* strains, proteins have been functionally characterized for the human isolates *L. plantarum* WCFS, *L. johnsonii* NCC 533, *L. acidophilus* NCFM and *L. salivarius* UCC118. The majority of research has focused on bacterium-host interaction, mainly with regard to the adhesive properties of these probiotic organisms and their ability to deal with *in vivo* environmental stresses such as bile acids. Table 3.2 lists all functional characterized genes of sequenced lactobacilli which are discussed in this review.

Adhesion factors

The interaction of an organism with its environment is often mediated by surface proteins. There are several mechanisms by which a surface protein can be anchored to the Gram-positive bacterial cell wall. Proteins can be linked to the peptidoglycan via a LysM-binding domain (Steen *et al.*, 2003), a choline-binding domain (Garcia *et al.*, 1988), GW-repeats (Marino *et al.*, 2002) or proteins can be maintained in the cell membrane by a so-called signal anchor (von Heijne, 1988). As previously mentioned, certain proteins can be covalently linked to the peptidoglycan by sortase (Fischetti *et al.*, 1990). The distribution of sortase-dependent proteins varies considerably between different species. For example, *L. plantarum* WCFS is predicted to encode 27 proteins of this family (Boekhorst *et al.*, 2005; Boekhorst *et al.*, 2006b) whereas the predicted secretomes of *L. sakei* 23K (Chaillou *et al.*, 2005) and *L. reuteri* ATCC 55730 (Bath *et al.*, 2005) reveal only four and five putative sortase-dependent proteins, respectively. Based on primary annotation, the majority of sortase-dependent proteins are of unknown function whereas a small subset of sortase-dependent proteins might have an enzymatic function or might play a role in adhesion (Boekhorst *et al.*, 2005).

L. acidophilus is predicted to secrete 173 proteins, of which 56 are predicted to be cleaved by a signal peptidase (van Pijkeren *et al.*, 2006). Proteins have been identified with an LPXTG-motif (12), a lipoprotein anchor (5), GW-repeats (3) and one protein is predicted to be anchored via an LysM-binding domain (van Pijkeren *et al.*, 2006). Based on sequence similarity to previous functional characterized adhesins, Buck *et al.* functionally characterized by insertional inactivation and *in vitro* cell-line adhesion a set of five genes encoding putative adhesion factors in *L. acidophilus* (Buck *et al.*, 2005). Two proteins in *L. acidophilus* are homologous to the functionally characterized R28-protein of *Streptococcus pyogenes*, which is involved in epithelial cell adhesion (Stalhammar-Carlemalm *et al.*, 1999). Inactivation of the R28-homologues in *L. acidophilus* caused a reduction in adhesion to Caco-2 cells but was not deemed significant, mainly because of experimental variation (Buck *et al.*, 2005). Inactivation of a surface-layer protein in *L. acidophilus* had a significant effect on the adhesion ability of this strain using the Caco-2 adhesion model (Buck *et al.*, 2005). Surface-layer proteins are present in many genera of Gram-positive bacteria and have been linked with

Table 3.2 Overview of functionally characterized genes in the sequenced lactobacilli

Organism	Locus	Annotation[†]	Function[‡]	Reference
L. acidophilus NCFM	LBA0395	Formyl-CoA transferase	Catabolism of oxalate	(Azcarate-Peril *et al*., 2006)
	LBA0892	Bile salt hydrolase (*bsh*A)	Hydrolyses TCDCA and GCDCA	(McAuliffe *et al*., 2005)
	LBA1012	Trehalose PTSII ABC	Trehalose transporter	(Duong *et al*., 2006)
	LBA1014	Trehalose 6-P hydrolase	Catabolism trehalose	
	LBA1078	Bile salt hydrolase (*bsh*B)	Hydrolyses TDCA, TCA and TCDCA	(McAuliffe *et al*., 2005)
	LBA1148	Fibronectin-binding protein	Binds Caco-2 cells	(Buck *et al*., 2005)
	LBA1377	Mucus binding protein	Binds Caco-2 cells	(Buck *et al*., 2005)
	LBA1392	Mucus binding protein	Binds Caco-2 cells	(Buck *et al*., 2005)
	LBA1524 LBA1525	Two-component system	Confers acid resistance	(Azcarate-Peril *et al*., 2005)
	LBA1633	Cell surface protein	Undefined: mutation does not significantly reduce binding to Caco-2 cells	(Buck *et al*., 2005)
	LBA1634	Cell surface protein	Undefined: mutation does not significantly reduce binding to Caco-2 cells	(Buck *et al*., 2005)
L. johnsonii NCC 533	LJ0464	GroEL	Recombinant GroEL binds mucins and HT29 cells; induces IL8-production by macrophages; aggregates *H. pylori*	(Bergonzelli *et al*., 2006)
	LJ1009	Elongation factor Tu	Binds epithelial cell lines Caco-2 and HT29, recombinant protein binds mucins and induces IL8-production by HT29	(Granato *et al*., 2004)
L. plantarum WCFS1	lp_1229	Cell surface protein	Mannose-specific adhesin	(Pretzer *et al*., 2005)
	lp_3580 lp_3581 lp_3582	Two-component system	Involved in adhesion and biofilm formation on glass surface	(Sturme *et al*., 2005)
L. salivarius UCC118	LSL_0311	Mucus-binding protein	Binds epithelial cell lines HT29 and Caco-2	(van Pijkeren *et al*., 2006)
	LSL_1085	Hypothetical surface protein	Undefined: mutant showed no significant reduction in adhesion to epithelial cells	(van Pijkeren *et al*., 2006)
	LSL_1335	Mucus binding protein	Undefined: mutant not expressed *in vitro*	(van Pijkeren *et al*., 2006)
	LSL_1606	Sortase	Sortase-deletion significantly reduced adhesion to epithelial cell lines HT29 and Caco-2	(van Pijkeren *et al*., 2006)
	LSL_1838	Hypothetical protein	Undefined: mutant showed no significant reduction in adhesion to epithelial cells	(van Pijkeren *et al*., 2006)

[†]PTSII, phosphotransferase system II; [‡]TCDCA, taurochenodeoxycholic acid; GCDCA, glycogenodeoxycholic acid; TDCA, taurodeoxycholic acid; TCA, taurocholic acid.

adhesive properties in several studies (Toba *et al.*, 1995; Sillanpaa *et al.*, 2000; Calabi *et al.*, 2002; Avall-Jaaskelainen *et al.*, 2003). As suggested by Buck *et al.* it is plausible that the observed phenotype of the *L. acidophilus* strain lacking the surface layer protein is not directly attributable to the surface-layer protein (2005). Removal of the surface-layer could affect the accessibility or conformation of adhesins or could possibly change the charge of the cell which subsequently affects the adhesive ability of the bacterium.

In all sequenced bacterial genomes of the lactic acid bacteria group, a gene is present that is annotated as a fibrinogen/fibronectin-binding protein. Fibronectin is a mammalian glycoprotein that is present as a soluble dimer in plasma. In addition, fibronectin can be found as a less soluble multimer in the extracellular matrix (such as the epithelium) (Ruoslahti, 1988) which could be a receptor for binding by intestinal organisms. Some lactobacilli bind epithelial cells which can be correlated with the binding to fibronectin (Kapczynski *et al.*, 2000). Also, binding to immobilized fibronectin has been shown for some lactobacilli (Lorca *et al.*, 2002). Inactivation of the fibronectin-binding protein in *L. acidophilus* resulted in a significant decrease in the ability to adhere to Caco-2 cells (Buck *et al.*, 2005). It remains to be determined, however, if the specific receptor for this protein in *L. acidophilus* is fibronectin.

In the small intestine and colon, the epithelium is lined with a thick layer of mucus which acts as a barrier between the mucosa and the luminal content. Within the human gut, bacteria reside on or in the mucus and binding of the bacteria to the mucus (or epithelium) is a prerequisite for persistence (Berg, 1996). Roos *et al.* identified and functionally characterized a large (358 kDa) mucus-binding protein of *L. reuteri* (Roos and Jonsson, 2002) which is predicted to be anchored to the cell wall by an LPXTG-motif. This protein has two types of repeats: type one repeats are divergent in length and similarity whereas type 2 repeats are highly conserved. Both type of repeats have been shown to interact with mucus or mucus components (Roos and Jonsson, 2002). *In silico* analyses of several genomes of lactic acid bacteria revealed that the majority of strains have mucus-binding domains (Boekhorst *et al.*, 2006a). The length of these domains and their sequence are variable and their presence seems to be linked with the ecological niche of the bacterium. This is supported by the finding that the majority of mucus-binding domains are found in bacteria that are associated with the gastrointestinal tract (Boekhorst *et al.*, 2006a).

In *L. acidophilus*, 12 ORFs are present with that encode proteins with one or more mucus-binding domains. Five of these proteins have an LPXTG signal anchor (Boekhorst *et al.*, 2006a). One of these five proteins has one of the largest open reading frames (>4,000 amino acids) in *L. acidophilus* and encodes a mucus-binding protein, homologous to the functionally characterized mucus-binding protein in *L. reuteri* (Roos and Jonsson, 2002; Buck *et al.*, 2005). Inactivation of this gene reduced the adherence to Caco-2 cells significantly. The authors suggested that the interaction of the protein with the cells is other than to mucus since no mucus could be detected on the Caco-2 cells (Buck *et al.*, 2005). In this particular study it is shown that this protein is involved in binding to Caco-2 cells, but the question remains with what component(s) this mucus-binding protein interacts. Since this family of proteins are conserved in intestinal lactobacilli, and are considered important with regard to the interaction of the bacterium with the host, more in depth studies are required to clarify this question. Also, four proteins were identified in *L. salivarius* that contain mucus-binding domains and all four are members of a group of 10 sortase-dependent proteins (SDPs) (Fig. 3.3) (van Pijkeren *et al.*, 2006). Five genes coding for SDPs are putatively encoded by the chromosome, four are putatively encoded by pMP118 and one is putatively encoded by pSF118–44 (Fig. 3.3). The SDPs in *L. salivarius* are not clustered (Fig. 3.3), in contrast to some SDPs in *L. plantarum* (Siezen *et al.*, 2006). As in *L. acidophilus*, one of the largest ORFs is annotated as a mucus-binding protein (*mbp*1) but is a pseudogene, whereas a second mucus-binding protein (*lsp*C) is not expressed *in vitro* in *L. salivarius* (van Pijkeren *et al.*, 2006). A small gene fragment with mucus-binding domains and a LPXTG-motif is present on pMP118 but is lacking a signal peptide (Fig. 3.3). The fourth gene encoding a mucus-binding protein (*lsp*A)

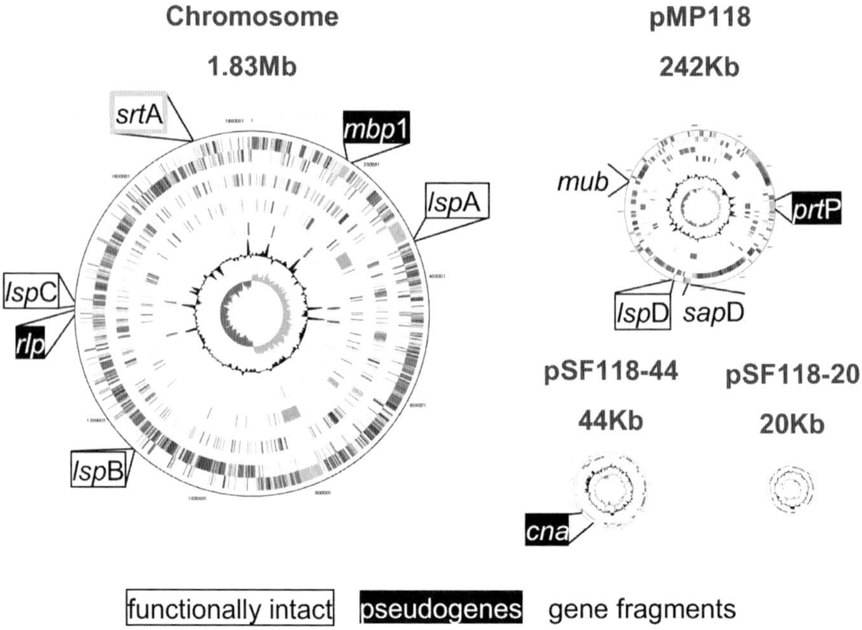

Figure 3.3 Genome atlas of *L. salivarius* UCC118 in which the gene coding for sortase (*srt*A), and the ten genes which putatively encode sortase-dependent proteins are depicted. Genes in black boxes are pseudogenes, indicative that the coding sequence is interrupted by an internal stop codon, or in the case of *prt*P, the coding sequence has a frame-shift. Green boxes indicate gene fragments corresponding to truncated proteins with an LPXTG-motif. The four transcriptionally and translationally intact genes encoding sortase-dependent proteins are named *Lactobacillus surface protein* A, B, C and D (*lsp*A, *lsp*B, *lsp*C and *lsp*D, respectively) of which only *lsp*D is encoded by pMP118. The gene coding for sortase is located on the chromosome and marked by a green box.

*mbp*1: mucus-binding protein 1; *rlp*: R28-like protein; *prt*P: protease P; *sap*D: salivary-agglutinin protein D; *mub*: mucus-binding protein; *cna*: collagen adhesin; *srt*A: sortase. This figure is adapted from Claesson *et al.* (Claesson *et al.*, 2006).

was inactivated by insertional inactivation and the mutant displayed a significant reduction in adhesion to the epithelial cell lines HT29 and Caco-C2. We suggested that the interaction of this protein with HT29 cells is plausibly with mucins, since mucin production is reported in HT29 cells when cultured on a glucose-containing medium (Mack and Hollingsworth, 1994). Mutants lacking production of other sortase-dependent proteins (*lsp*B and *lsp*D) did not show a significant reduction to HT29 and Caco-C2 cells. Moreover, the gene coding for sortase (*srt*A) was deleted in *L. salivarius* by double cross-over which resulted in a significant reduction in adherence to these cell lines at levels comparable to the mucus-binding mutant, suggesting the importance of this mucus-binding protein in this model (van Pijkeren *et al.*, 2006).

Since the sortase deletion did not completely abolish the adhesive ability of this strain, it is plausible that proteins other than sortase-dependent ones, are involved in adhesion. These proteins could be anchorless adhesions, such as the fibronectin-binding protein (Chhatwal, 2002; Buck *et al.*, 2005) which may, or may not, be exported via a non-typical secretion pathway (Bendtsen *et al.*, 2005). Other examples of functionally characterized adhesins that are non-sortase-dependent are the elongation factor Tu (Granato *et al.*, 2004) and the heat shock protein GroEL (Bergonzelli *et al.*, 2006), as characterized in *L. johnsonii* NCC 533. The elongation factor Tu lacks a typical secretion sequence and anchor mechanism. Despite this, it is displayed on the surface of the *L. johnsonii* cells and shows specific interaction with human intestinal cells and mucins (Granato *et al.*, 2004). Thus it is possibly another adhesion mechanism for intestinal lactobacilli since it is encoded by all sequenced intestinal lactobacilli. GroEL is a heat-shock protein involved in mediating protein folding within the cell, and is highly conserved among many organisms (Craig *et al.*, 1993; Gupta, 1995; Braig, 1998). Although GroEL is predicted to be intracellular, there have been several reports that this protein is displayed on the surface and might play a role in adhesion (Frisk *et al.*, 1998; Hennequin *et al.*, 2001; Skar *et al.*, 2003). A study by Bergonzelli *et al.* showed that the GroEL protein of *L. johnsonii* is displayed on the cell surface and is also found in the culture supernatant (Bergonzelli *et al.*, 2006). Recombinant GroEL adhered to mucus and the epithelial cell line HT29 in a pH-dependent manner, i.e. good binding at pH 5 whereas little binding was observed at pH 7.2. A four-fold induction of IL8 production was reported upon addition of recombinant GroEL to HT29 cells. Similar induction levels were obtained with recombinant GroEL proteins of *L. helveticus*, *Bacillus subtilis*, *Lactococcus lactis* and *Helicobacter pylori* (Bergonzelli *et al.*, 2006). Moreover, recombinant GroEL induced aggregation of the pathogen *H. pylori*, whereas decreasing aggregation levels were observed with recombinant GroEL of *L. helveticus*, *B. subtilis* and *L. lactis*, respectively, when compared to recombinant GroEL of *L. johnsonii*. To date it is unclear whether there is an export mechanism for GroEL proteins or that these proteins are located on the cell surface because of autolysis, as has been reported for the gastric pathogen *H. pylori* (Phadnis *et al.*, 1996).

As binding to mucus and epithelial cells are considered important features in colonization, Adlerberth *et al.* described the ability of *L. plantarum* to adhere to mannose containing sugar-moieties, which can be part of mucus, via a mannose-specific adherence mechanism (Adlerberth *et al.*, 1996). This mannose-specific adhesion mechanism might be involved in the colonization of the bacterium in the intestine and thus could possibly compete with pathogens for mannose-receptors at the epithelium (Adlerberth *et al.*, 1996; Pretzer *et al.*, 2005). *Saccharomyces cerevisiae* was used as a model to test the mannose-specific adhesion phenotype based on agglutination of the yeast cells. The basis for this is the presence of mannose-containing polysaccharides in the cell wall of *Saccharomyces cervisiae* (Pretzer *et al.*, 2005). Several *L. plantarum* strains were tested but only some of these aggregated the yeast cells (Pretzer *et al.*, 2005). *L. plantarum* encodes 27 predicted sortase-dependent proteins (Boekhorst *et al.*, 2005) and inactivation of sortase resulted in loss of aggregating phenotype. By *in silico* matching of genotypic and phenotypic traits of a set of *L. plantarum* strains, a sortase-dependent protein was identified that was responsible for this trait (Pretzer *et al.*, 2005). This was further substantiated by gene inactivation and over-expression of the mannose-specific adhesin.

Environmental influences on *Lactobacillus* gene expression

Both *in vivo* and *ex vivo* bacteria, can encounter some harsh conditions requiring alteration of gene expression patterns to enable survival. Intestinal organisms may encounter gastric acids, bile acids and scarcity of nutritional resources. Apart from scarce nutrients, organisms living *ex vivo* might encounter stresses such as low temperature, high salt concentrations, and fluctuation of oxygen levels. When a genome sequence is available, genes encoding proteins central to some survival mechanisms can be identified *in silico*. This predictive capacity has limitations since many proteins are of unknown function or have no homologue based on the proteins in the databases. Alternatively, chromosomal fragments can be cloned in a vector with a marker which can be used to identify promoters which are conditionally expressed. Several promoter-probe expression systems have been developed based on antibiotic markers, hydrolases such as β-galactosidases, and light emission using luciferase as a promoter probe. Disadvantages of these systems are lack of complete genome coverage and the fact they are very laborious to construct. An addition, and possibly an alternative to these established systems, are microarrays. Microarrays are being used as a resource for identification of differential expressed genes. This allows genome wide identification of differential gene expression and has also been proven a useful tool in genome wide comparative analyses.

Bile acid response

In humans, and most other mammals, bile salts are synthesized in the liver from cholesterol (Hofmann and Mysels, 1992). The produced and concentrated bile salts are stored in the gallbladder and upon ingestion of food, bile salts are secreted via the bile duct into the duodenum where they play a role in fat digestion throughout the small intestine (Hofmann, 1989; Hofmann and Mysels, 1992). Some bacteria have the ability to deconjugate bile salts, including several species of the genus *Lactobacillus* (Bateup *et al.*, 1995; De Smet *et al.*, 1995; De Boever and Verstraete, 1999; Grill *et al.*, 2000; Elkins *et al.*, 2001; Pereira *et al.*, 2003; Bron *et al.*, 2004c; Azcarate-Peril *et al.*, 2005; McAuliffe *et al.*, 2005). Bile salts are deconjugated by bile salt hydrolases, encoded by *bsh*. Homologues of this protein are found in all strains associated with the GIT (Begley *et al.*, 2006), whereas some organisms, such as *L. johnsonii* NCC533 and *L. plantarum*, WCFS1 have multiple *bsh* genes. The exact *in vivo* function of bile salt hydrolases are currently unknown but several roles have been suggested, such as a nutritional role, change of bacterial membrane characteristics resulting in a stronger membrane, bile detoxification in the host, and increased bacterial survival (reviewed in (Begley *et al.*, 2006)).

In *L. acidophilus* NCFM, two bile salt hydrolases have been identified and functionally characterized (McAuliffe *et al.*, 2005). Inactivation of one or other of the bile salt hydrolases did not affect the ability to grow in the presence of bile. However, hydrolysis of different bile salts was shown whereby each of the mutants displayed different substrate specificities. The bile salt response was also investigated in *L. plantarum* WCFS1 (Bron *et al.*, 2004c). Growth of this organism at low concentrations of porcine bile (0.05%) caused cells to clump together and their membrane surfaces appeared less smooth. Higher concentrations of bile (0.1–0.15%) exaggerated these phenotypic changes. Using an *alr*-complementation library strategy (Bron *et al.*, 2004b) 4000 colonies were screened for induced promoter activity when grown on MRS plates containing bile. Thirty-one unique loci were identified based on clones which displayed conditional growth. These clones have a predicted role in membrane function (8), cell wall function (3), redox reactions (5), regulation (5) and other functions (4) and loci annotated as (conserved) hypothetical function (6) (Bron *et al.*, 2004c). Also, a clone was identified which showed conditional expression corresponding to a homologue of the bile acid 7α-dehydratase. Although this homologue is found in an operon structure in other intestinal organisms (Mallonee *et al.*, 1990; Wells and Hylemon, 2000), in *L. plantarum* this gene is not part of an operon structure, yet it is conditionally expressed in the presence of bile (Bron *et al.*, 2004c). Two genes encoding an integral membrane protein and an argininosuccinate synthase, were subjected to further investigation. Bacterial RNA was isolated from the duodenum of mice which were fed with *L.* plantarum and expression analysis confirmed up-regulation of those two genes *in vivo* (Bron *et al.*, 2004c). Since bile in the duodenum is released via the bile duct, these *in vivo* finding could be directly correlated with the observed *in vitro* findings and thus exemplify the usefulness of the *alr*-complementation system in *L. plantarum*, and potentially in other lactic acid bacteria. Another approach to identify bile-responsive genes in *L. plantarum* is based on DNA microarray analysis. Transcriptional profiling based on growth of *L. plantarum* with and without bile showed that bile exposure resulted in the up-regulation of 28 genes and down-regulation of 68 genes (Bron *et al.*, 2006). Differential gene expression was observed in genes related to stress, such as increased expression of glutathionine reductase which could protect the cells against oxidative stress, as previously shown in *L. lactis* (Li *et al.*, 2003; Bron *et al.*, 2006). One of the four genes encoding a choloylglycine hydrolase, annotated as *bsh*1, was up-regulated 5-fold whereas a second choloylglycine hydrolase was down-regulated (9.8- and 3.0-fold in experiment 1 and 2, respectively) which does not add to the clarification of the exact role of bile salt hydrolases. Many genes that were differentially expressed were related to cytoplasmic membrane and cell-wall-associated functions, as was noted in a previous study (Li *et al.*, 2003; Bron *et al.*, 2004c; Bron *et al.*, 2006) which would be suggestive that the bacterium is adapting to a new environment resulting in a better chance for survival. This could possibly be supported by the fact that expression of some bile salt hydrolases is up-regulated upon exposure to bile.

Two-component regulatory systems (2CS)

Changing levels of gene expression are often preceded by sensing a change in environmental conditions. In many bacteria, including lactic acid bacteria, one of the mechanisms is the family of two-component systems (2CS) (Kleerebezem *et al.*, 1997; Hellingwerf *et al.*, 1998). This system includes a membrane-associated histidine protein kinase (HPK) which 'senses' the environmental change, and a so-called response-regulator (RR) which is located in the cytoplasm of the cell and has a direct effect on the gene regulation (Hellingwerf *et al.*, 1998; Kleerebezem *et al.*, 1997). Both the HPK and the RR consist of two domains. The N-terminal domain and the C-terminal domain of the HPK hold the 'sensing domain' and 'transmitter domain', respectively. The N-terminal domain and the C-terminal domain of the RR hold the 'receiver domain' and 'output domain', respectively. Upon sensing an environmental change by the HPK, autophosphorylation occurs at the conserved histidine residue at the C-terminus. The phosphate is transported to the RR and is covalently bound at the conserved aspartate residue, located at the N-terminus, resulting in a change of activity at the output domain which eventually results in an adaptive response. For example, in *L. lactis* MG1363, six 2CS have been identified, involved in sensing change in pH, osmotic and oxidative conditioning, and regulation of phosphatase activity (O'Connell-Motherway *et al.*, 2000). Another example is the ability of *B. subtilis* to sense a change in temperature, also mediated by a 2CS (Aguilar *et al.*, 2001).

Based on its genome sequence, a total of nine putative 2CS have been identified in *L. acidophilus*. One of these 2CS in *L. acidophilus* is homologous to a 2CS of *L. monocytogenes* which is responsible for acid tolerance and virulence (Cotter *et al.*, 1999). The corresponding locus in *L. acidophilus* was inactivated by plasmid insertion and survival of these strains was assessed when grown in an acidified environment (pH 3.5) (Azcarate-Peril *et al.*, 2005). As was observed in *L. monocytogenes*, the mutant was more susceptible to acid then the control strain, reflected in a 2-log reduction in viability of the mutant over a 2.5 hour period. Interestingly, when bacteria were pre-exposed to medium at pH 5.5, followed by incubation at pH 3.5, an adaptive response was observed whereby the mutant showed similar levels of sensitivity as the control. This is indicative that the adaptive response to acid is regulated by mechanism(s) other than this 2CS (Azcarate-Peril *et al.*, 2005). Examination of the expression profiles of the control strain and the mutant when subjected to various acid conditions identified over 80 genes which were differentially expressed, of which the majority were up-regulated by acid exposure. Over 40 of these genes are predicted to encode membrane proteins. Highest levels of expression were observed in genes coding for proteolytic enzymes. This was further evaluated by assessing the ability to acidify skimmed milk. The mutant was not able to acidify the medium below pH 5 whereas the pH of medium containing the wild type strain dropped to almost 4.0. The mutant was able to reach similar pH levels when the medium was supplemented with yeast extract, thus suggesting that the mutant was lacking proteolytic activity (Azcarate-Peril *et al.*, 2005). The authors argued that it is plausible that some amino acid transporters were expressed at lower levels, resulting in low intracellular levels of amino acids. Subsequently, proteolytic enzymes and peptide transporters are up-regulated for compensation of this deficiency. Because the majority of genes were up-regulated, it was proposed that this investigated response regulator protein acts as a repressor (Azcarate-Peril *et al.*, 2005).

Gene regulation can also be influenced by two-component systems whereby the trigger to a response is cell density, also referred to as 'quorum sensing' (Waters and Bassler, 2005). This type of regulation has been identified in Gram-negative as well as Gram-positive bacteria. Quorum sensing involves a small signalling molecule that interacts with the response regulator resulting in a change in gene regulation. In many Gram-negative bacteria this signalling molecule is N-acyl homoserine lactone (Eberl, 1999), whereas in Gram-positive bacteria signalling occurs mostly via a small peptide, also known as autoinducing peptides (AIPs) (Kleerebezem *et al.*, 1997; Lazazzera and Grossman, 1998).

In *L. plantarum* an *agr*-like two-component system was identified and named *lam*BDCA (Sturme *et al.*, 2005). This 2CS was shown to be growth-phase dependent and the mutant strain, in which the predicted response regulator (*lam*A) was deleted, displayed a significant reduction in adherence to a glass surface. Also, biofilm was no longer produced on a glass surface by the *lam*A-deficient strain (Sturme *et al.*, 2005). Only a limited number of loci were differentially expressed in the mutant, as determined by microarray analysis. The authors argued that the observed transcriptional effects could be secondary, rather than primary effects regulated by *lam*A, since no specific *cis*-acting elements upstream of the regulated genes could be identified (Sturme *et al.*, 2005). All in all, this 2CS might play a role in bacteria–host interaction because of its role in adhesion.

In vivo bacterial gene regulation

In vivo gene regulation, i.e. when bacteria are in the GIT, has been investigated using a range of methods, such as signature tagged mutagenesis (STM) (Autret *et al.*, 2003; Lestrate *et al.*, 2003; Paik *et al.*, 2005; Cuccui *et al.*, 2006), selective capture of transcribed sequences (SCOTS) (Graham *et al.*, 2002; Dozois *et al.*, 2003; Faucher *et al.*, 2006) and *in vivo* expression technology (IVET), originally described for *Salmonella* (Mahan *et al.*, 1993) and later exploited in other organisms (Walter *et al.*, 2003; Fedhila *et al.*, 2006). If the IVET system is applied using antibiotic markers, dual reporters or auxotrophic markers to trap promoters, weakly expressed genes *in vitro* might be identified as strongly expressed promoters *in vivo* (Bron *et al.*, 2004a).

In *L. plantarum* an alternative to the IVET system was applied, namely recombinant-IVET (R-IVET) (Bron *et al.*, 2004a). An erythromycin marker flanked by two *lox*P-sites was integrated in the chromosome. A library was constructed using 1–2 kb chromosomal fragments which were cloned upstream of the promoter-less *cre*-resolvase gene. Promoter activity results in expression of the *cre*-resolvase leading to excision of the chromosomal located erythromycin marker by recognition of the LoxP sites. Using this technology, 72 genes were identified which were induced *in vivo* (Bron *et al.*, 2004a). These genes were categorized in the following groups: sugar transport and utilization (9 genes), acquisition of non-sugars and amino acids (9 genes), stress (4 genes), extracellular proteins (4 genes), regulators (3 genes), hypothetical function (24 genes) and genes involved in diverse pathways, such as DNA and energy metabolism, fermentation and synthesis (19 genes) (Bron *et al.*, 2004a). Although this study revealed which proteins are required for adaptation, maintenance or survival *in vivo*, no information could be deduced with regards to the specific site of gene expression when *in transit*. This was addressed in a separate study whereby different parts of the GIT were investigated upon feeding of *L. plantarum* to mice (Marco *et al.*, 2007). Based on the previously performed R-IVET study (Bron *et al.*, 2004a) 15 genes were selected and their expression profiles were investigated at different time points and at different sites *in vivo* (Marco *et al.*, 2007). High induction of expression was detected for a putative copper-transporting ATPase gene (*copA*) and a putative bi-functional alcohol dehydrogenase (*adhE*) in all intestinal compartments and at all times examined. A gene annotated as a haemolysin (*hem*) was consistently down-regulated throughout transit. Modest levels of differential expression were detected for the remaining 12 genes which in some cases were dependent on the *in vivo* location (Marco *et al.*, 2007). In terms of localization, the majority of *L. plantarum* cells present were identified in the caecum and colon, whereas over an 8-hour period bacteria were shed faster from the stomach and small intestine. After 24-hours, the number of *L. plantarum* cells detected was comparable to control levels, suggestive that the majority of bacteria ingested were being shed (Marco *et al.*, 2007). The use of R-IVET in a previous study made it possible to select a subset of genes for further investigation. In this particular study, RT-PCR proved to be a critically important tool for the identification of differentially expressed genes in the different compartments of the GIT of the mouse. Transit of *L. plantarum* through the GIT of a mouse is nearly complete after 24 hours and thus microarray analysis would be hard to implement since relatively large amounts of RNA are required. The combination of R-IVET and directed RT-PCR analysis are useful and powerful tools to yield a better understanding of *in vivo* expression profiles of bacteria.

Metabolism and catabolism

The ability of an organism to metabolize, catabolize and transport substrates e.g. carbohydrates and amino acids, is a reflection of its dependence on nutrient sources in its environment, and thus how versatile an organism is. In environments with limited nutritional resources, amino acid catabolism could play an important role in obtaining energy. Sugar catabolism is another energy source, and for lactic acid bacteria it is not an uncommon feature to have the ability to break down several sugars. Still, little is known with regard to expression profiles of pathways responsible for carbohydrate utilization.

Transcriptional profiles of *L. acidophilus* NCFM were obtained when grown on a set of eight different carbohydrates, namely glucose, fructose, sucrose, lactose, galactose, trehalose, raffinose and fructo-oligosaccharides (Barrangou *et al.*, 2006), causing 63 genes (3% of the total genome) to be up-regulated. Based on transcription profiles, the phosphotransferase system (PTS) transporters were predicted to transport monosaccharides (fructose and sucrose) and a disaccharide (trehalose). Fructo-oligosaccharides and raffinose were predicted to be transported by ABC-transporters. Galactoside-pentose hexuronide (GPH) translocators were responsible for transport of galactosides. Most carbon sources induced expression of genes encoding proteins related to their own transport and hydrolysis. This was not the case for glucose and fructose since gene clusters corresponding to transport and hydrolysis of these carbohydrates were constitutively highly expressed. This suggests that this organism preferably uses these carbon sources for growth, although it has the ability to use others (Barrangou *et al.*, 2006). In an analogous *in silico* study, highly expressed genes were predicted using the codon adaptation index (CAI) in the genome of *L. plantarum* WCFS1 (Karlin *et al.*, 2001; Kleerebezem *et al.*, 2003). Many genes were predicted to be highly expressed related to phosphotransferase systems (PTSs) and thus involved in sugar catabolism, especially the mannose and fructose PTSs (Kleerebezem *et al.*, 2003).

The disaccharide trehalose has been linked with cryoprotection, although little is known with regard to the exact mechanism behind this (Israeli *et al.*, 1993; Hincha and Hagemann, 2004; Li and Tian, 2006). The *tre* locus has been examined in *L. acidophilus* NCFM (Duong *et al.*, 2006). This locus consists of a transcriptional regulator (*treR*), a trehalose PTS transporter (*treB*) and a trehalose-6-phospahte-hydrolase (*treC*). Separate mutants were constructed by plasmid insertion mutagenesis, targeting the *treB* and *treC* genes. Both mutants were unable to grow on trehalose, showing that both transport and metabolism of this carbohydrate is required for growth. Both intracellular and extracellular trehalose contributed to a cryoprotective effect, but extracellular trehalose appeared to be less effective for protection compared to intracellular trehalose (Duong *et al.*, 2006).

The ability to break down toxic intestinal substances would be a desirable trait for bacteria that will be ingested. This could either be bacteria used in fermentation processes or bacteria that have been specifically selected for their probiotic properties. An example of a toxic substance is oxalate, which may be consumed in the diet and is found in cereals (Siener *et al.*, 2006), tea (Savage *et al.*, 2003) and many vegetables (Brinkley *et al.*, 1990). Oxalate can complex with calcium and is believed to play a role in the formation of urolithiasis (kidney stones) (Micali *et al.*, 2006). *Oxalobacter formigenes* can be found in the intestine (Allison *et al.*, 1985) and it has been shown to break down oxalate via a oxalyl-coenzyme A decarboxylase, catalysing the decarboxylation of oxalyl-CoA to formyl-CoA (Baetz and Allison, 1989). A correlation was found between the absence of this bacterium in particular individuals and the formation of kidney stones (Kumar *et al.*, 2002; Kwak *et al.*, 2003). The enzyme is not common in LAB acid bacteria but has been identified in *B. lactis* DSM 10140, *L. gasseri* NCK334 and *L. acidophilus* NCFM (Azcarate-Peril *et al.*, 2006). The enzyme has been functionally characterized in *B. lactis* (Federici *et al.*, 2004) and more recently in *L. acidophilus* NCFM (Azcarate-Peril *et al.*, 2006).

An operon in *L. acidophilus* potentially involved in oxalate catabolism was identified, consisting of two genes, namely an oxalyl coenzyme A decarboxylase gene (*oxc*) and a formyl coenzyme A transferase gene (*frc*). Cells in the logarithmic phase of growth were exposed to MRS (pH 6.8) and acidified MRS (pH 4.5 and 5.5). It was noted that at pH 5.5, the *oxc* and *frc* genes were up-regulated over three fold. When cells were adapted to ammonium oxalate (0.05%) for three passages and then exposed to MRS (pH 5.5) containing 0.5% ammonium oxalate, 3–4-fold up-regulation was observed for *frc* and *oxc*. However, when non-adapted cells were exposed to MRS (pH 5.5) containing ammonium oxalate, only 4-fold up-regulation was observed for *oxc* whereas little change in expression was observed for *frc* (Azcarate-Peril *et al.*, 2006). Lower amounts of oxalate were detected in the supernatant of the wild type compared to an *frc*-mutant, suggesting Frc is involved in oxalate catabolism. It was noted by the authors that these genes were possibly obtained by horizontal gene transfer, because of their divergent G+C% (Azcarate-Peril *et al.*, 2006). Although several questions remain to be answered with regard to, for example, the observation of differential expression within the same operon and how oxalate can be actively transported into the cell, this clearly shows the potential of *L. acidophilus* to catabolize oxalate. The use of probiotic bacteria to break down toxic compounds could have prophylactic or therapeutic potential, and might provide new opportunities for exploiting additional health-promoting properties of commensal lactobacilli.

Bacteriophages

Bacteriophages (or phage) are bacterial viruses which often exhibit strain specificity. In the dairy industry phages are undesirable, as they may lyse starter cultures. In contrast, bacteriophages, or bacteriophage products, can be useful for therapeutic applications and may aid in curing bacterial infections (Stone, 2002; Clark and March, 2006). Also, prophages have been linked with pathogenicity in *S. aureus* (Bae *et al.*, 2006) and *S. agalactiae* (van der Mee-Marquet *et al.*, 2006) since genes related to virulence were bacteriophage related. A recent survey screening 302 completed bacterial genomes identified 403 putative prophage regions, corresponding to 2.7% of the combined bacterial genome size (Fouts, 2006). However, larger prophage contents have been reported for *Streptococcus pyogenes* with four to six prophages encompassing 12% of the total genome content (Canchaya *et al.*, 2003).

Prophages or prophage elements have been detected in most sequenced *Lactobacillus* genomes and have been functionally characterized in *L. plantarum* WCFS1 (Ventura *et al.*, 2003), *L. johnsonii* NCC 533 (Ventura *et al.*, 2004) and in the *Lactobacillus* species *gasseri*, *salivarius*, and *casei* (Ventura *et al.*, 2006).

Integrases can be correlated with mobile DNA elements, such as integrative plasmids, pathogenicity islands and prophages (Brugger *et al.*, 2002). A total of nine integrases were identified in the genome of *L. plantarum* WCFS1 and led to the identification of one complete prophage (Lp1) and a prophage which is likely to be defective (Lp2), both sharing 88% DNA similarity over 10 kb. In addition, two prophage remnants were detected (R-Lp3 and R-Lp4) (Ventura *et al.*, 2003). Phages could not be induced by mitomycin C treatment. Transcription was examined and transcripts were obtained for the lysogeny module of Lp1, whereas no transcripts were obtained for early lytic genes, DNA replication genes, transcription regulator genes, structural genes and phage holin genes (Ventura *et al.*, 2003). By searching for integrase genes, two prophages were identified in *L. johnsonii* NCC 533 of 38 kb (Lj928) and 40 kb (Lj965). One integrase homologue identified an additional region DNA sequence with weak phage relatedness (Ventura *et al.*, 2004). Induction by mitomycin C did not result in excised phage, and transcriptional analysis showed that the majority of genes were not transcribed (Ventura *et al.*, 2004). Four prophages were identified in the genome of *L. salivarius* UCC118 (Sal1, Sal2, Sal3 and Sal4) of approximately 49 kb, 40 kb, 13 kb and 11 kb, respectively. Sal1 and Sal2 appeared to be complete whereas Sal3 and Sal4 were remnants. The lysin genes present in Sal1 and Sal2 were shown to be transcribed. Induction by mitomycin excised Sal2 and Sal4 prophages and these are the only inducible prophages in lactobacilli sequenced so far. Genome-based characterization of phages may impact on selection of strains for commercial application, or technological processes for of probiotic products.

Conclusions and perspectives

Access to genome sequences of a range of lactobacilli from different niches has provided a better understanding of this complex family. However, one of the largest groups of *Lactobacillus*-specific genes encodes hypothetical proteins, emphasizing the need for functional genomics. The use of microarrays allows us to compare closely related species from different environments which will improve our limited understanding of the evolutionary process underlying this complex taxonomy. Although information is expanding for individual organisms, relatively little is still known about how lactobacilli co-exist with other organisms within complex communities. The range of tools already available to investigate the metabolome and transcriptome of bacteria will inevitably lead us to new exciting areas. With novel higher-throughput sequencing technology now being developed, the application of metagenomic analysis to the gastrointestinal microbiota will undoubtedly provide novel insights into this diverse genus.

Acknowledgements

Work in our laboratory is supported by Enterprise Ireland, and by Science Foundation Ireland through a Centre for Science, Engineering and Technology award to the Alimentary Pharmabiotic Centre, UCC. We thank Carlos Canchaya for advice on performing whole-genome alignments, and Michael Mangan for critically reading the manuscript.

References

Adlerberth, I., Ahrne, S., Johansson, M.L., Molin, G., Hanson, L.A., and Wold, A.E. (1996). A mannose-specific adherence mechanism in *Lactobacillus plantarum* conferring binding to the human colonic cell line HT-29. Appl. Environ. Microbiol. *62*, 2244–2251.

Aguilar, P.S., Hernandez-Arriaga, A.M., Cybulski, L.E., Erazo, A.C., and de Mendoza, D. (2001). Molecular basis of thermosensing: a two-component signal transduction thermometer in *Bacillus subtilis*. Embo J. *20*, 1681–1691.

Ahrne, S., Nobaek, S., Jeppsson, B., Adlerberth, I., Wold, A.E., and Molin, G. (1998). The normal *Lactobacillus* flora of healthy human rectal and oral mucosa. J. Appl. Microbiol. *85*, 88–94.

Allison, M.J., Dawson, K.A., Mayberry, W.R., and Foss, J.G. (1985). *Oxalobacter formigenes* gen. nov., sp. nov.: oxalate-degrading anaerobes that inhabit the gastrointestinal tract. Arch. Microbiol. *141*, 1–7.

Altermann, E., Russell, W.M., Azcarate-Peril, M.A., Barrangou, R., Buck, B.L., McAuliffe, O., Souther, N., Dobson, A., Duong, T., Callanan, M., Lick, S., Hamrick, A., Cano, R., and Klaenhammer, T.R. (2005). Complete genome sequence of the probiotic lactic acid bacterium *Lactobacillus acidophilus* NCFM. Proc. Natl. Acad. Sci. USA *102*, 3906–3912.

Autret, N., Raynaud, C., Dubail, I., Berche, P., and Charbit, A. (2003). Identification of the agr locus of *Listeria monocytogenes*: role in bacterial virulence. Infect. Immun. *71*, 4463–4471.

Avall-Jaaskelainen, S., Lindholm, A., and Palva, A. (2003). Surface display of the receptor-binding region of the *Lactobacillus brevis* S-layer protein in *Lactococcus lactis* provides nonadhesive lactococci with the ability to adhere to intestinal epithelial cells. Appl. Environ. Microbiol. *69*, 2230–2236.

Azcarate-Peril, M.A., Bruno-Barcena, J.M., Hassan, H.M., and Klaenhammer, T.R. (2006). Transcriptional and functional analysis of oxalyl-coenzyme A (CoA) decarboxylase and formyl-CoA transferase genes from *Lactobacillus acidophilus*. Appl. Environ. Microbiol. *72*, 1891–1899.

Azcarate-Peril, M.A., McAuliffe, O., Altermann, E., Lick, S., Russell, W.M., and Klaenhammer, T.R. (2005). Microarray analysis of a two-component regulatory system involved in acid resistance and proteolytic activity in *Lactobacillus acidophilus*. Appl. Environ. Microbiol. *71*, 5794–5804.

Bae, T., Baba, T., Hiramatsu, K., and Schneewind, O. (2006). Prophages of *Staphylococcus aureus* Newman and their contribution to virulence. Mol. Microbiol. *62*, 1035–1047.

Baetz, A.L., and Allison, M.J. (1989). Purification and characterization of oxalyl-coenzyme A decarboxylase from *Oxalobacter formigenes*. J. Bacteriol. *171*, 2605–2608.

Barrangou, R., Azcarate-Peril, M.A., Duong, T., Conners, S.B., Kelly, R.M., and Klaenhammer, T.R. (2006). Global analysis of carbohydrate utilization by *Lactobacillus acidophilus* using cDNA microarrays. Proc. Natl. Acad. Sci. USA *103*, 3816–3821.

Bateup, J.M., McConnell, M.A., Jenkinson, H.F., and Tannock, G.W. (1995). Comparison of *Lactobacillus* strains with respect to bile salt hydrolase activity, colonization of the gastrointestinal tract, and growth rate of the murine host. Appl. Environ. Microbiol. *61*, 1147–1149.

Bath, K., Roos, S., Wall, T., and Jonsson, H. (2005). The cell surface of *Lactobacillus reuteri* ATCC 55730 highlighted by identification of 126 extracellular proteins from the genome sequence. FEMS Microbiol. Lett. *253*, 75–82.

Begley, M., Hill, C., and Gahan, C.G. (2006). Bile salt hydrolase activity in probiotics. Appl. Environ. Microbiol. *72*, 1729–1738.

Bendtsen, J.D., Kiemer, L., Fausboll, A., and Brunak, S. (2005). Non-classical protein secretion in bacteria. BMC Microbiol. 5, 58.

Benson, D.A., Karsch-Mizrachi, I., Lipman, D.J., Ostell, J., Rapp, B.A., and Wheeler, D.L. (2000). GenBank. Nucleic Acids Res. *28*, 15–18.

Berg, R.D. (1996). The indigenous gastrointestinal microflora. Trends Microbiol. *4*, 430–435.

Berger, B., Pridmore, R.D., Barretto, C., Delmas-Julien, F., Schreiber, K., Arigoni, F., and Brussow, H. (2007). Similarity and Differences in the *Lactobacillus acidophilus* Group identified by Polyphasic Analysis and Comparative Genomics. J. Bacteriol. *189*, 1311–1321.

Bergonzelli, G.E., Granato, D., Pridmore, R.D., Marvin-Guy, L.F., Donnicola, D., and Corthesy-Theulaz, I.E. (2006). GroEL of *Lactobacillus johnsonii* La1 (NCC 533) is cell surface associated: potential role in interactions with the host and the gastric pathogen *Helicobacter pylori*. Infect. Immun. *74*, 425–434.

Boekhorst, J., de Been, M.W., Kleerebezem, M., and Siezen, R.J. (2005). Genome-wide detection and analysis of cell wall-bound proteins with LPxTG-like sorting motifs. J. Bacteriol. *187*, 4928–4934.

Boekhorst, J., Helmer, Q., Kleerebezem, M., and Siezen, R.J. (2006a). Comparative analysis of proteins with a mucus-binding domain found exclusively in lactic acid bacteria. Microbiology *152*, 273–280.

Boekhorst, J., Siezen, R.J., Zwahlen, M.C., Vilanova, D., Pridmore, R.D., Mercenier, A., Kleerebezem, M., de Vos, W.M., Brussow, H., and Desiere, F. (2004). The complete genomes of *Lactobacillus plantarum* and *Lactobacillus johnsonii* reveal extensive differences in chromosome organization and gene content. Microbiology *150*, 3601–3611.

Boekhorst, J., Wels, M., Kleerebezem, M., and Siezen, R.J. (2006b). The predicted secretome of *Lactobacillus plantarum* WCFS1 sheds light on interactions with its environment. Microbiology *152*, 3175–3183.

Bolotin, A., Quinquis, B., Renault, P., Sorokin, A., Ehrlich, S.D., Kulakauskas, S., Lapidus, A., Goltsman, E., Mazur, M., Pusch, G.D., Fonstein, M., Overbeek, R., Kyprides, N., Purnelle, B., Prozzi, D., Ngui, K., Masuy, D., Hancy, F., Burteau, S., Boutry, M., Delcour, J., Goffeau, A., and Hols, P. (2004). Complete sequence and comparative genome analysis of the dairy bacterium *Streptococcus thermophilus*. Nat. Biotechnol. *22*, 1554–1558.

Braig, K. (1998). Chaperonins. Curr. Opin. Struct. Biol. *8*, 159–165.

Brinkley, L.J., Gregory, J., and Pak, C.Y. (1990). A further study of oxalate bioavailability in foods. J. Urol. *144*, 94–96.

Bron, P.A., Grangette, C., Mercenier, A., de Vos, W.M., and Kleerebezem, M. (2004a). Identification of *Lactobacillus plantarum* genes that are induced in the gastrointestinal tract of mice. J. Bacteriol. *186*, 5721–5729.

Bron, P.A., Hoffer, S.M., Van, S., II, De Vos, W.M., and Kleerebezem, M. (2004b). Selection and characterization of conditionally active promoters in *Lactobacillus plantarum*, using alanine racemase as a promoter probe. Appl. Environ. Microbiol. *70*, 310–317.

Bron, P.A., Marco, M., Hoffer, S.M., Van Mullekom, E., de Vos, W.M., and Kleerebezem, M. (2004c). Genetic characterization of the bile salt response in *Lactobacillus plantarum* and analysis of responsive promoters in vitro and in situ in the gastrointestinal tract. J. Bacteriol. *186*, 7829–7835.

Bron, P.A., Molenaar, D., de Vos, W.M., and Kleerebezem, M. (2006). DNA micro-array-based identification of bile-responsive genes in *Lactobacillus plantarum*. J. Appl. Microbiol. *100*, 728–738.

Brugger, K., Redder, P., She, Q., Confalonieri, F., Zivanovic, Y., and Garrett, R.A. (2002). Mobile elements in archaeal genomes. FEMS Microbiol. Lett. *206*, 131–141.

Buck, B.L., Altermann, E., Svingerud, T., and Klaenhammer, T.R. (2005). Functional analysis of putative adhesion factors in *Lactobacillus acidophilus* NCFM. Appl. Environ. Microbiol. *71*, 8344–8351.

Calabi, E., Calabi, F., Phillips, A.D., and Fairweather, N.F. (2002). Binding of *Clostridium difficile* surface layer proteins to gastrointestinal tissues. Infect. Immun. *70*, 5770–5778.

Canchaya, C., Claesson, M.J., Fitzgerald, G.F., van Sinderen, D., and O'Toole, P.W. (2006). Diversity of the genus *Lactobacillus* revealed by comparative genomics of five species. Microbiology *152*, 3185–3196.

Canchaya, C., Proux, C., Fournous, G., Bruttin, A., and Brussow, H. (2003). Prophage genomics. Microbiol. Mol. Biol. Rev. *67*, 238–276.

Chaillou, S., Champomier-Verges, M.C., Cornet, M., Crutz-Le Coq, A.M., Dudez, A.M., Martin, V., Beaufils, S., Darbon-Rongere, E., Bossy, R., Loux, V., and Zagorec, M. (2005). The complete genome sequence of the meat-borne lactic acid bacterium *Lactobacillus sakei* 23K. Nat. Biotechnol. *23*, 1527–1533.

Chhatwal, G.S. (2002). Anchorless adhesins and invasins of Gram-positive bacteria: a new class of virulence factors. Trends Microbiol. *10*, 205–208.

Claesson, M.J., Li, Y., Leahy, S., Canchaya, C., van Pijkeren, J.P., Cerdeno-Tarraga, A.M., Parkhill, J., Flynn, S., O'Sullivan, G.C., Collins, J.K., Higgins, D., Shanahan, F., Fitzgerald, G.F., van Sinderen, D., and O'Toole, P.W. (2006). Multireplicon genome architecture of *Lactobacillus salivarius*. Proc. Natl. Acad. Sci. USA *103*, 6718–6723.

Clark, J.R., and March, J.B. (2006). Bacteriophages and biotechnology: vaccines, gene therapy and antibacterials. Trends Biotechnol. *24*, 212–218.

Cossart, P., and Jonquieres, R. (2000). Sortase, a universal target for therapeutic agents against Gram-positive bacteria? Proc. Natl. Acad. Sci. USA *97*, 5013–5015.

Cotter, P.D., Emerson, N., Gahan, C.G., and Hill, C. (1999). Identification and disruption of lisRK, a genetic locus encoding a two-component signal transduction system involved in stress tolerance and virulence in *Listeria monocytogenes*. J. Bacteriol. *181*, 6840–6843.

Craig, E.A., Gambill, B.D., and Nelson, R.J. (1993). Heat shock proteins: molecular chaperones of protein biogenesis. Microbiol. Rev. *57*, 402–414.

Cuccui, J., Easton, A., Chu, K.K., Bancroft, G.J., Oyston, P.C., Titball, R.W., and Wren, B.W. (2006). Development of signature tagged mutagenesis in

Burkholderia pseudomallei to identify genes important in survival and pathogenesis. Infect. Immun.

De Boever, P., and Verstraete, W. (1999). Bile salt deconjugation by *Lactobacillus plantarum* 80 and its implication for bacterial toxicity. J. Appl. Microbiol. *87*, 345–352.

De Smet, I., Van Hoorde, L., Vande Woestyne, M., Christiaens, H., and Verstraete, W. (1995). Significance of bile salt hydrolytic activities of lactobacilli. J. Appl. Bacteriol. *79*, 292–301.

Delcher, A.L., Phillippy, A., Carlton, J., and Salzberg, S.L. (2002). Fast algorithms for large-scale genome alignment and comparison. Nucleic Acids Res. *30*, 2478–2483.

Dozois, C.M., Daigle, F., and Curtiss, R.,3rd (2003). Identification of pathogen-specific and conserved genes expressed in vivo by an avian pathogenic *Escherichia coli* strain. Proc. Natl. Acad. Sci. USA *100*, 247–252.

Duong, T., Barrangou, R., Russell, W.M., and Klaenhammer, T.R. (2006). Characterization of the tre locus and analysis of trehalose cryoprotection in *Lactobacillus acidophilus* NCFM. Appl. Environ. Microbiol. *72*, 1218–1225.

Eberl, L. (1999). N-acyl homoserinelactone-mediated gene regulation in Gram-negative bacteria. Syst. Appl. Microbiol. *22*, 493–506.

Elkins, C.A., Moser, S.A., and Savage, D.C. (2001). Genes encoding bile salt hydrolases and conjugated bile salt transporters in *Lactobacillus johnsonii* 100–100 and other *Lactobacillus* species. Microbiology *147*, 3403–3412.

Ennahar, S., Cai, Y., and Fujita, Y. (2003). Phylogenetic diversity of lactic acid bacteria associated with paddy rice silage as determined by 16S ribosomal DNA analysis. Appl. Environ. Microbiol. *69*, 444–451.

Faucher, S.P., Porwollik, S., Dozois, C.M., McClelland, M., and Daigle, F. (2006). Transcriptome of *Salmonella enterica* serovar Typhi within macrophages revealed through the selective capture of transcribed sequences. Proc. Natl. Acad. Sci. USA *103*, 1906–1911.

Federici, F., Vitali, B., Gotti, R., Pasca, M.R., Gobbi, S., Peck, A.B., and Brigidi, P. (2004). Characterization and heterologous expression of the oxalyl coenzyme A decarboxylase gene from *Bifidobacterium lactis*. Appl. Environ. Microbiol. *70*, 5066–5073.

Fedhila, S., Daou, N., Lereclus, D., and Nielsen-LeRoux, C. (2006). Identification of *Bacillus cereus* internalin and other candidate virulence genes specifically induced during oral infection in insects. Mol. Microbiol. *62*, 339–355.

Fischetti, V.A., Pancholi, V., and Schneewind, O. (1990). Conservation of a hexapeptide sequence in the anchor region of surface proteins from Gram-positive cocci. Mol. Microbiol. *4*, 1603–1605.

Fitch, W.M. (1970). Distinguishing homologous from analogous proteins. Syst. Zool. *19*, 99–113.

Fouts, D.E. (2006). Phage_Finder: automated identification and classification of prophage regions in complete bacterial genome sequences. Nucleic Acids Res. *34*, 5839–5851.

Frisk, A., Ison, C.A., and Lagergard, T. (1998). GroEL heat shock protein of *Haemophilus ducreyi*: association with cell surface and capacity to bind to eukaryotic cells. Infect. Immun. *66*, 1252–1257.

Garcia, E., Garcia, J.L., Garcia, P., Arraras, A., Sanchez-Puelles, J.M., and Lopez, R. (1988). Molecular evolution of lytic enzymes of *Streptococcus pneumoniae* and its bacteriophages. Proc. Natl. Acad. Sci. USA *85*, 914–918.

Graham, J.E., Peek, R.M., Jr., Krishna, U., and Cover, T.L. (2002). Global analysis of *Helicobacter pylori* gene expression in human gastric mucosa. Gastroenterol. *123*, 1637–1648.

Granato, D., Bergonzelli, G.E., Pridmore, R.D., Marvin, L., Rouvet, M., and Corthesy-Theulaz, I.E. (2004). Cell surface-associated elongation factor Tu mediates the attachment of *Lactobacillus johnsonii* NCC533 (La1) to human intestinal cells and mucins. Infect. Immun. *72*, 2160–2169.

Grill, J.P., Cayuela, C., Antoine, J.M., and Schneider, F. (2000). Isolation and characterization of a *Lactobacillus amylovorus* mutant depleted in conjugated bile salt hydrolase activity: relation between activity and bile salt resistance. J. Appl. Microbiol. *89*, 553–563.

Gupta, R.S. (1995). Evolution of the chaperonin families (Hsp60, Hsp10 and Tcp-1) of proteins and the origin of eukaryotic cells. Mol. Microbiol. *15*, 1–11.

Hammes, W.P., and Vogel, R.F. (1995). The genus *Lactobacillus*. In The lactic acid bacteria, B. Wood, and H. Holzapfel, eds. (London, Blackie Academic and Professional), pp. 19–54.

Hellingwerf, K.J., Crielaard, W.C., Joost Teixeira de Mattos, M., Hoff, W.D., Kort, R., Verhamme, D.T., and Avignone-Rossa, C. (1998). Current topics in signal transduction in bacteria. Antonie Van Leeuwenhoek *74*, 211–227.

Hennequin, C., Porcheray, F., Waligora-Dupriet, A., Collignon, A., Barc, M., Bourlioux, P., and Karjalainen, T. (2001). GroEL (Hsp60) of *Clostridium difficile* is involved in cell adherence. Microbiology *147*, 87–96.

Hincha, D.K., and Hagemann, M. (2004). Stabilization of model membranes during drying by compatible solutes involved in the stress tolerance of plants and microorganisms. Biochem. J. *383*, 277–283.

Hofmann, A.F. (1989). Current concepts of biliary secretion. Dig. Dis. Sci. *34*, 16S-20S.

Hofmann, A.F., and Mysels, K.J. (1992). Bile acid solubility and precipitation in vitro and in vivo: the role of conjugation, pH, and Ca2+ ions. J. Lipid Res. *33*, 617–626.

Israeli, E., Shaffer, B.T., and Lighthart, B. (1993). Protection of freeze-dried *Escherichia coli* by trehalose upon exposure to environmental conditions. Cryobiology *30*, 519–523.

Kandler, O., and Weiss, N. (1986). Regular, nonsporing Gram-positive rods. (Baltimore, Williams and Wilkins).

Kapczynski, D.R., Meinersmann, R.J., and Lee, M.D. (2000). Adherence of *Lactobacillus* to intestinal 407 cells in culture correlates with fibronectin binding. Curr. Microbiol. *41*, 136–141.

Karlin, S., Mrazek, J., Campbell, A., and Kaiser, D. (2001). Characterizations of highly expressed genes of four fast-growing bacteria. J. Bacteriol. *183*, 5025–5040.

Kleerebezem, M., Boekhorst, J., van Kranenburg, R., Molenaar, D., Kuipers, O.P., Leer, R., Tarchini, R., Peters, S.A., Sandbrink, H.M., Fiers, M.W., Stiekema, W., Lankhorst, R.M., Bron, P.A., Hoffer, S.M., Groot, M.N., Kerkhoven, R., de Vries, M., Ursing, B., de Vos, W.M., and Siezen, R.J. (2003). Complete genome sequence of *Lactobacillus plantarum* WCFS1. Proc. Natl. Acad. Sci. USA *100*, 1990–1995.

Kleerebezem, M., Quadri, L.E., Kuipers, O.P., and de Vos, W.M. (1997). Quorum sensing by peptide pheromones and two-component signal-transduction systems in Gram-positive bacteria. Mol. Microbiol. *24*, 895–904.

Konstantinidis, K.T., and Tiedje, J.M. (2004). Trends between gene content and genome size in prokaryotic species with larger genomes. Proc. Natl. Acad. Sci. USA *101*, 3160–3165.

Kumar, R., Mukherjee, M., Bhandari, M., Kumar, A., Sidhu, H., and Mittal, R.D. (2002). Role of *Oxalobacter formigenes* in calcium oxalate stone disease: a study from North India. Eur. Urol. *41*, 318–322.

Kwak, C., Kim, H.K., Kim, E.C., Choi, M.S., and Kim, H.H. (2003). Urinary oxalate levels and the enteric bacterium *Oxalobacter formigenes* in patients with calcium oxalate urolithiasis. Eur. Urol. *44*, 475–481.

Lazazzera, B.A., and Grossman, A.D. (1998). The ins and outs of peptide signaling. Trends Microbiol. *6*, 288–294.

Lestrate, P., Dricot, A., Delrue, R.M., Lambert, C., Martinelli, V., De Bolle, X., Letesson, J.J., and Tibor, A. (2003). Attenuated signature-tagged mutagenesis mutants of *Brucella melitensis* identified during the acute phase of infection in mice. Infect. Immun. *71*, 7053–7060.

Li, B.Q., and Tian, S.P. (2006). Effects of trehalose on stress tolerance and biocontrol efficacy of *Cryptococcus laurentii*. J. Appl. Microbiol. *100*, 854–861.

Li, Y., Hugenholtz, J., Abee, T., and Molenaar, D. (2003). Glutathione protects *Lactococcus lactis* against oxidative stress. Appl. Environ. Microbiol. *69*, 5739–5745.

Lorca, G., Torino, M.I., Font de Valdez, G., and Ljungh, A. (2002). Lactobacilli express cell surface proteins which mediate binding of immobilized collagen and fibronectin. FEMS Microbiol. Lett. *206*, 31–37.

Mack, D.R., and Hollingsworth, M.A. (1994). Alteration in expression of MUC2 and MUC3 mRNA levels in HT29 colonic carcinoma cells. Biochem. Biophys. Res. Commun. *199*, 1012–1018.

Mahan, M.J., Slauch, J.M., and Mekalanos, J.J. (1993). Selection of bacterial virulence genes that are specifically induced in host tissues. Science *259*, 686–688.

Makarova, K., Slesarev, A., Wolf, Y., Sorokin, A., Mirkin, B., Koonin, E., Pavlov, A., Pavlova, N., Karamychev, V., Polouchine, N., Shakhova, V., Grigoriev, I., Lou, Y., Rohksar, D., Lucas, S., Huang, K., Goodstein, D.M., Hawkins, T., Plengvidhya, V., Welker, D., Hughes, J., Goh, Y., Benson, A., Baldwin, K., Lee, J.H., Diaz-Muniz, I., Dosti, B., Smeianov, V., Wechter, W., Barabote, R., Lorca, G., Altermann, E., Barrangou, R., Ganesan, B., Xie, Y., Rawsthorne, H., Tamir, D., Parker, C., Breidt, F., Broadbent, J., Hutkins, R., O'Sullivan, D., Steele, J., Unlu, G., Saier, M., Klaenhammer, T., Richardson, P., Kozyavkin, S., Weimer, B., and Mills, D. (2006). Comparative genomics of the lactic acid bacteria. Proc. Natl. Acad. Sci. USA *103*, 15611–15616.

Makarova, K.S, and Koonin, E.V. (2007). Evolutionary genomics of lactic acid bacteria. J. Bacteriol. *189*, 1199–208.

Mallonee, D.H., White, W.B., and Hylemon, P.B. (1990). Cloning and sequencing of a bile acid-inducible operon from *Eubacterium* sp. strain VPI 12708. J. Bacteriol. *172*, 7011–7019.

Marco, M.L., Bongers, R.S., de Vos, W.M., and Kleerebezem, M. (2007). Spatial and Temporal Expression of *Lactobacillus plantarum* Genes in the Gastrointestinal Tracts of Mice. Appl. Environ. Microbiol. *73*, 124–132.

Marino, M., Banerjee, M., Jonquieres, R., Cossart, P., and Ghosh, P. (2002). GW domains of the *Listeria monocytogenes* invasion protein InlB are SH3-like and mediate binding to host ligands. EMBO J. *21*, 5623–5634.

McAuliffe, O., Cano, R.J., and Klaenhammer, T.R. (2005). Genetic analysis of two bile salt hydrolase activities in *Lactobacillus acidophilus* NCFM. Appl. Environ. Microbiol. *71*, 4925–4929.

Metchnikoff, E. (1907). Lactic acid as inhibiting intestinal putrefaction. In The prolongation of life: optimistic studies, P. Chalmers Mitchell, ed. (London, UK: Butterworth-Heinemann), pp. 161–183.

Micali, S., Grande, M., Sighinolfi, M.C., De Carne, C., De Stefani, S., and Bianchi, G. (2006). Medical therapy of urolithiasis. J. Endourol. *20*, 841–847.

Mohamadzadeh, M., Olson, S., Kalina, W.V., Ruthel, G., Demmin, G.L., Warfield, K.L. Bavari, S., and Klaenhammer, T.R. (2005). Lactobacilli activate human dendritic cells that skew T cells toward T helper 1 polarization. Proc. Natl. Acad. Sci. USA *102*, 2880–2885.

Mohr, K.I., and Tebbe, C.C. (2006). Diversity and phylotype consistency of bacteria in the guts of three bee species (Apoidea) at an oilseed rape field. Environ. Microbiol. *8*, 258–272.

Molenaar, D., Bringel, F., Schuren, F.H., de Vos, W.M., Siezen, R.J., and Kleerebezem, M. (2005). Exploring *Lactobacillus plantarum* genome diversity by using microarrays. J. Bacteriol. *187*, 6119–6127.

Nakamura, C.E., and Whited, G.M. (2003). Metabolic engineering for the microbial production of 1,3-propanediol. Curr. Opin.Biotechnol. *14*, 454–459.

Navarre, W.W., and Schneewind, O. (1994). Proteolytic cleavage and cell wall anchoring at the LPXTG motif of surface proteins in Gram-positive bacteria. Mol. Microbiol. *14*, 115–121.

Nelson, K.E., Paulsen, I.T., Heidelberg, J.F., and Fraser, C.M. (2000). Status of genome projects for nonpathogenic bacteria and archaea. Nat. Biotechnol. *18*, 1049–1054.

O'Connell-Motherway, M., van Sinderen, D., Morel-Deville, F., Fitzgerald, G.F., Ehrlich, S.D., and Morel, P. (2000). Six putative two-component regulatory systems isolated from *Lactococcus lactis* subsp. *cremoris* MG1363. Microbiology *146 (Pt 4)*, 935–947.

O'Hara, A.M., O'Regan, P., Fanning, A., O'Mahony, C., Macsharry, J., Lyons, A., Bienenstock, J., O'Mahony, L., and Shanahan, F. (2006). Functional modulation of human intestinal epithelial cell responses by *Bifidobacterium infantis* and *Lactobacillus salivarius*. Immunology *118*, 202–215.

Ochman, H., Lawrence, J.G., and Groisman, E.A. (2000). Lateral gene transfer and the nature of bacterial innovation. Nature *405*, 299–304.

Paik, S., Senty, L., Das, S., Noe, J.C., Munro, C.L., and Kitten, T. (2005). Identification of virulence determinants for endocarditis in *Streptococcus sanguinis* by signature-tagged mutagenesis. Infect. Immun. *73*, 6064–6074.

Paludan-Muller, C., Huss, H.H., and Gram, L. (1999). Characterization of lactic acid bacteria isolated from a Thai low-salt fermented fish product and the role of garlic as substrate for fermentation. Int. J. Food Microbiol. *46*, 219–229.

Pepe, O., Blaiotta, G., Anastasio, M., Moschetti, G., Ercolini, D., and Villani, F. (2004). Technological and molecular diversity of *Lactobacillus plantarum* strains isolated from naturally fermented sourdoughs. Syst. Appl. Microbiol. *27*, 443–453.

Pereira, D.I., McCartney, A.L., and Gibson, G.R. (2003). An *in vitro* study of the probiotic potential of a bile-salt-hydrolyzing *Lactobacillus fermentum* strain, and determination of its cholesterol-lowering properties. Appl. Environ. Microbiol. *69*, 4743–4752.

Phadnis, S.H., Parlow, M.H., Levy, M., Ilver, D., Caulkins, C.M., Connors, J.B., and Dunn, B.E. (1996). Surface localization of *Helicobacter pylori* urease and a heat shock protein homolog requires bacterial autolysis. Infect. Immun. *64*, 905–912.

Pretzer, G., Snel, J., Molenaar, D., Wiersma, A., Bron, P.A., Lambert, J., de Vos, W.M., van der Meer, R., Smits, M.A., and Kleerebezem, M. (2005). Biodiversity-based identification and functional characterization of the mannose-specific adhesin of *Lactobacillus plantarum*. J. Bacteriol. *187*, 6128–6136.

Pridmore, R.D., Berger, B., Desiere, F., Vilanova, D., Barretto, C., Pittet, A.C., Zwahlen, M.C., Rouvet, M., Altermann, E., Barrangou, R., Mollet, B., Mercenier, A., Klaenhammer, T.R., Arigoni, F., and Schell, M.A. (2004). The genome sequence of the probiotic intestinal bacterium *Lactobacillus johnsonii* NCC 533. Proc. Natl. Acad. Sci. USA *101*, 2512–2517.

Reeson, A.F., Jankovic, T., Kasper, M.L., Rogers, S., and Austin, A.D. (2003). Application of 16S rDNA-DGGE to examine the microbial ecology associated with a social wasp *Vespula germanica*. Insect Mol. Biol. *12*, 85–91.

Reid, G., Sanders, M.E., Gaskins, H.R., Gibson, G.R., Mercenier, A., Rastall, R., Roberfroid, M., Rowland, I., Cherbut, C., and Klaenhammer, T.R. (2003). New scientific paradigms for probiotics and prebiotics. J. Clin. Gastroenterol. *37*, 105–118.

Roos, S., and Jonsson, H. (2002). A high-molecular-mass cell-surface protein from *Lactobacillus reuteri* 1063 adheres to mucus components. Microbiology *148*, 433–442.

Ruoslahti, E. (1988). Fibronectin and its receptors. Annu. Rev. Biochem. *57*, 375–413.

Savage, G.P., Charrier, M.J., and Vanhanen, L. (2003). Bioavailability of soluble oxalate from tea and the effect of consuming milk with the tea. Eur. J. Clin. Nutr. *57*, 415–419.

Schleifer, K.H., and Ludwig, W. (1995). Phylogeny of the genus *Lactobacillus* and related genera. Syst. Appl. Microbiol. *18*, 461–467.

Siener, R., Honow, R., Voss, S., Seidler, A., and Hesse, A. (2006). Oxalate content of cereals and cereal products. J. Agric. Food Chem. *54*, 3008–3011.

Siezen, R., Boekhorst, J., Muscariello, L., Molenaar, D., Renckens, B., and Kleerebezem, M. (2006). *Lactobacillus plantarum* gene clusters encoding putative cell-surface protein complexes for carbohydrate utilization are conserved in specific Gram-positive bacteria. BMC Genomics *7*, 126.

Sillanpaa, J., Martinez, B., Antikainen, J., Toba, T., Kalkkinen, N., Tankka, S., Lounatmaa, K., Keranen, J., Hook, M., Westerlund-Wikstrom, B., Pouwels, P.H., and Korhonen, T.K. (2000). Characterization of the collagen-binding S-layer protein CbsA of *Lactobacillus crispatus*. J. Bacteriol. *182*, 6440–6450.

Skar, C.K., Kruger, P.G., and Bakken, V. (2003). Characterisation and subcellular localisation of the GroEL-like and DnaK-like proteins isolated from *Fusobacterium nucleatum* ATCC 10953. Anaerobe *9*, 305–312.

Stalhammar-Carlemalm, M., Areschoug, T., Larsson, C., and Lindahl, G. (1999). The R28 protein of *Streptococcus pyogenes* is related to several group B streptococcal surface proteins, confers protective immunity and promotes binding to human epithelial cells. Mol. Microbiol. *33*, 208–219.

Steen, A., Buist, G., Leenhouts, K.J., El Khattabi, M., Grijpstra, F., Zomer, A.L., Venema, G., Kuipers, O.P., and Kok, J. (2003). Cell wall attachment of a widely distributed peptidoglycan binding domain is hindered by cell wall constituents. J. Biol. Chem. *278*, 23874–23881.

Stone, R. (2002). Bacteriophage therapy. Stalin's forgotten cure. Science *298*, 728–731.

Stoyancheva, G.D., Danova, S.T., and Boudakov, I.Y. (2006). Molecular identification of vaginal lactobacilli isolated from Bulgarian women. Antonie Van Leeuwenhoek *90*, 201–210.

Sturme, M.H., Nakayama, J., Molenaar, D., Murakami, Y., Kunugi, R., Fujii, T., Vaughan, E.E., Kleerebezem, M., and de Vos, W.M. (2005). An *agr*-like two-component regulatory system in *Lactobacillus plantarum* is involved in production of a novel cyclic peptide and regulation of adherence. J. Bacteriol. *187*, 5224–5235.

Szajewska, H., Kotowska, M., Mrukowicz, J.Z., Armanska, M., and Mikolajczyk, W. (2001). Efficacy of *Lactobacillus* GG in prevention of nosocomial diarrhea in infants. J. Pediatr. *138*, 361–365.

Szajewska, H., Ruszczynski, M., and Radzikowski, A. (2006). Probiotics in the prevention of antibiotic-associated diarrhea in children: a meta-analysis of randomized controlled trials. J. Pediatr. *149*, 367–372.

Tamames, J. (2001). Evolution of gene order conservation in prokaryotes. Genome Biol. *2*, 1–11.

Toba, T., Virkola, R., Westerlund, B., Bjorkman, Y., Sillanpaa, J., Vartio, T., Kalkkinen, N., and Korhonen, T.K. (1995). A Collagen-Binding S-Layer Protein in *Lactobacillus crispatus*. Appl. Environ. Microbiol. *61*, 2467–2471.

Ton-That, H., Liu, G., Mazmanian, S.K., Faull, K.F., and Schneewind, O. (1999). Purification and characterization of sortase, the transpeptidase that cleaves surface proteins of *Staphylococcus aureus* at the LPXTG motif. Proc.Natl. Acad. Sci. USA *96*, 12424–12429.

Underdahl, N.R. (1983). The effect of feeding *Streptococcus faecium* upon *Escherichia coli* induced diarrhea in gnotobiotic pigs. Prog. Food Nutr. Sci. *7*, 5–12.

Underdahl, N.R., Torres-Medina, A., and Dosten, A.R. (1982). Effect of *Streptococcus faecium* C-68 in control of *Escherichia coli*-induced diarrhea in gnotobiotic pigs. Am. J. Vet. Res. *43*, 2227–2232.

van de Guchte, M., Penaud, S., Grimaldi, C., Barbe, V., Bryson, K., Nicolas, P., Robert, C., Oztas, S., Mangenot, S., Couloux, A., Loux, V., Dervyn, R., Bossy, R., Bolotin, A., Batto, J.M., Walunas, T., Gibrat, J.F., Bessieres, P., Weissenbach, J., Ehrlich, S.D., and Maguin, E. (2006). The complete genome sequence of *Lactobacillus bulgaricus* reveals extensive and ongoing reductive evolution. Proc. Natl. Acad. Sci. USA *103*, 9274–9279.

van der Mee-Marquet, N., Domelier, A.S., Mereghetti, L., Lanotte, P., Rosenau, A., van Leeuwen, W., and Quentin, R. (2006). Prophagic DNA fragments in *Streptococcus agalactiae* strains and association with neonatal meningitis. J. Clin. Microbiol. *44*, 1049–1058.

van Pijkeren, J.P., Canchaya, C., Ryan, K.A., Li, Y., Claesson, M.J., Sheil, B., Steidler, L., O'Mahony, L., Fitzgerald, G.F., van Sinderen, D., and O'Toole, P.W. (2006). Comparative and functional analysis of sortase-dependent proteins in the predicted secretome of *Lactobacillus salivarius* UCC118. Appl. Environ. Microbiol. *72*, 4143–4153.

Ventura, M., Canchaya, C., Bernini, V., Altermann, E., Barrangou, R., McGrath, S., Claesson, M.J., Li, Y., Leahy, S., Walker, C.D., Zink, R., Neviani, E., Steele, J., Broadbent, J., Klaenhammer, T.R., Fitzgerald, G.F., O'Toole P.W., and van Sinderen, D. (2006). Comparative genomics and transcriptional analysis of prophages identified in the genomes of *Lactobacillus gasseri*, *Lactobacillus salivarius*, and *Lactobacillus casei*. Appl. Environ. Microbiol. *72*, 3130–3146.

Ventura, M., Canchaya, C., Kleerebezem, M., de Vos, W.M., Siezen, R.J., and Brussow, H. (2003). The prophage sequences of *Lactobacillus plantarum* strain WCFS1. Virology *316*, 245–255.

Ventura, M., Canchaya, C., Pridmore, R.D., and Brussow, H. (2004). The prophages of *Lactobacillus johnsonii* NCC 533: comparative genomics and transcription analysis. Virology *320*, 229–242.

von Heijne, G. (1988). Transcending the impenetrable: how proteins come to terms with membranes. Biochim. Biophys. Acta *947*, 307–333.

Walter, J., Heng, N.C., Hammes, W.P., Loach, D.M., Tannock, G.W., and Hertel, C. (2003). Identification of *Lactobacillus reuteri* genes specifically induced in the mouse gastrointestinal tract. Appl. Environ. Microbiol. *69*, 2044–2051.

Waters, C.M., and Bassler, B.L. (2005). Quorum sensing: cell-to-cell communication in bacteria. Annu Rev. Cell Dev. Biol. *21*, 319–346.

Weiser, J.N., Bae, D., Fasching, C., Scamurra, R.W., Ratner, A.J., and Janoff, E.N. (2003). Antibody-enhanced pneumococcal adherence requires IgA1 protease. Proc. Natl. Acad. Sci. USA *100*, 4215–4220.

Wells, J.E., and Hylemon, P.B. (2000). Identification and characterization of a bile acid 7alpha-dehydroxylation operon in *Clostridium* sp. strain TO-931, a highly active 7alpha-dehydroxylating strain isolated from human feces. Appl. Environ. Microbiol. *66*, 1107–1113.

Wheeler, D.L., Chappey, C., Lash, A.E., Leipe, D.D., Madden, T.L., Schuler, G.D., Tatusova, T.A., and Rapp, B.A. (2000). Database resources of the National Center for Biotechnology Information. Nucleic Acids Res. *28*, 10–14.

Studies of the Intestinal Microflora by Traditional, Functional and Molecular Techniques

4

Elisabeth Norin, Cecilia Jernberg, Hans-Olof Nilsson and Lars Engstrand

Abstract

To resolve the complexity of a bacterial population sophisticated analytical methods and alternative genetic techniques are required. Traditional cultivation, microscopy and determination of fermentation/degradation products are still important. Future studies of the microbe-microbe and the microbe-host cross-talk will strengthen our knowledge about the composition and function of the microflora. We will discuss functional studies including the GAC-MAC concept, and the development of different molecular techniques, which provide microbiologists complementary methods to study complex microbial communities. The most commonly utilized targets are the genes coding for small subunit ribosomal (r) RNAs. The 16S and 23S rRNA genes contain conserved, variable and hypervariable sequences, making them suitable for studies of bacterial evolution and microbial community composition. Denaturing gradient gel electrophoresis and temperature gradient gel electrophoresis are gel-based techniques that separate double-stranded DNA molecules of identical size that vary in sequence and composition. Terminal-restriction fragment length polymorphism also provides a reproducible fingerprint of complex microbial communities. The recent development of sequence based methods including 454-pyrosequencing has improved sequencing technology and enabled genomics to take a quantum leap in terms of high-throughput DNA sequence analysis facilitating completely new types of investigations. Finally, metagenomics, i.e. analyses of microbial populations by the use of genetic/molecular techniques as well as functional analyses, will be discussed.

Introduction

In man, the GI tract is divided into three major regions with regard to flora composition and anatomy, namely the stomach, small intestine (duodenum, jejunum and ileum) and large intestine or colon. A rapid transit time and acidic conditions in the stomach restrict the number of microbes in this region, but at least some aciduric microbes as *Helicobacter* spp. are detected (10^2–10^4 cells/ml). In the upper small intestine, the flow is somewhat slower and in the lower part, the number of microbes increase and is estimated to be 10^6–10^8 CFU/ml and with a high diversity – both facultative and strict anaerobe microbes. As the digesta reaches the colon, the passage time is even slower, the pH is higher and the microbial community is estimated to extend to around 10^{10}–10^{12} CFU/g intestinal content, comprising more than 1000 different species, of which only a limited number of dominating species so far are identified (Ben-Amor *et al.*, 2005).

Traditional cultivation techniques

Much of today's knowledge about the composition and function of the intestinal flora and influences from external factors as, for example, intake of drugs and different dietary components originate mainly from traditional cultivating, microscopy and determination of fermentation/

degradation products of known reagents, an expensive and time-consuming technique. These methods are still important, but are to-day complemented with new techniques (Zoetendal *et al.*, 2006). You can cultivate on selective media under defined conditions, where you exclude unwanted groups of bacteria, and check, for example, for increases of different probiotic strains, but this type of studies may fail to detect more subtle changes at species level, and other species are not detected at all. The lack of knowledge of the total composition of an intestinal microflora and its alterations due to age, sex, diet and other factors as disturbances due to i.e. drugs as antimicrobial drugs and/or intake of pre- and probiotics, is due to the fact that almost only faecal samples are, so far investigated. Material from the upper intestine is difficult to obtain and you should also differentiate between luminal and mucosal microbes. Which microbes from the upper compartments that are detectable in faecal samples is difficult to say – one example is demonstrated in the *Helicobacter* field, where *Helicobacter* never were identified in faecal samples previously (Quesada-Perez *et al.*, 2005). However, the microbes could be identified by some invasive methods, biopsies or by ^{13}C-urea breath test (Roberts *et al.*, 2000). In this chapter, we will present some complementary methods suitable for studying the intestinal flora under influences of external factors, e.g. after ingestion of pre- or probiotics. The microbe-microbe and the microbe-host cross-talk are to-day evaluated in several ways and each of them will strengthen our knowledge about the function and composition of the intestinal flora.

Functional studies: the GAC-MAC concept

This concept is applicable to different structures and functions in man and animal under both physiological and pathophysiological conditions. You can search for metabolites which reflect a metabolic or catabolic microbial activity – MACs (microflora-associated characteristics), a process that might have occurred in the upper GI tract, maybe by subdominant microbes, thus from locations where samples are almost impossible to investigate by traditional cultivating techniques. A MAC has been defined as the recording of any anatomical structure, physiological, biochemical or immunological function in an organism, that has been influenced by the microflora in either an anabolic or catabolic way (Midtvedt *et al.*, 1986). In the absence of functionally active bacteria the opposite structure or function is described as a GAC (germ-free animal characteristic). The GACs are at hand in germ-free animals, newborn babies and sometimes developed in relation to antimicrobial therapy (Table 4.1). Some of these functions will be further commented upon in the light of probiotic use.

Degradation of β-aspartylglycine

The presence of β-aspartylglycine indicates a reduction of the intestinal colonization resistance – the resistance of the ecosystem to colonization by establishing microorganisms. This dipeptide has been found in germ-free animals as lambs, piglets, rats and mice. It is shown that host derived intestinal proteolytic enzymes are capable to break down dietary proteins to β-aspartylglycine, but its β-carboxyl dipeptide binding is only broken down by microbial derived proteases (Welling *et al.*, 1988). Thus, the presence/absence of β-aspartylglycine represents one GAC/MAC function which is depending on presence of a dietary precursor; host derived proteolytic enzymes and microbial derived proteolytic enzymes. As intestinal contents from germ-free animals contain the β-aspartylglycine you can follow microbes establishing in the intestine, as the amount of the β-aspartylglycine diminishes, and the conventional animals are totally devoid of this dipeptide – but it may re-occur after antimicrobial treatment. At present, we have never found any effect of any of the probiotics given to germ-free animals – it appears that this function is influenced by the interaction of increasing amounts of different microbes rather than just one single strain.

Conversion of bilirubin to urobilins

Bilirubin is the end product of catabolism of haemoglobin and some other heme-containing substances, and bilirubin metabolism involves both a microbial deconjugation of bilirubin glucuronide, conjugated and excreted from the liver, and a conversion to urobilins. The capacity to transform bilirubin to urobilins seems however

Table 4.1 Some major intestinal structures and functions influenced by the microflora

Function	MAC	GAC	Microbes involved
Biochemical			
β-Aspartylglycine	Absent	Present	Unknown (Areano *et al.*, 1996)
Bile acid metabolism	Deconjugation	No deconjugation	Many species (Hylemon, 1985)
	Dehydrogenation	No dehydrogenation	Many species (Midtvedt, 1974)
	Dehydroxylation	No dehydroxylation	Few species (Gustafsson *et al.*, 1966)
Bilirubin metabolism	Deconjugation	Little deconjugation	Many species (Gadelle *et al.*, 1985)
	Urobilin	No urobilin	One species (Midtvedt *et al.*, 1981)
Cholesterol	Coprostanol	No coprostanol	Few species (Sadzikowski *et al.*, 1977)
Intestinal gases	Carbon dioxide	Some carbon dioxide	Many species (Midtvedt, 1999)
	Hydrogen	No hydrogen	Some species (Midtvedt, 1999)
	Methane	No methane	Few species (Midtvedt, 1999)
Mucin	Degraded	No degradation	Several species (Carlstedt-Duke *et al.*, 1986)
Short-chain fatty acids	Large amounts	Far less	Many species (Midtvedt, 1999)
Tryptic activity	Little or absent	High activity	Few species (Ramare *et al.*, 1996)
Anatomical/physiological			
Intestinal wall	Thicker	Thinner	Unknown (Savage, 1977)
Cell kinetics	Fast	Slower	Unknown (Gustafsson, 1982)
Migration motor complexes	Normal	Fewer	Unknown (Midtvedt, 1999)
Production of peptides	Normal	Altered	Unknown (Midtvedt, 1999)
Sensitivity to peptides	Normal	Reduced	Unknown (Midtvedt, 1999)
Caecum size (rodents)	Normal	Enlarged	Partly known (Gustafsson *et al.*, 1970)
Osmolarity	Normal	Reduced	Unknown (Midtvedt, 1999)
Colloid osmotic pressure	Normal	Increased	Unknown (Midtvedt, 1999)
Oxygen tension	Low	High as in tissue	Several species (Midtvedt, 1999)
Electropotential (mV)	Under 100 mV	Above 100 mv	Unknown (Midtvedt, 1999)

MAC, microflora-associated characteristic; GAC, germ-free animal characteristic.

to be a rare event among intestinal microbes – so far only very few strains of *Clostridium* and one *Bacteroides* have been associated to this function (Vitek., 2005) – germ-free animals and newborns are devoid of this function. Because of the implication of bilirubin metabolism in neonatal jaundice, it is especially of importance to investigate the possible microbial metabolism in very young children. Moreover, in studies performed within our group, we have never shown any conversion of bilirubin to urobilins when inoculating probiotic strains into germ-free mice (Cardona *et al.*, 2002).

Metabolism of bile acids

The bile acids are synthesized in the liver from cholesterol, and in the liver, they are conjugated mainly with taurine or glycine prior to secretion from the gall bladder into the intestine, where they undergo microbial transformations including deconjugation, dehydrogenation and dehydroxylation (Midtvedt, 1974). The deconjugation is performed only by bacteria in the distal ileum and colon and therefore, bile acids are excreted only as conjugates under germ-free conditions. However, most bile acids under conventional conditions are deconjugated and reabsorbed into an enterohepatic circulation, and this function could be responsible for a cholesterol-lowering activity. We have shown that 12 probiotics tested out of 14 were able to split bile acid conjugates *in vitro*, and probably this process is the same *in vivo*, however, not so far confirmed (Cardona *et al.*, 2000). Thus, a cholesterol-lowering effect by several probiotic strains has often been claimed, but our group has never been able to establish such a function.

Conversion of cholesterol to coprostanol

Cholesterol conversion to coprostanol is a microbial route for the body to eliminate or reduce the host cholesterol pool – cholesterol deriving from both endogenous and exogenous sources. Endogenous cholesterol is synthesized mainly in the liver and the small intestine, and the exogenous is mainly of dietary origin. Cholesterol in the intestine can be reabsorbed, but there are also some microbes known which are able to convert the cholesterol to coprostanol, unabsorbable, and thus reducing the host cholesterol pool. Some studies have shown an association between high excretion of coprostanol and plasma cholesterol lowering effects after feeding *Eubacterium coprostanoliences* to rabbits and mice. However, when the same strain was given to hens, no plasma cholesterol lowering effect was seen (Cardona *et al.*, 2002). None of 18 probiotic strains tested in our laboratory as mono-contaminants to germ-free mice were able to convert cholesterol to coprostanol – rather maybe one function could be that probiotics are able to de-conjugate bile acids, and by that way reduce the intestinal cholesterol pool.

In man, we have found that most often this microbial function is established before the age of two years, and that it can be disturbed by, for example, an ordinary antimicrobial treatment. In adults we have arbitrary divided a group of 633 healthy volunteers into three groups – <36, 36–50 and >50 years of age – males and females, with a conversion capacity of non-, low and high capacity. In males there were a higher percentage of non-converters in the younger group, compared to the two other groups, whereas the percentage of high converters was increased in the oldest group. In females the differences were not that pronounced (Benno *et al.*, 2005). Probiotics given to healthy elderly did not significantly alter the conversion capacity (E. Norin, unpublished results).

Degradation of mucin

Mucin is produced in the intestine, mainly by the goblet cells, and it constitutes the major organic component of the defences barrier covering intestinal epithelial cells. At the same time it acts as nutrient for the intestinal micro-organisms. There are growing evidences that mucin and the intestinal flora play a significant role in the pathophysiological role of some intestinal disturbances, e.g. Crohn's disease and intestinal ulceration. In the same study as referred to earlier regarding inoculation of different probiotic strains to germ-free animals, no effect on the faecal mucin pattern was seen (Cardona *et al.*, 2002) – the mucin pattern was similar both qualitatively and quantitatively, as in the germ-free animals.

Intestinal gases

The main gases occurring in the large intestine are carbon dioxide, hydrogen and methane, and they are nearly all products by an anaerobic metabolism of carbohydrates and proteins. The CO_2 can be of both host and microbial origin, and methane is detected in almost half of the world's human population. Intestinal gases can sometimes develop big problems for some individuals, thus persons suffering from intestinal unbalances of the flora such as in inflammatory bowel syndrome and 'balloon-stomach' patients. Different pre- and probiotics are given to these patients with both positive and negative results.

Production of short-chain fatty acids

The short chain fatty acids (SCFAs) (C_2–C_6) are products after mainly an anaerobic metabolism of carbohydrates and proteins of both dietary and endogenous origin. One important role of these acids is that they provide energy to the gut epithelium and facilitate the absorption of Na^+ and water. Dietary components escaping digestion or absorption in the small intestine are fermented in the colon and give rise to these acids. The amount produced depends on the kind of microflora present, the substrate and gut transit time. In adults, they supply approximately two-thirds of the colonic anion concentration (Siigur *et al.*, 1994) mainly in the form of acetate, propionate and butyrate. However, in other species quite different values are found, due to different dietary intake and of different physiology (herbivores/omnivores). The SCFA production and absorption is closely related to the nourishment of the colonic mucosa. Much of the acids are readily absorbed; acetate enters the peripheral circulation to be metabolized by peripheral tissues. Butyric acid is the major energy source for the colonocytes, and propionic acid is taken up by the liver. As the main anions of the colon and the major source of energy for colonocytes, the fatty acids, and especially butyric acid, enhance the growth of lactobacilli and bifidobacteria, and the effect of pre- and probiotics on cell proliferation, differentiation, apoptosis, mucin production, immune function, mineral absorption, lipid metabolism, and intestinal peptides is well documented.

Molecular techniques

As mentioned previously, isolation of bacterial strains by microbial culture is the golden standard for diagnosis and characterization of pathogens as well as commensal bacteria. However, the importance to analyse bacterial communities, rather than examination of a bacterial strain or members of a bacterial genus, has become apparent during the last decade. It is for example desirable to study the effects of potentially probiotic bacteria, as well as the effect(s) of prebiotics, on the existing gastrointestinal (GI) tract bacterial assemblage over time. Additionally, several genera of micro-organisms have been linked to inflammatory bowel disease (IBD) and ulcerative colitis (UC), but none has conclusively been established as an etiological agent. Hence, a need to study the population dynamics of the intestinal microflora in its entirety has emerged. For this reason, standard microbial culture techniques have inherent limitations.

The development of molecular techniques has provided the microbiologist with an array of new methods for the study of microbial community diversity (Furri, 2006). The most commonly utilized target for such analysis is the ubiquitous gene coding for small subunit ribosomal RNA. The 16S rRNA gene (16S rDNA), as well as the 23S rRNA gene, has historically proved to be the most accurate genes for studies of bacterial diversity, evolution, as well as phylogenetic analysis (Woese, 1987). The 16S rDNA consists of consensus sequences, universal for all prokaryotes, and variable sequences that are specific for particular groups or species of bacteria. Hypervariable sequences, that may be unique for certain strains within a species, are contained within the variable areas.

Denaturing and temperature gradient gel electrophoresis

Denaturing gradient gel electrophoresis (DGGE) and temperature gradient gel electrophoresis (TGGE) are gel-based techniques that separate double-stranded (ds) DNA molecules of identical size that vary in sequence and composition according to their melting properties. The DGGE method was first developed to detect single base-pair gene substitutions. It was shown that wild-type and single base substituted DNA molecules that migrated into a gradient of ascending concentrations of denaturants (urea and formamide) separated as single DNA-species in focused sharp bands (Fischer and Lerman, 1983).

The separation is based on the decreased electrophoretic mobility of partially denaturized dsDNA. PCR-fragments with varying sequence have different melting properties and will therefore stop migrating at different gel positions. TGGE is based on the same principle, but instead of chemical denaturants a linear temperature gradient is applied to the gel. The DGGE technique is further improved by the addition of a so called GC-clamp, a guanine cytosine-rich 40–70 base-pair DNA sequence, at the 5′ end of one of the PCR-primers. This modification leads to increased resolving efficiency and nearly all base substitutions, both in high- and low-temperature melting domains of a DNA molecule can be detected.

The first report describing profiling of complex microbial populations using this technique was published in 1993 (Muyzer and Waal, 1993).

DNA is amplified using one or more sets of so called universal or broad range primers, which are constructed in 16S rDNA consensus regions found in most or all bacteria that are known to inhabit the sample to be analysed. PCR with genus- or group-specific primers is often used in combination with universal primers. The PCR-products are run through the denaturing gel and a band pattern visualized. To compare DGGE/TGGE profiles, the gels are scanned and similarity indices calculated. Cluster analysis can also be applied to scanned gel patterns to construct dendrograms that demonstrate the relationships between DGGE profiles (Zoetendal *et al.*, 2002; Saarela *et al.*, 2007; Zhang *et al.*, 2007).

An advantageous feature of DGGE/TGGE, compared with some other fingerprinting methods, is the possibility to excise separated PCR-products from gels to identify members of the analysed bacterial community by DNA sequencing and/or probe hybridization. DGGE and TGGE have also been combined with other methods to increase the analytical power. The bacterial diversity contained in different 16S rDNA clone libraries has been displayed using DGGE (Burr *et al.*, 2006).

Temporal temperature gradient gel electrophoresis (TTGE) is a further development of the DGGE/TGGE concept that combines the effect of urea and a temperature gradient to denature and separate DNA molecules. This technique has been applied to characterize bifidobacteria in intestinal samples and competitive PCR was then used to estimate the quantity of bifidobacteria (Mangin *et al.*, 2006). TTGE has been used for bacterial profiling in colon and rectum biopsies. To increase the sensitivity of the analysis, total genomic DNA was amplified using multiple displacement amplification (MDA) followed by 16S rDNA PCR and TTGE (Monstein *et al.*, 2005).

Terminal restriction fragment length polymorphism (T-RFLP)

Terminal restriction fragment length polymorphism (T-RFLP) is a PCR based method that provides a fingerprint of complex microbial communities like those that occur in the gastrointestinal tract. Characterization of the richness and evenness of the dominant members of the specific microbial community can be done and compared between different samples. Changes in community compositions due to treatment effects or spatial variations in an environment can also be assessed.

After PCR amplification of the 16S rRNA genes the amplicons are digested with a restriction enzyme (Fig. 4.1). A fluorescently labelled primer is used during PCR amplification and this enables detection of only the terminal restriction fragments (TRFs). Different species most often have restriction sites at different locations in their 16S rRNA genes. This will result in TRFs of different sizes from different organisms. The restricted fragments are separated either by polyacrylamide gel or by capillary gel electrophoresis. The lengths, in base pairs, and the amounts (peak height and area) of the fluorescent TRFs are calculated by an automated fragment analysis program. This is achieved by the addition of a size standard in every lane which allows the technique to be reproducible.

The data output from a T-RFLP run is in the form of an electropherogram. In order to account for the effect of loading different amounts of DNA, the output data must first be standardized for comparison purposes. This is done by calculating the relative abundances of the individual TRFs, so that the area of each peak in the electropherogram is normalized to the total amounts of DNA analysed in that specific sample. A threshold value is often set so that TRFs with very low relative abundance values (<0.5%) can be excluded from further statistical analysis to reduce background.

Today the most common way to analyse T-RFLP data is by multivariate statistical analyses methods such as principal components analysis (PCA), correspondence analysis (COA) (Wang *et al.*, 2004; Jernberg *et al.*, 2007) or cluster analysis (Jernberg *et al.*, 2005). A few studies have used only information of presence and absence data when analyzing the community profile (Sakata *et al.*, 2005). However, use of relative abundance data is more sensitive and provides more information about subtle changes in community compositions.

T-RFLP can also be used to monitor specific bacteria in a complex community. This can be done either by using general 16S rRNA gene

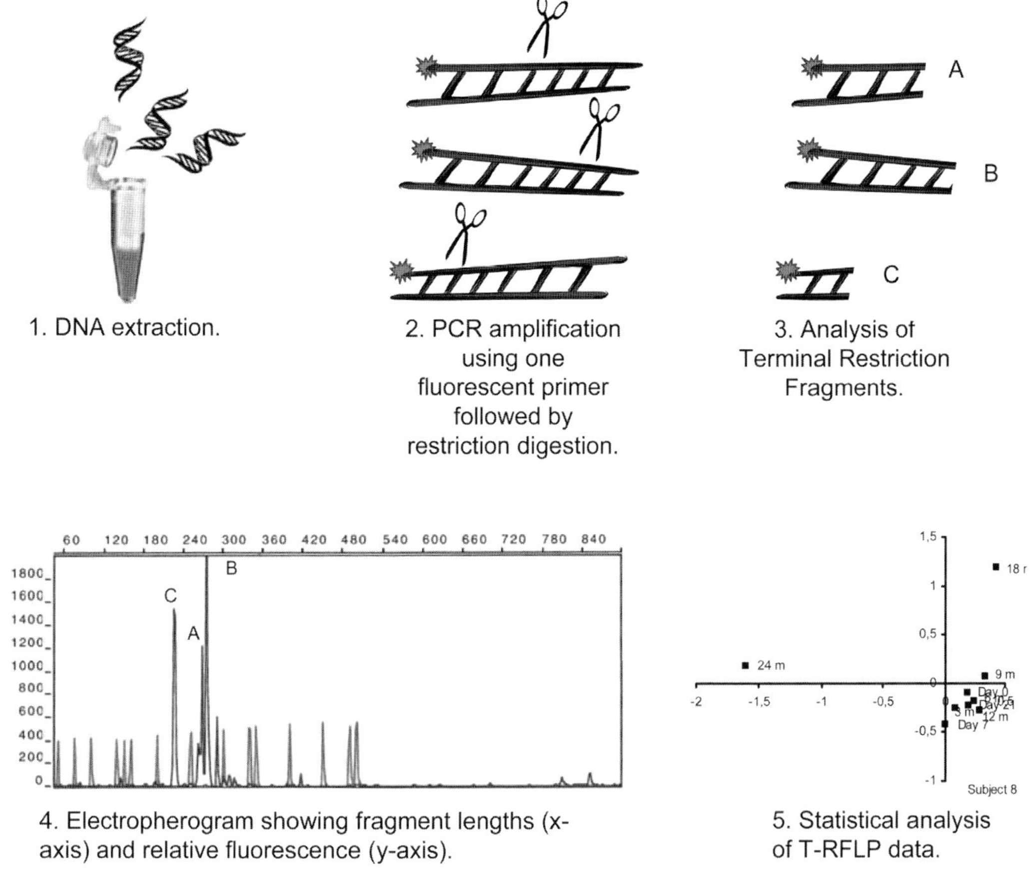

Figure 4.1 The T-RFLP technique.

primers or by using more genus/species specific primers. When using general primers the bacterium must belong to one of the dominant species in the sample analysed. Otherwise the bacterium will fall under the detection limit of the method. Furthermore, the TRF of the specific bacterium must have a length that is unique so that only the bacterium of interest is monitored. The potential beneficial effects of a probiotic product during clindamycin administration was investigated by T-RFLP (Jernberg *et al.*, 2005). PCA analysis using T-RFLP data with general primers indicated that the disturbances in the microflora were more prolonged in subjects not ingesting the probiotic. Furthermore, by using a *Lactobacillus*-specific primer it was shown that the probiotic lactobacilli strains dominated this more defined community during the ingestion of the probiotic.

Putative identity matches for the different TRFs can be done by comparison with a clone library with sequenced clones from the same material. Furthermore, the sizes of TRFs can be matched to databases where *in silico* digestion of existing 16S rRNA gene sequences can be made (Maidak *et al.*, 1999). However, the strength with T-RFLP analysis does not lie in identification purposes. The technique is better suited for rapid screening of multiple samples for comparative purposes.

Sequence based methods including 454-pyrosequencing

There are several sequencing based methods that can be used to study the diversity of the gastrointestinal microflora. The traditional Sanger DNA sequencing technique is well established with a low cost per sample. However, when thousands of samples from one individual need to be sequenced to achieve information about the diversity of the microflora the cost per individual sample will be high. The recent development of improved sequencing technology has enabled the field of genomics to take a quantum leap in terms of high throughput DNA sequence analysis facilitating completely new types of investigations in all types of biological niches. By adopting a massively parallel sequencing strategy the complexity of bacterial communities can be studied down to the species level. The complexity of a bacterial population requires extreme throughput of the analytical methods, and therefore its structure and dynamics has only recently started to be unravelled.

A novel DNA sequencing instrument, built up on the pyrosequencing technology which originally was developed at the Royal Institute of Technology, Stockholm (KTH), has truly revolutionized the field of genomics. The instrument is applied using a micro-fluidics platform that facilitates 300,000 parallel sequencing reactions in less than 3 hours. In the pioneering work published in Nature by Margulies and co-workers (Marguiles *et al.*, 2005) they demonstrate use of this technology by sequencing a bacterial genome in a single experiment, an experiment that otherwise would take weeks or months with conventional capillary sequencing technology

A novel application of the picotitre platform for characterization of microflora by sequencing of the corresponding 16S rDNA populations has recently been described (Sogin *et al.*, 2006). For microflora characterization, DNA is extracted from the biopsy/faecal sample and the 16S rDNA content amplified by PCR using universal 16S primers. Subsequently, these PCR products will be individually amplified and sequenced using the picotitre platform. Universal primers targeting conserved sites on the 16S gene are used for amplification. As for the other molecular techniques described above, this allows a single primer pair to be used for amplification of 16S from virtually any bacteria. The amplicon includes one or several variable sites enabling identification of bacterial species or phylotypes by sequencing. Sequencing of 100 bases in a variable region will in most cases be enough for identification at the species or genus level.

This technique enables parallel in-depth analysis of several individual samples with limited sample processing and without potential cloning bias. The method correctly describes the microbiota down to operational taxonomic units (OTUs) beyond the genus level. We applied this technique to analyse the microbial content in throat, stomach and faecal samples, and obtained 51,869 annotated reads representing 609 OTUs belonging to 14 phyla. The microbiota clustered according to their sampling local, demonstrating important habitat-selection pressure (Anderson *et al.*, 2008).

Metagenomics

Metagenomics, defined as the science of biological diversity, consists of genomic analysis of a microbial population with similar but not identical members, by the use of genetic and molecular analysis. A comprehensive metagenomic study provides understanding of the dynamics of a microbial population and includes analysis of nucleotide sequence, structure, regulation and function (Handelsman, 2004).

The metagenomics approach starts with the isolation of total genomic DNA from an environmental or human sample. Heterologous DNA from the sample is ligated with a vector and transformed into a host bacterium; the pool of resulting transformants is defined as a metagenomic library (Fig. 4.2). The library of clones may then be screened for biologically active compounds, such as enzymatic or antibiotic activity. PCR and/or hybridization can be applied to detect conserved phylogenetic marker genes, such as the 16S rRNA, recA, rpoB or others. Genes that flank phylogenetic markers may also be sequenced and the information gained may give insights into the physiology and ecology of that clone. It is also common that the metagenome library is randomly sequenced for identification purposes.

Metagenomic approaches have been applied to study human intestinal microorganism communities. Gill *et al.* revealed a microbial genomic and genetic diversity and identified some of the distinctive functional attributes found in the human distal gut microbiome. Whole-genome shotgun sequencing, sequence identification of phylotypes, and comparative functional analysis of metagenomic library clones from two healthy subjects defined part of the gene content and encoded functional traits of the gut microbial

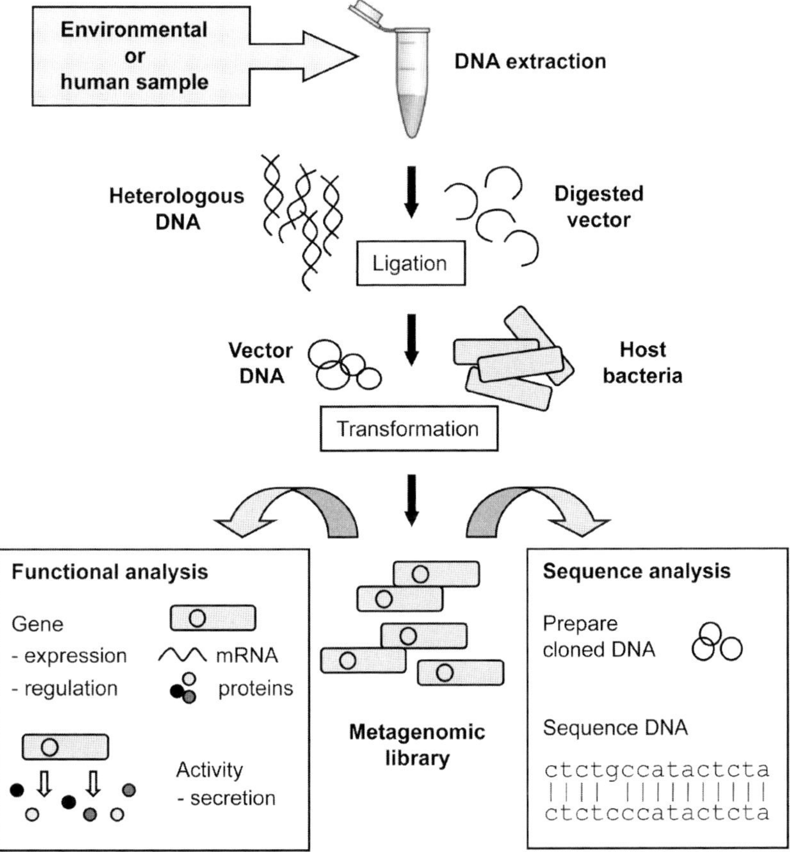

Figure 4.2 Schematic representation of construction and screening of a metagenomic library. Adapted from Handelsman, Microbiol. Mol. Biol. Rev. 2004;68:669–685.

community (Gill *et al.*, 2006). Manichanh *et al.* constructed two metagenomic libraries (25,000 clones each) from pooled genomic DNA of healthy individuals and patients with Crohn's disease (CD), respectively. Each library was used to construct a macro array that were analysed for microbial diversity by hybridization screening of 16S rDNA, sequencing and phylogenetic analysis. Reduced diversity of the faecal microbiota in CD was revealed by this approach (Manichanh *et al.*, 2006).

Next generation sequencing techniques

Sequencing technology has improved dramatically since the human genome project. The next generation sequencing techniques (Table 4.2) will require computational treatment of massive sequencing including large-scale data handling, automated quality control and annotation, phylogenetic analysis of selected genes and gene families, and ultimately separate genome assembly of many metagenomically sequenced organisms in the intestinal microflora.

Fully automated genome assembly of many microorganisms sequenced simultaneously will thus become a major computational challenge. There will be a need for modified genome assembly methods, and *de novo* algorithm development. Examples are binary codes for third generation sequencing where the four-digit genome sequence will be converted to a binary code (0s, 1s) with the purpose to simplify the read-out. Novel sample preparation methods will also be needed such as magnetic bead technology to facilitate enrichment of genome regions through hybridization capture for targeted analysis.

Table 4.2 'Next generation' sequencing techniques

Parameter	Instrument/technique		
	454	Solexa	SOLiD
Average length	200–300 nt	25–35 nt	25–35 nt
Reads	400,000	30 million	90 million
Data	100 Mb	1 Gb	3 Gb
Cost/run	$5000		$ 3000
Cost instrument	~$0.5million	~$0.5million	~$0.5million
Scale-up of # reads	+	+++	+++
Future increase of read length	+++	+	+
Access to instruments	+++	++	+
Drawbacks	High error rate for homopolymers	Error rate increases with read length	Error rate increases with read length
Advantages	Long read length	Easy to scale up	Easy to scale up

nt, nucleotide; Mb, megabyte, Gb, gigabyte.

References

Andersson, A.F., Lindberg, M., Jakobsson, H., Bäckhed, F., Nyrén, P., and Engstrand, L. (2008). Comparative analysis of human gut microbiota by barcoded pyrosequencing. PLoS ONE. *3*, e2836.

Araneo, B.A, Cebra, J.J, Beuth, J., Fuller, R., Heidt, P.J., and Midvedt, T. (1996). Problems and priorities for controlling opportunistic pathogens with new antimicrobial strategies; an overview of current literature. Zentralbl. Bakteriol. *283*, 431–465.

Ben-Amor, K., Heilig, H., Smidt, H., Vaughan, E.E, Abee, T., and de Vos, W.M. (2005). Genetic diversity of viable, injured, and dead fecal bacteria assessed by fluorescence-activated cell sorting and 16S rRNA gene analysis. Appl. Environ. Microbiol. *71*, 4679–4689.

Benno, P., Midtvedt, K., Alam, M., Collinder, E., Norin, E., and Midtvedt, T. (2005). Examination of intestinal conversion of cholesterol to coprostanol in 633 healthy subjects reveals an age- and sex-dependent pattern. Microb. Ecol. Health Dis. *17*, 200–204.

Burr, M.D, Clark, S.J, Spear, C.R, and Camper, A.K. (2006). Denaturing gradient gel electrophoresis can rapidly display the bacterial diversity contained in 16S rDNA clone libraries. Microb. Ecol. *51*, 479–486.

Cardona, M., de Vibe Vanay, V., Midtvedt, T., and Norin, E. (2000). Probiotics in gnotobiotic mice – Conversion of cholesterol to coprostanol in vitro and in vivo and bile acid deconjugation in vitro. Microb. Ecol. Health Dis. *12*, 219–224.

Cardona, M., de V Vanay, V., Midtvedt, T., and Norin, E. (2002). Effect of probiotics on five biochemical microflora-associated characteristics, in vitro and in vivo. Scand. J. Nutr. *46*, 73–79.

Carlstedt-Duke, B., Midtvedt, T., Nord, C.-E., and Gustafsson, B.E. (1986). Isolation and characterization of a mucin-degrading strain of Peptostreptococcus from rat intestinal tract. Acta Path. Microbiol. Immunol. Scand. *94*, 293–300.

Fischer, S.G, and Lerman, L.S. (1983). DNA fragments differing by single base pair substitutions are separated in denaturing gradient gels: Correspondence with melting theory. Proc. Nat. Acad. Sci. USA *80*, 1579–1583.

Furrie, E. (2006). A molecular revolution in the study of intestinal microflora. Gut. 55, 141–143.

Gadelle, D., Raibaud, P., and Sacquet, E. (1985). Beta-Glucuronidase activities of intestinal bacteria determined both in vitro and in vivo in gnotobiotic rats. Appl. Environ. Microbiol. *49*, 682–685.

Gill, S.R, Pop, M., Deboy, R.T, Eckburg, P.B, Turnbaugh, P.J, Samuel, B.S, Gordon, J.I, Relman, D.A, Fraser-Liggett, C.M, and Nelson, K.E. (2006). Metagenomic analysis of the human distal gut microbiome. Science *312*, 1355–1359.

Gustafsson, B.E, Midtvedt, T., and Strandberg, K. (1970). Effects of microbial contamination on the cecum enlargement of germ-free rats. Scand. J. Gastroent. 5, 309–314.

Gustafsson, B. (1982). Introduction to the ecology of the intestinal microflora and its general characteristics. XV Symposium Swedish Nutrition Fund, Stockholm, pp. 11–16.

Gustafsson, B., Midtvedt, T., and Norman, A. (1966). Isolated fecal microorganisms capable of 7-alpha-dehydroxylation bile acids. J. Exp. Med. *123*, 413–432.

Handelsman, J. (2004). Metagenomics: application of genomics to uncultured microorganisms. Microb. Mol. Biol. Rev. *68*, 669–685.

Hylemon, P., ed. (1985). Metabolism of Bile Acids In Intestinal Microflora. Amsterdam: Elsevier.

Jernberg, C., Sullivan, A., Edlund, C., and Jansson, J.K. (2005). Monitoring of antibiotic-induced alterations in the human intestinal microflora and detection of probiotic strains by use of terminal restriction fragment length polymorphism. Appl. Environ. Microbiol. *71*, 501–506.

Jernberg, C., Löfmark, S., Edlund, C., and Jansson, J.K. (2007). Long-term ecological impacts of antibiotic administration on the human intestinal microbiota. ISME J. *1*, 56–66.

Mangin, I., Suau, A., Magne, F., Garrido, D., Gotteland, M., Neut, C., and Pochart, P. (2006). Characterization of human intestinal bifidobacteria using competitive PCR and PCR-TTGE. FEMS Microbiol. Ecol. 55, 28–37.

Margulies, M., Egholm, M., Altman W.E, Attiya, S., Bader, J.S, Bemben, L.A, Berka, J., Braverman, M.S, Chen, Y.J, Chen, Z., Dewell, S.B, Du, L., Fierro, J.M, Gomes, X.V, Godwin, B.C, He, W., Helgesen, S., Ho, C.H, Irzyk, G.P, Jando, S.C, Alenquer, M.L, Jarvie, T.P, Jirage, K.B, Kim, J.B, Knight, J.R, Lanza, J.R, Leamon, J.H, Lefkowitz, S.M, Lei M., Li J., Lohman, K.L, Lu, H., Makhijani, V.B, McDade, K.E, McKenna, M.P, Myers, E.W, Nickerson E., Nobile, J.R, Plant, R., Puc, B.P, Ronan, M.T, Roth, G.T, Sarkis, G.J, Simons, J.F, Simpson, J.W, Srinivasan, M., Tartaro, K.R, Tomasz, A., Vogt, K.A, Volkmer, G.A, Wang, S.H, Wang, Y., Weiner, M.P, Yu, P., Begley, R.F., and Rothberg, J.M. (2005). Genome sequencing in microfabricated high-density picolitre reactors. Nature *437*, 376–380.

Maidak, B.L, Cole, J.R, Parker, C.T, Garrity, G.M, Larsen, N., Li, B., Lilburn, T.G, McCaughey, M.J, Olsen, G.J, Overbeek, R., Pramanik, S., Schmidt, T.M, Tiedje, J.M., and Woese, C.R. (1999). A new version of the RDP (Ribosomal Database Project). Nucleic Acids Res. *27*, 171–173.

Manichanh, C., Rigottier-Gois, L., Bonnaud, E., Gloux, K., Pelletier, E., Frangeul, L., Nalin, R., Jarrin, C., Chardon, P., Marteau, P., Roca, J., and Dore, J. (2006). Reduced diversity of faecal microbiota in Crohn's disease revealed by a metagenomic approach. Gut 55, 205–211.

Midtvedt, T. (1974). Microbial bile acid transformation. Am. J. Clin. Nutr. *27*, 1341–1347.

Midtvedt, T., and Gustafsson, B.E. (1981). Microbial conversion of bilirubin to urobilins in vitro and in vivo. Acta Pathol. Microbiol. Scand. *89*, 57–60.

Midtvedt, T., Carlstedt-Duke, B., Hoverstad, T., Lingaas, E., Norin, E., Saxerholt, H., Steinbakk, M. (1986). Influence of peroral antibiotics upon the biotransformatory activity of the intestinal microflora in healthy subjects. Europ. J. Clin. Invest. *16*, 11–17.

Midtvedt, T. (1999). Microbial functional activities. In: Hansson L., Yolken, RH, ed. Probiotics, other Nutritional Factors, and Intestinal Microflora. Philadelphia: Lippincott-Raven, pp. 79–96.

Monstein, H.J, Olsson, C., Nilsson, I., Grahn, N., Benoni, C., Ahrné, S. (2005). Multiple displacement amplification of DNA from human colon and rectum biopsies: bacterial profiling and identification of *Helicobacter pylori*-DNA by means of 16S rDNA-based TTGE and pyrosequencing analysis. J. Microbiol. Methods *63*, 239–247.

Muyzer, G., de Waal, E.C, and Uitterlinden, A.G. (1993). Profiling of complex microbial populations by denaturing gradient gel electrophoresis analysis of polymerase chain reaction-amplified genes encoding for 16S rRNA. Appl. Environ. Microbiol. *59*, 695–700.

Quesada-Perez, M., Martin-Molina, A., and Hidalgo-Alvarez, R. (2005). Simulation of electric double layers undergoing charge inversion: mixtures of mono- and multivalent ions. Langmuir *21*, 9231–9237.

Ramare, F., Hautefort, I., Verhe, F., Raibaud, P., and Iovanna, J. (1996). Inactivation of tryptic activity by a human-derived strain of *Bacteroides distasonis* in the large intestines of gnotobiotic rats and mice. Appl. Environ. Microbiol. *62*, 1434–1436.

Roberts, A.P, Childs, S.M, Rubin, G., and de Wit, N.J. (2000). Tests for *Helicobacter pylori* infection: a critical appraisal from primary care. Family Practice *17*, S12–S20.

Saarela, M., Maukonen, J., von Saarela, M., Maukonen, J., von Wright, A., Vilpponen-Salmela, T., Patterson, A.J, Scott, K.P, Hamynen, H., and Matto, J. (2007). Tetracycline susceptibility of the ingested *Lactobacillus acidophilus* LaCH-5 and *Bifidobacterium animalis* subsp. *lactis* Bb-12 strains during antibiotic/probiotic intervention. Int. J. Antimicrob. Agents. *29*, 271–280.

Sadzikowski, M.R, Sperry, J.F, and Wilkins, T.D. (1977). Cholesterol-reducing bacterium from human feces. Appl. Environ. Microbiol. *34*, 355–362.

Sakata, S., Tonooka, T., Ishizeki, S., Takada, M., Sakamoto, M., Fukuyama, M., and Benno, Y. (2005). Culture-independent analysis of fecal microbiota in infants, with special reference to *Bifidobacterium* species. FEMS Microbiol. Lett. *243*, 417–423.

Savage, D.C. (1977). Microbial ecology of the gastrointestinal tract. Ann. Rev. Microbiol. *31*, 107–133.

Siigur, U., Norin, K.E, Allgood, G., Schlagheck, T., and Midtvedt, T. (1994). Concentrations and correlations of faecal short-chain fatty acids and faecal water content in man. Microb. Ecol. Health. Dis. *7*, 287–294.

Sogin, M.L, Morrison, H.G, Huber, J.A, Welch, D.M, Huse, S.M, Neal, P.R, Arrieta, J.M, and Herndl, G.J. (2006). Microbial diversity in the deep sea and the underexplored 'rare biosphere'. Proc. Nat. Acad. Sci. USA *103*, 12115–12120.

Wang, M., Ahrne, S., Antonsson, M., and Molin, G. (2004). T-RFLP combined with principal component analysis and 16S rRNA gene sequencing: an effective strategy for comparison of fecal microbiota in infants of different ages. J. Microb. Methods 59, 53–69.

Welling, G.W., and Meijer-Severs, G.J. (1988). Beta-aspartylglycinease activity, a microflora-associated characteristic (MAC); its presence in different strains of anaerobic bacteria. Microb. Ecol. Health Dis. *1*, 45–49.

Vitek, L., Zelenka, J., Zadinova, M., and Malina, J. (2005). The impact of intestinal microflora on serum bilirubin levels. J. Hepatol. *42*, 238–243.

Woese, C. (1987). Bacterial evolution. Microb. Rev. *51*, 221–271.

Zhang, M., Liu, B., Zhang, Y., Wei, H., Lei, Y., Zhao, L. (2007). Structural shifts of mucosa-associated lactobacilli and *Clostridium leptum* subgroup in patients with ulcerative colitis. J. Clin. Microbiol. *45*, 496–500.

Zoetendal, E.G, von Wright, A., Vilpponen-Salmela, T., Ben-Amor, K., Akkermans, A.D, de Vos, W.M. (2002). Mucosa-associated bacteria in the human gastrointestinal tract are uniformly distributed along the colon and differ from the community recovered from feces. Appl. Environ. Microbiol. *68*, 3401–3407.

Zoetendal, E.G, Vaughan, E.E., and de Vos, W.M. (2006). A microbial world within us. Mol. Microbiol. *59*, 1639–1650.

Surface Proteins of *Lactobacillus* Involved in Host Interactions

5

Jenni Antikainen, Timo K. Korhonen, Veera Kuparinen, Takahiro Toba and Stefan Roos

Abstract

Specific recognition of host components is central in bacterial adhesion and colonization at host surfaces as well as in bacterial interaction with physiological and immunological processes of the host. Isolates of *Lactobacillus* express a variety of adhesive surface proteins, many of which are multifunctional adhesins or also involved in physiological processes in the bacteria. These adhesins can be grouped as S-layer proteins, proteins with the LPXTG surface-anchoring motif, surface-localized housekeeping proteins, as well as transporter proteins. Recognized targets for lactobacillar adhesins include epithelial and phagocytic cells, extracellular matrices, mucins, and circulating components. A more detailed, mechanistic knowledge of lactobacillar adhesion proteins will help to understand their role in colonization and to develop their probiotic use.

Introduction

Bacterial species that infect eukaryotic organisms show a repertoire of interactions with their hosts; on the host side, these interactions target specific tissues as well as tissue domains and components, epithelial and nonepithelial cell types, as well as physiologically important circulating components. The interactions demonstrate a high degree of molecular specificity and serve different aspects of bacterial colonization within the host. Adhesion to host tissues is the first step in bacterial colonization and also influences subsequent phases leading to infectious disease or commensalism, e.g. by targeting bacteria into the extracellular milieu within the host (Westerlund and Korhonen, 1993) or by affecting responses in epithelial or phagocytic cells (Vaarala, 2003; Merk *et al.*, 2005). Recent studies have demonstrated the presence of several distinct surface components in lactobacilli that function in host interactions. This Chapter addresses their identity, structure, as well as functions and possible biological importance.

As in other prokaryotes, surface proteins in lactobacilli make up the most numerous adhesion molecules and will be the focus of this Chapter. Initially, the role of proteinaceous surface molecules in adhesion was indicated in several studies (Greene and Klaenhammer, 1994; Tuomola *et al.*, 2000; Lorca *et al.*, 2002) that addressed lactobacillar adhesion to mucus, epithelial cells and the extracellular matrix, but also lipoteichoic acids (LTAs) were reported to mediate adhesion of *Lactobacillus* to epithelial cells (Granato *et al.*, 1999). To date, several lactobacillar surface proteins have been identified, and it is becoming obvious that lactobacilli express a wealth of recognition processes with their hosts. The lactobacillar surface proteins functioning in these processes include the surface layer (S-layer) proteins; the functionally diverse group of proteins anchored to the cell wall via the so-called LPXTG motif; the so-called anchorless housekeeping proteins which are classically considered cytoplasmic and whose secretion and surface-anchoring mechanisms have remained poorly known; transporter proteins; as well as proteins not belonging to any of the recognized protein groups (Table 5.1).

Table 5.1 Proposed or identified adhesive surface proteins of *Lactobacillus*

Adhesin	Target	Species/strain	Reference
Surface layer proteins			
S-layer protein	Avian intestinal epithelial cells	*Lactobacillus acidophilus* spp.	Schneitz *et al.*, 1993
CbsA	Collagens, laminin, chicken intestinal subepithelial tissue	*Lactobacillus crispatus* JCM 5810	Toba *et al.*, 1995
SlpA	Fibronectin, human epithelial cell line	*Lactobacillus brevis* ATCC8287	Hynönen *et al.*, 2002
S-layer protein	Red blood cells	*Lactobacillus kefir* CIDCA 8321, *Lactobacillus parakefir* CIDCA 8328	Garrote *et al.*, 2004
SlpA	Murine ileal epithelial cells	*Lactobacillus acidophilus* M92	Frece *et al.*, 2005
LPXTG-motif proteins			
Mub	Hen intestinal mucus, pig mucin	*Lactobacillus reuteri* 1063	Roos and Jonsson, 2002
Mub (LBA1392)	Human intestinal epithelial cell line	*Lactobacillus acidophilus* NCFM	Buck *et al.*, 2005
Lsp	Murine gut epithelium	*Lactobacillus reuteri* 100–23	Walter *et al.*, 2005
Msa (LP1229)	Mannosides	*Lactobacillus plantarum* WCFS1	Pretzer *et al.*, 2005
LspA	Human intestinal epithelial cell line	*Lactobacillus salivarius* UCC118	van Pijkeren *et al.*, 2006
Anchorless housekeeping proteins			
EF-Tu	Human intestinal epithelial cell line, mucin	*Lactobacillus johnsonii* NCC533	Granato *et al.*, 2004
GroEL	Human intestinal epithelial cell line, mucin	*Lactobacillus johnsonii* NCC533	Bergonzelli *et al.*, 2006
Enolase	Laminin, fibronectin, plasminogen	*Lactobacillus crispatus* ST1	Hurmalainen *et al.*, 2007; Antikainen *et al.*, 2007b
GAPDH	Plasminogen	*Lactobacillus crispatus* ST1	Hurmalainen *et al.*, 2007
Transporter proteins			
CnBP	Type I collagen	*Lactobacillus reuteri* NCIB 11951	Roos *et al.*, 1996
MapA	Porcine intestinal mucus, human intestinal epithelial cell line	*Lactobacillus reuteri*104R	Rojas *et al.*, 2002; Miyoshi *et al.*, 2006
Others			
FbpA	Human intestinal epithelial cell line	*Lactobacillus acidophilus* NCFM	Buck *et al.*, 2005

Adhesive surface layer proteins

S-layers are periodic crystalline arrays that are composed of protein or glycoprotein subunits, which self-assemble to cover up to 70% of the bacterial cell surface. The S-layer is not impermeable and has pores between the identical lattice units (Sára and Sleytr, 2000). Lactobacillar S-layers have a relatively high isoelectric point and are homopolymers of 25 to 71-kDa subunits (reviewed by Åvall-Jääskeläinen and Palva, 2005). S-layer proteins represent 10–15% of the total amount of proteins in *Lactobacillus* cells (Boot and Pouwels, 1996), and their transcription and secretion mechanisms must be efficient and tightly regulated. Multiple promoters precede several S-layer genes (Boot and Pouwels, 1996), including S-layer genes of *Lactobacillus acidophilus* (Boot *et al.*, 1996a) and *L. brevis* (Vidgren *et al.*, 1992; Kahala *et al.*, 1997) and are likely to ensure the efficient transcription of these genes. Also, the half-lives of mRNA encoding lactobacillar S-layer proteins are relatively high, approximately 15 min, which enables efficient protein translation (Boot *et al.*, 1996a; Kahala *et al.*, 1997). The predicted lactobacillar S-layer proteins contain a conserved N-terminal signal sequence of 25–30 amino acids, which indicates that their secretion occurs via the general Sec-pathway. The highly efficient lactobacillar promoter regions and signal sequences have been utilized in various heterologous protein expression systems (Savijoki *et al.*, 1997; Kahala and Palva, 1999; Åvall-Jääskeläinen *et al.*, 2002), for instance, in expression of the adhesive S-layer peptides of *L. crispatus* JCM 5810 (Martínez *et al.*, 2000).

Several lactobacillar S-layers have been proposed or identified as adhesins (Table 5.1). Treatment of *L. kefir* and *L. parakefir* cells with lithium chloride – which is a routine method to release S-layer proteins from the bacterial surface – abolished the haemagglutination ability of the bacteria (Garrote *et al.*, 2004). Haemagglutination, however, apparently is not a common characteristic in species of *Lactobacillus* (Ocaña *et al.*, 1999; Colloca *et al.*, 2000). Schneitz *et al.* (1993) observed that treatment of *L. acidophilus* cells with LiCl abolished bacterial adhesiveness to chicken intestinal cells and proposed that S-layer of *L. acidophilus* mediates the bacterial binding. Similarly, Frece *et al.*, (2005) found that treatment of *L. acidophilus* M92 cells with LiCl abolished bacterial adhesiveness to mouse ileal epithelial cells and Chen *et al.* (2007) reported that depletion of S-layer of *L. crispatus* ZJ001 abolished adhesiveness to HeLa cells. Removal of S-layer proteins with LiCl may, however, simultaneously remove other cell-wall proteins important in adhesion, and therefore the observations described above remain suggestive only. Along this line, deletion of the S-layer gene *slpA* in *L. acidophilus* NCFM abolished the bacterial adherence to a human intestinal epithelial cell line, but the authors suggested that the nonadhesive phenotype probably resulted from loss of other surface proteins bound onto the S-layer (Buck *et al.*, 2005).

So far, the adhesive functions of two lactobacillar S-layer proteins have been confirmed by genetic means. *L. crispatus* JCM 5810 adheres very efficiently to collagens, laminin, and fibronectin (Fn), which are major components of mammalian extracellular matrices (ECM), and the S-layer protein CbsA (Collagen binding S-layer protein A) extracted from *L. crispatus* JCM5810 cell surface bound to solubilized as well as to immobilized type IV collagen (Toba *et al.*, 1995; Sillanpää *et al.*, 2000). Inhibition studies indicated that type I and type V collagens are also recognized by CbsA (Toba *et al.*, 1995). Interestingly, bacteria expressing CbsA do not bind solubilized laminin but adhere to immobilized laminin (Toba *et al.*, 1995; Antikainen *et al.*, 2002), such a conformation- or receptor-density dependent recognition of an ECM protein has been reported for other bacteria and adhesins as well (Westerlund and Korhonen, 1993).

The *cbsA* gene was cloned into *Escherichia coli*, where CbsA was expressed as a His-tagged fusion protein that exhibited collagen binding (Sillanpää *et al.*, 2000). *cbsA* encodes a mature protein of 410 amino acids with typical features of lactobacillar S-layer proteins, such as a high content of basic amino acids. CbsA has a conserved C-terminal region with pI 10.0 and a nonconserved N-terminal region with a pI 6.8. It is interesting that the strain JCM 5810 has another S-layer gene, termed *cbsB*, which was not expressed by JCM 5810 cells and whose recombinant form did not bind to collagens

(Sillanpää *et al.*, 2000). An extensive mutational analysis with His-tagged proteins revealed that the N-terminal region of CbsA is responsible for binding to collagens and laminin, whereas the basic C-terminus anchors the S-layer onto negatively charged LTAs on the bacterial surface (Sillanpää *et al.*, 2000; Antikainen *et al.*, 2002). That CbsA indeed is an ECM-specific adhesin was further supported by the finding that recombinant *L. casei* expressing the N-terminal part of CbsA specifically adhered to collagen-containing areas of chicken colon (Sillanpää *et al.*, 2000; Antikainen *et al.*, 2002). The N-terminal region also is responsible for formation of the periodic surface layer, and analysis of the recombinant His_6-CbsA peptides indicated that the polymeric form of CbsA is important for collagen-binding (Antikainen *et al.*, 2002). Analysis of intra- and intermolecular forces that drive the folding and assembly of CbsA have been initiated by atomic force microscopy (Verbelen *et al.*, 2007). The domain structure and structure/function relationships in CbsA are depicted in Fig. 5.1.

Treatment of *L. brevis* ATCC 8287 cell with GnHCl – another common method to remove S-layer from cell surface – abolished binding of this strain to human nonpolarized Intestine 407 cells, which suggested the involvement of S-layer. Expression of fragments of the *L. brevis* S-layer protein SlpA as a genetic fusion in flagellar FliC subunits in *E. coli* conferred binding of chimeric flagella to Intestine 407 cells and to Fn, which confirmed the adhesive characteristics of the *L. brevis* SlpA. By testing hybrid flagella expressing different *slpA* regions, the receptor-binding region in SlpA was mapped to the 81 amino acids in the N-terminal part of the protein (Hynönen *et al.*, 2002). Analysis of antibody binding to ATCC 8287 and to chimeric flagellas suggested that the fibronectin-binding fragment in SlpA is located in a groove inaccessible to antibodies, a reminiscent of the canyon hypothesis in viral adhesion proteins (Rossmann, 1989). It is interesting to note that adherence of *Lactobacillus* isolates to human intestine 407 cells has earlier been correlated with binding to Fn (Kapczynski *et al.*, 2000). Fn is a large, multifunctional ECM protein that occurs in a circulating form, plasma Fn, as well as in a cellular form (Pankov and Yamada, 2002). Fn has adhesive and immunostimulating functions in tissues and it is possible that it mediates bacterial adherence to epithelial cells by a bridging mechanism, i.e. by binding to bacteria as well as to epithelial cells (Schwarz-Linek *et al.*, 2004), or by a direct mechanism through cellular Fn (Sarén *et al.*, 1999).

The two-domain structure resembling that in CbsA (Fig. 5.1) has also been detected in SlpA protein of *L. acidophilus* ATCC 4356. The two S-layer proteins are related in primary sequence but adhesive functions have not been described for SlpA. A fragment containing the N-terminal two-thirds of the SlpA protein (SAN) crystallized into a layer and was proposed to be composed of two sub-domains with a surface-exposed loop (Smit *et al.*, 2002). The C-terminal one-third of the S-layer protein from *L. acidophilus* (SAC) was shown to bind to bacterial cells depleted of the S-layer, and SAC binds to LiCl-extracted cell surface of *L. crispatus* and *Lactobacillus helveticus*, which have a closely related S-layer protein (Smit *et al.*, 2001). Further, the cell-wall binding site was localized to the N-terminal region of 65 amino acids in the SAC domain, and a preliminary analysis by selective extraction of cell wall components suggested that SAC binds to teichoic acids (Smit and Pouwels, 2002).

The predicted amino acid sequences of *L. brevis*, *L. crispatus*, and *L. acidophilus* S-layer proteins are not identical and, hence, the two-domain structure cannot be extended to *L. brevis*. The S-layer proteins from *L. brevis* and *L. buchneri* were proposed to bind to a neutral polysaccharide moiety in the cell wall, but not to PG or teichoic acids (Masuda and Kawata, 1980; 1981). It thus remains open how conserved the domain architecture and molecular interactions within the cell wall are in lactobacillar S-layer proteins.

S-layer proteins represent a common and abundant class of surface proteins in lactobacilli, and given the common occurrence of ECM-adherence in lactobacilli (Aleljung *et al.*, 1991; Styriak *et al.*, 2001), it is somewhat surprising that another collagen-binding S-layer protein in lactobacilli has not been reported, not even in closely related isolates of *L. crispatus*. This indicates that CbsA is rather an exception in its functions and that functions of S-proteins

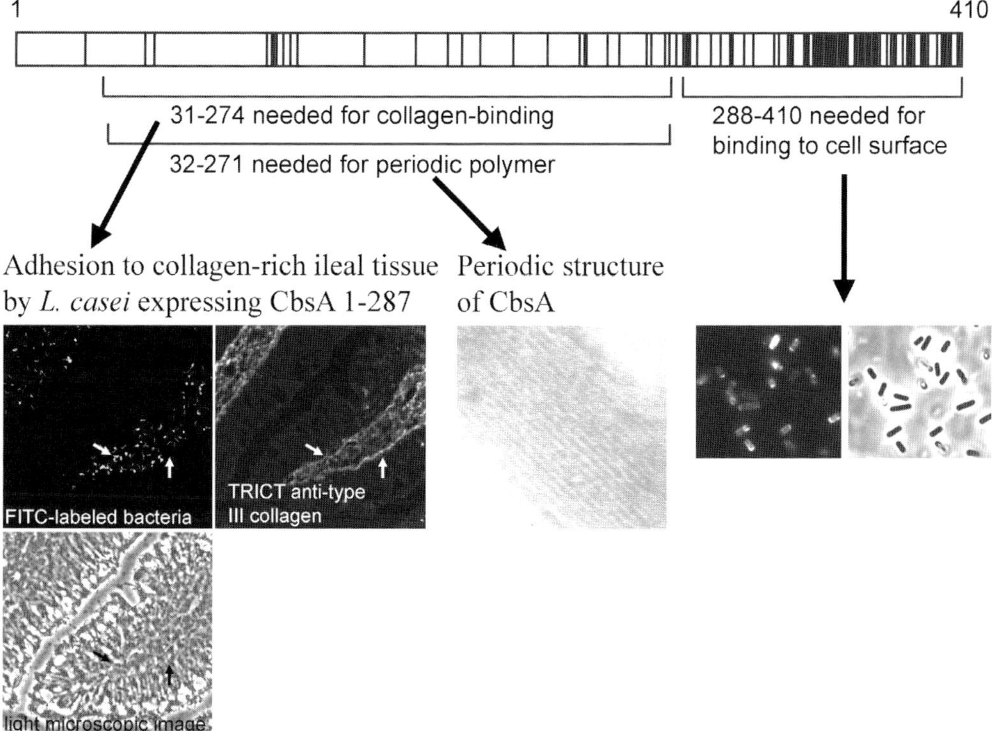

Figure 5.1 Domain structure of the collagen-binding S-layer protein CbsA of *Lactobacillus crispatus*. The bar represents CbsA with 410 amino acid residues, and residues conserved in eight S-layer proteins of *Lactobacillus crispatus*, *Lactobacillus acidophilus* and *Lactobacillus helveticus* are represented by a line. Functional regions are indicated below the bar. Adhesion to frozen section of chicken ileum by a recombinant *L. casei* strain expressing N-terminal 287 amino acids of CbsA is shown in the left-bottom panels. The same microscopic field is shown for adherent fluorescein isothiocyanate (FITC)-labelled bacteria, for double staining of the tissue section with anti-type III- collagen-specific immunoglobulins conjugated to tetramethylrhodamine isothiocyanate (TRITC), the bottom panel is the light microscopic image of the same field. An electron microscopic picture of the periodic structure of His$_6$-CbsA 32–271 is shown in the middle. On the right, binding of His$_6$-CbsA 288–410 to the cell wall of *L. crispatus* cells from which S-layer had been extracted, the peptide binding was detected anti-His$_6$ antibody. The figure is modified from Antikainen *et al.*, (2002) with permission of Blackwell Publishing Ltd.

vary between species of *Lactobacillus* and even within a species. S-layer genes and proteins have been cloned and characterized from *L. crispatus* (Sillanpää *et al.*, 2000; Antikainen *et al.*, 2002), *L. acidophilus* (Boot *et al.*, 1993), *L. gallinarum* (Hagen *et al.*, 2005), *L. helveticus* (Callegari *et al.*, 1998; Gatti *et al.*, 2005) and from *L. brevis* (Vidgren *et al.*, 1992; Jakava-Viljanen *et al.*, 2002), and multiple S-layer genes have been identified in the genomes of *L. acidophilus, L. amylovorus, L. gallinarum, L. crispatus, L. brevis, L. gasseri and L. johnsonii* (Boot *et al.*, 1996b; Jakava-Viljanen *et al.*, 2002; Ventura *et al.*, 2002). Formerly, *L. johnsonii* and *L. gasseri* were regarded lacking an S-layer (Boot *et al.*, 1996b), but recently, (Ventura *et al.*, 2002) identified a protein called aggregation-promoting factor in these species as an S-layer-like protein, whose amino acid composition and physical properties are similar to those in lactobacillar S-layers, which indicates that their presence in lactobacilli is more common than presently assumed. The S-protein primary sequences are conserved only in closely related species of *Lactobacillus* (Åvall-Jääskeläinen and Palva, 2005), and our hypothesis is that variability in primary sequences, as well as in the domain architecture are related to the functional variations in lactobacillar S-layer proteins and perhaps, on a broader basis, to their functional adaptation to enhance bacterial colonization in different environments.

LPXTG adhesin proteins in lactobacilli

Among Gram-positive bacteria a common mechanism for anchoring proteins to the cell envelope is the sortase-dependent anchoring via an LPXTG motif. The proteins anchored through this mechanism have an N-terminal secretion signal sequence and, at the C-terminus, a sorting signal with the sequence LPXTG (sometimes slightly modified), a hydrophobic membrane-spanning region as well as a tail of charged amino acids. The LPXTG sequence is recognized by a membrane-associated sortase enzyme, which cleaves the peptide bond between the threonine and glycine residues in the motif and thereafter covalently links the protein to a peptide cross-bridge of peptidoglycan (Paterson and Mitchell, 2004; Marraffini *et al.*, 2006). In Gram-positive pathogens such as streptococci, staphylococci, and listeria, proteins anchored in this way often are involved in adhesion and important for the virulence of the bacteria (Paterson and Mitchell, 2004).

During the last years it has become evident that also lactobacilli harbours surface proteins anchored via an LPXTG motif. The first example of a sortase-dependent adhesin from lactobacilli was found in *L. reuteri* 1063 isolated from pig intestine (Roos and Jonsson, 2002). The gene encoding the adhesin was found by screening a lambda genomic library with antibodies directed against cell surface proteins of the bacteria. These antibodies were known to inhibit the adhesion of the bacteria to Caco-2 cells and mucus material. Functional studies of recombinant proteins from two lambda clones performed with an ELISA showed that both adhered to mucus. After sequencing of the gene, the deduced protein was found to be very large (358 kDa) and extremely repetitive. The protein, named Mub, contained two types of repeats of approximately 200 amino acids and the typical features of an LPXTG envelope anchored protein (Fig. 5.2A). After subcloning, the two types of repeats were produced separately and in an ELISA they were shown to adhere to both porcine and hen intestinal mucus. Although the exact ligand not was determined, the authors suggested that the repeats adhered to carbohydrate moieties of mucus components with molecular masses from approximately 15 kDa to more than 2 MDa (Fig. 5.2B). The binding was enhanced at pH values around 4–5 (Fig. 5.2C). At the time for publication of the protein Mub, no *Lactobacillus* genomic sequence was available and the closest related protein was an unknown protein from *Lactococcus lactis* containing repeats with similarities to Mub and a C-terminal sorting signal (Roos and Jonsson, 2002). However, today when several *Lactobacillus* species has been sequenced it is clear that protein Mub represents a class of proteins found in most of them (Boekhorst *et al.*, 2006; van Pijkeren *et al.*, 2006). The Mub repeat sequences cluster into the pfam06458 (MucBP), and a search in Genbank reveals that MucBP also is found in a few proteins from *L. lactis*, *Pediococcus pentosaceus*, *Streptococcus thermophilus* and *Listeria monocytogenes*.

Besides protein Mub from *L. reuteri*, the function of two other MucBP proteins has been studied. The determinant mediating the mannose-specific adhesion earlier described in *L. plantarum* (Adlerberth *et al.*, 1996) was recently identified by Pretzer *et al.* (2005) using a biodiversity-based identification strategy. By correlating the genotype (assessed by DNA microarray-based genotyping) of 14 *L. plantarum* strains with agglutination of yeast in a mannose-specific manner, four candidate genes, whereof two encoded LPXTG proteins, were identified. Since a sortase mutant of *L. plantarum* WCFS1 lost the capacity to agglutinate yeast the conclusion was that one or both of these LPXTG proteins were responsible for this feature. Inactivation of the corresponding genes indeed revealed that one of them mediated the mannoside-adhesion property. This gene was named *msa* (mannose-specific adhesin) and *in silico* analysis showed that the protein harboured the N-terminal signal sequence, the C-terminal LPXTG anchor sequence, two MucBP repeats, and a domain with similarities to SasA, a lectin from *Staphylococcus aureus*. The authors found it likely that those domains are relevant for the mannose adhesion and are involved in the interaction of *L. plantarum* with its host. During the

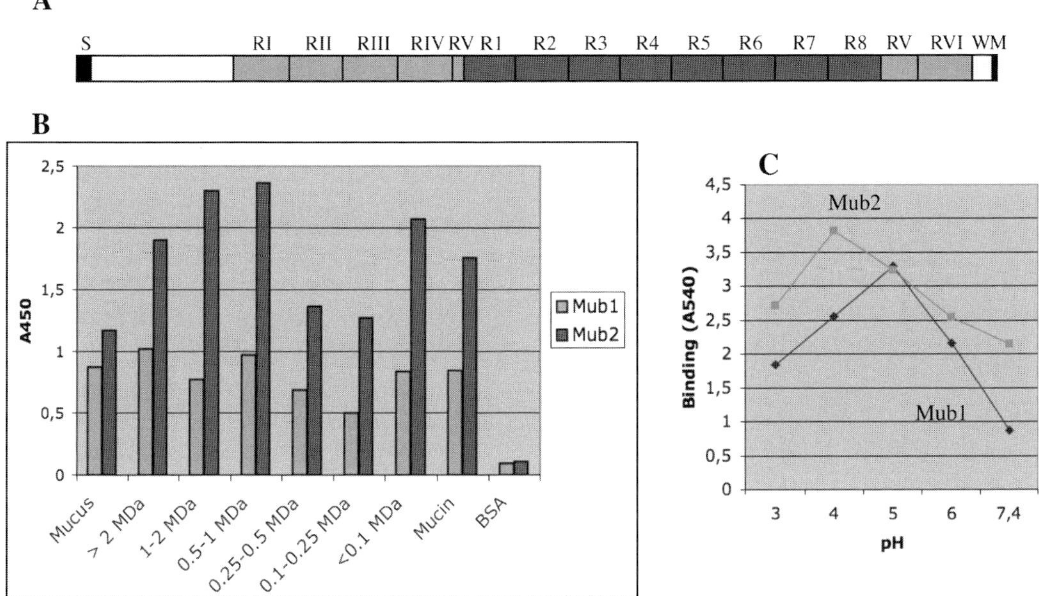

Figure 5.2 Map and adhesion of protein Mub.A. Map of the 358-kDa mucus binding protein Mub. Regions and sites: S, Signal sequence; RI-VI, Type 1 repeats; R1–8, Type 2 repeats; W, Cell wall-spanning region; M, LPXTG motif, membrane-spanning region and membrane anchor; B. Binding of Mub1 and Mub2 repeats to fractionated pig intestinal mucus and mucin (purified from pig stomach mucus) as detected by an ELISA; the size range of the mucus fractions are given at the bottom. C. Binding of Mub repeats to mucus at different pH. A originates from Roos and Jonsson (2002) and is published with permission of Society for General Microbiology.

same period, the role of a number of cell surface proteins of *L. acidophilus* NCFM in attachment to Caco-2 cells were investigated by Buck *et al.* (2005). They searched the previously obtained genome sequence for genes encoding putative adhesion proteins and found one gene encoding a MucBP protein. Inactivation of this gene gave a 65% reduction in the adhesion to Caco-2 cells.

The fact that all MucBP proteins investigated so far are for bacterial adhesion to cells or mucus argues for that proteins belonging to this group are involved in binding of the bacteria to different ligands on mucosal surfaces. The group of proteins containing MucBP domains is very heterogenic, and the members show sequence differences in different bacterial species (Boekhorst *et al.*, 2006). Also, there are indications that there are large sequence variations within a species. Neither the protein Mub from *L. reuteri* 1063 (isolated from pig intestine) nor the large MucBP protein Mlp from *L. reuteri* (formerly *L. fermentum*) BR11 (Turner *et al.*, 2003; isolate from guinea pig vagina) are present in *L. reuteri* ATCC 55730 (isolate from mother's milk). Instead, this strain has a so-called GW protein (contains repeats with glycine-trypthophane [GW] dipeptide motifs) harbouring two MucBP domains (Accession nr: AAY86887; Båth *et al.*, 2005). The large variation of the MucBP adhesins opens for speculations that they are important for adhesion to diverse complex carbohydrates present on mucosal surfaces and contribute to specific colonization of different ecological niches. This and the importance for probiotic properties has to be addressed in the future research on these proteins.

A second type of lactobacillar LPXTG adhesin was found by Walter *et al.* (2005). In an attempt to find genes of importance for the interaction between *L. reuteri* 100–23 (isolated from rat) and the host, a gene encoding an unknown large cell surface protein was partly cloned and sequenced. The encoded protein, named Lsp, showed similarities with uncharacterized LPXTG proteins in *L. johnsonii* and *L. gasseri*, but more interestingly there were also similarities with proteins from Gram-positive cocci involved in biofilm formation or adherence to epithelial cells. After inactivation of the *lsp* gene in the strain 100–23, a mutant with reduced ability to colonize mice and adhere to forestomachs *ex vivo* was obtained, showing that Lsp are important for the interaction with the mucosal surface. In the study of Buck *et al.* (2005) two genes encoding proteins with similarities to Lsp were mutated and their importance for bacterial binding to Caco-2 cells were evaluated. In contrast to the Mub mutant, only a slight and not significant reduction in the adhesion could be seen. In common with these two proteins, Lsp contains Rib repeats (pfam08428) that also are present in the related adhesins from Gram-positive cocci. Rib proteins have also been found in two other *L. reuteri* strains (Båth *et al.*, 2005; Turner *et al.*, 2003). Searches in Genbank reveals that this second group of putative LPXTG adhesins in lactobacilli only are present in *L. reuteri* and members of the *L. acidophilus* group.

Finally, LPXTG adhesins have also been described in *L. salivarius*. Four complete sortase-dependent LPXTG protein sequences (LspA-D) of the strain UCC118 were identified from the genome sequence (van Pijkeren *et al.*, 2006). Importance of these proteins in adhesion was demonstrated by that a sortase mutant had a decreased binding to HT-29 and Caco-2 cells. Investigation of the expression of the LPXTG protein genes revealed that *lspA*, *lspB* and *lspD* were transcribed during the conditions tested, and the genes were inactivated. The *lspA* mutant showed a significant reduction in adhesion to HT-29 cells, but since the decrease only was around 30%, the authors concluded that other unknown determinants on the cell surface also contribute to the adhesiveness of strain UCC118. LspA shows similarities with many LPXTG proteins from Gram-positive bacteria with unknown functions, and might be a representative for a third class of LPXTG adhesins in lactobacilli.

Anchorless, housekeeping proteins secreted by lactobacilli

Recently, several proteins with essential intracellular roles in bacterial growth and metabolism ('housekeeping proteins') have also been detected on the bacterial surface or in the extracellular proteome. The studies were initiated with pathogenic bacterial species, in which these proteins enhance virulence by mediating adhesion or expressing proteolytic or immuno-stimulating activities (Chhatwal, 2002; Pancholi and Chhatwal, 2003; Bergmann *et al.*, 2005). These surface proteins are called 'anchorless', since no established signal sequence or anchoring motif is present in their predicted sequences. Recently, such anchorless proteins have also been identified in lactobacilli, where they include GroEL and EF-Tu, as well as the glycolytic enzymes enolase and glyceraldehyde-3-phosphate dehydrogenase (GAPDH) (Table 5.1).

Surface association of GroEL and EF-Tu has been described in the probiotic strain *L. johnsonii* La1 (NCC 533), where they were identified by immunological methods on the cell surface and GroEL in the culture medium also (Granato *et al.*, 2004; Bergonzelli *et al.*, 2006). Both proteins have been earlier detected on the surface of various other bacterial species. GroEL (also called Hsp60) is a chaperone and essential in intracellular protein folding. The recombinant La1 GroEL protein from *E. coli* binds to mucins from the human adenocarcinoma cell line HT29 as well as to HT29 cells at acidic pH (Bergonzelli *et al.*, 2006). In addition, recombinant GroEL stimulates interleukin 8 secretion in HT29 cells as well as in blood macrophages; the stimulation of human macrophages was dependent on the presence of the CD14 molecule and was also seen with recombinant GroEL proteins originating from *L. helveticus* and *L. lactis*. EF-Tu is an elongation factor, a guanosine binding protein, important in intracellular protein synthesis. Recombinant La1 EF-Tu from *E. coli* bound to mucins isolated from normal human colon as well as to cultured human intestinal Caco-2 and HT29 epithelial cells, it also stimulated release of IL-8 by HT29 cells in the presence of soluble CD14 (Granato *et al.*, 2004). Thus, the adhesive functions and the induction of a proinflammatory immune response in human cells by GroEL and EF-Tu of *L. johnsonii* La1 appear very similar. Their functional similarity is underlined by the observations that the adhesive functions of both proteins are pH-dependent and higher at pH 5 than at pH 7.2 (Granato *et al.*, 2004; Bergonzelli *et al.*, 2006).

Lactobacilli are obligate fermentative organisms and secrete lactic acid as a primary metabolite, which rapidly reduces the pH of the environment to 4. It is therefore not surprising that the adhesive and the host interactions by lactobacilli are sensitive to pH changes. Adhesiveness of lactobacilli to Caco-2 cells and intestinal mucus (Blum *et al.*, 1999) as well as to proteins of the extracellular matrix (Harty *et al.*, 1994) is higher at low pH. This results from two distinct properties of the lactobacillar adhesins. First, as described above for the GroEL and EF-Tu proteins of *L. johnsonii* La1 as well as the Mub proteins (Fig. 5.2C), the individual adhesion proteins may have higher affinity for their ligands at low pH (Granato *et al.*, 2004; Bergonzelli *et al.*, 2006). Second, the surface-association mechanisms of the 'anchorless' proteins may be based on ionic interactions and thus sensitive to pH or salts (Antikainen *et al.*, 2007a; Hurmalainen *et al.*, 2007). This is exemplified in Fig. 5.3, where the release of enolase and GAPDH from the surface of *L. crispatus* strain ST1 are shown to take place at neutral pH. The ST1 enolase and GAPDH proteins have pIs of 4.8 and 5.2, and their release from cell surface becomes detectable at pH 5.2, i.e. close to or above their isoelectric points. The release does not involve *de novo* protein synthesis, and enolase and GAPDH appear to bind cell surface and LTA by ionic interactions at pH below their pIs. In contrast, the S-layer protein of ST1 with a high pI remains cell-bound at acidic and neutral pH (Fig. 5.3); also, binding of the S-layer protein to LTA is similar at pH 4.4 and pH 7 (Antikainen *et al.*, 2007a). This is the first report on surface association mechanism of 'anchorless proteins'. The pH-dependent release was also detected in *L. acidophilus*, *L. amylovorus*, *L. gallinarum*, *L. gasseri*, and *L. johnsonii* (Antikainen *et al.*, 2007a), but it remains to be established how common this mechanism is among other bacterial species as well as how many surface proteins are present in the extracellular proteome of lactobacilli. Taken together, these findings illustrate that environmental pH is an important factor affecting the structure of lactobacillar cell surface as well as the functions of lactobacillar adhesion proteins.

Enolase and GAPDH are essential intracellular glycolytic enzymes. GAPDH catalyses oxidation and phosphorylation of glyceraldehyde-3-phosphate to 1,3-biphosphoglycerate, whereas enolase catalyses dehydration of 2-phosphoglycerate (2-PGE) to phosphoenolpuryvate. Enolase also catalyses reverse reaction in gluconeogenesis. These enzymes have been detected on the surface of several Gram-positive and Gram-negative bacterial species as well as in fungi and other eukaryotic organisms (reviewed in Pancholi, 2001; Pancholi and Chhatwal, 2003; Lähteenmäki *et al.*, 2005). A major function detected for surface-exposed enolases and GAPDH in Gram-positive pathogens is immobilization of plasminogen (Plg) onto bacterial surface and subsequent enhancement of its activation to the serine protease plasmin (Pancholi, 2001; Pancholi and Chhatwal, 2003; Lähteenmäki *et al.*, 2005). Plasmin is a powerful protease involved in several physiological processes, such as fibrinolysis, degradation of ECM, enhancement of cell migration and activation of prohormones and growth factors (Mignatti and Rifkin, 1993; Lijnen and Collen, 1995; Plow *et al.*, 1999; Myöhänen and Vaheri, 2004). Plg activation and enolase are central in the pathogenesis of pneumococcal (Bergmann *et al.*, 2005) as well as group A streptococcal (Sun *et al.*, 2004) infectious diseases. Rather surprisingly, *L. crispatus* ST1 and several other species of the genus *Lactobacillus* were recently found to enhance Plg activation. Enolase and GAPDH were identified in the extracellular proteome obtained at pH 7, shown to bind Plg and to enhance its activation (Hurmalainen *et al.*, 2007). Another major function of surface enolase and GAPDH is in bacterial adhesion (Brassard *et al.*, 2004; Carneiro *et al.*, 2004; Ge *et al.*, 2004; Jin *et al.*, 2005; Pancholi and Fischetti, 1992; Seifert *et al.*, 2003), and enolase of *L. crispatus* was found to bind the ECM proteins laminin and Fn (Antikainen *et al.*, 2007b). Overall, it seems that lactobacilli and Gram-positive pathogens express enolases and GAPDHs with similar functional properties, a major difference being that in lactobacilli these proteins are cell-bound only at low pH.

The importance of extracellular proteome in *Lactobacillus*-host interactions is also indicated by the findings that cell-free culture medium of the probiotic *L. rhamnosus* GG inhibits expression of the proinflammatory cytokine TNF-α in murine macrophages as well as induces cytoprotective heat-shock protein expression and modulates

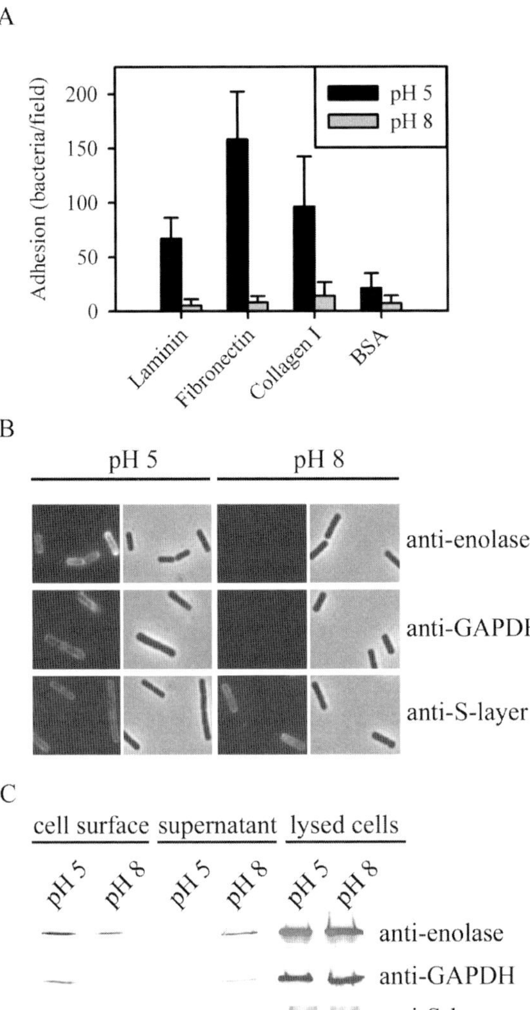

Figure 5.3 Modification of the adhesiveness and cell-wall architecture of *Lactobacillus crispatus* ST1 in response to change in pH. (A) Bacterial adhesiveness to immobilized laminin, fibronectin, type I collagen and bovine serum albumin (BSA) at pH 5 and pH 8 are shown. (B) Indirect immunofluorescence staining of ST1 cells suspended after overnight growth to pH 5 or pH 8, the stainings were done with anti-enolase, anti-GAPDH, or anti-S-layer protein immunoglobulins as primary antibodies. (C) Western blotting of enolase and GAPDH on ST1 cell surface, from supernatant and from lysed cells obtained after 1 h of incubation of the cells at pH 5 and pH 8. As controls, reactivity with the anti-S-layer and anti-RNA polymerase (pol) is shown. B and C originate from Antikainen *et al.* (2007a) and are published with permission of American Society for Microbiology.

signal transduction pathways in a mouse colonic cell line (Peña and Versalovic, 2003; Tao *et al.*, 2006). Further, Tao *et al.* (2006) showed that the factor responsible for heat-shock protein induction is a small-molecular-weight peptide whose activity is higher at pH 4 than at pH 7.

Transporter proteins as adhesins

CnBP is a collagen-binding protein of *L. reuteri* NCIB 11951^{T} whose predicted sequence shows high homology to amino acid binding subunit of an ABC-type transport system (Roos *et al.*, 1996). CnBP has at least two closely related orthologs in other *L. reuteri* strains, the BspA protein of *L. reuteri* BR11 (Turner *et al.*, 1999; formerly described as *Lactobacillus fermentum*) and the MapA protein of *L. reuteri* 104R (Rojas *et al.*, 2002; Miyoshi *et al.*, 2006). MapA (mucus adhesion promoting protein A) binds to mucins as well as to Caco-2 cells, where two putative receptor-active molecules were identified by ligand blotting (Miyoshi *et al.*, 2006). BspA, which functions in uptake of L-cystine, was found not to bind collagen or Fn and its adhesive properties remain open (Turner *et al.*, 1999). However, these results strongly suggest that members of the family III of bacterial solute binding proteins form a class of lactobacillar adhesion proteins, similar to what has been seen in, for example, streptococci (Spellerberg *et al.*, 1999).

The results cited above give a view of multiple adhesin types in lactobacilli and their multiple targets in the hosts. The available genomic sequences have given a possibility to analyse putative adhesin genes and to search for adhesin variants present in strains of *Lactobacillus*. For example, 10 LPXTG-type of adhesins were predicted to be present in the genome of *L. plantarum* strain WCFS1, most of which were predicted mucus-binding proteins (Boekhorst *et al.*, 2006). Such analyses give basis for functional genomic studies on lactobacillar interactions with their hosts. Along this line, Buck *et al.*, (2005) mutated *L. acidophilus* NCFM genes encoding a mucus-binding protein, a Fn-binding protein, an S-layer protein, as well as two LPXTG-type adhesins and tested the mutants for adherence to Caco-2 cell line. A significant decrease in adhesion was seen with Fn-binding protein mutant as well as the mucin-binding mutant. The gene encoding the fibronectin-binding protein (FbpA) of *L. acidophilus* NCFM was identified on basis of an N-terminal conserved domain in the predicted protein; the protein shows overall ca. 40% sequence identity with streptococcal Fn-binding proteins. Interestingly, orthologous genes were by the authors identified in the genomes of *L. gasseri* and *L. johnsonii* and can also be found in the other sequenced *Lactobacillus* species. In addition, FbpA also clusters perfectly into COG1293, predicted to be RNA-binding proteins, and was by Christie *et al.* (2002) suggested to play a regulatory role in the modulation of the fibronectin-binding protein CshA in *Streptococcus gordonii*. The role of FbpA in adhesion thus needs to be further investigated.

Application of adhesion proteins

The potential probiotic use of lactobacilli and the increasing number of their genomic sequences have given a rationale approach for understanding probiotic mechanisms of these bacteria. Adhesion is essential in bacterium-host interactions, and adhesion proteins are considered a determinant in the probiotic potential of lactobacilli (reviewed by Reid and Burton, 2002; Servin, 2004). Early studies addressed the inhibitory effects by probiotics and commensals on pathogen colonization and, in particular, on the inhibition of adhesion of enteric pathogens. Adhesion proteins also are important for the immunological responses that the strains of *Lactobacillus* exert on mammalian epithelial cells and phagocytes; further, the *in-vivo* use of genetically modified cytokine-secreting lactococci (Steidler, 2002) as well as the heterologous antigen-display strains of lactobacilli (Lee *et al.*, 2006; Liu *et al.*, 2006) may benefit from use of highly adherent and efficiently colonizing host strains.

Early studies by Reid and co-workers demonstrated that adhesive *Lactobacillus* strains isolated from urovaginal tract inhibited the adhesion of uropathogens to uroepithelial cells (Chan *et al.*, 1985). Later, Servin and his co-workers reported that adhesive *Lactobacillus* strains inhibited the *in-vitro* adhesion of enteroinvasive pathogens to human intestinal Caco-2 cell line (Chauvière *et al.*, 1992; Coconnier *et al.*, 1993b; Bernet *et al.*, 1993; Bernet *et al.*, 1994). Thereafter, inhibitory effects of lactobacillar strains against *in-vitro* adhesion or invasion by enteric pathogens to cultured intestinal epithelial cell lines, to frozen sections of intestinal tissue, as well as to intestinal mucus, uroepithelial cells, or basement membrane preparations have been described in numerous reports (Table 5.2). The studies have mainly addressed pathogen adherence or invasion into Caco-2 or HT-29 cell lines, and the lactobacillar adhesin(s) were often uncharacterized. The inhibition mechanisms appear to based on competition for the adhesion receptors – although utilization of same tissue receptors by the pathogen and the commensal/

Table 5.2 *In vitro* experiments showing pathogen exclusion

Adhesion/invasion target	*Lactobacillus* strain	Adhesin	Pathogen	Presumed or established inhibitory mechanism	Reference
Caco-2 cells	*L. acidophilus* LB	Unknown	*Escherichia coli* (ETEC)	Steric hindrance	Chauvière *et al.*, 1992
Caco-2 cells and HT29-MTX cells	*L. acidophilus* LB	Unknown	*Salmonella* Typhimurium, *Escherichia coli* (EPEC), *Yersinia pseudotuberculosis*, *Listeria monocytogenes*	Steric hindrance and/or antimicrobial substance	Coconnier *et al.*, 1993a,b
Caco-2 cells and HT29-MTX cells	*L. johnsonii* La1	Lipoteichoic acid and Carbohydrate-binding adhesin	*Escherichia coli* (ETEC, DAEC and EPEC), *Yersinia pseudotuberculosis*, *Salmonella* Typhimurium	Steric hindrance, antimicrobial substance and stimulation of sectretion of an antimicrobial substance	Bernet *et al.*, 1994; Bernet-Camard *et al.*, 1997; Granato *et al.*, 1999; Neeser *et al.*, 2000
Caco-2 cells, HT-29 cells, HT29-MTX cells	*L. acidophilus* HM107, *L. rhamnosus* DR20	Unknown	*Escherichia coli* O157:H7	Lactic acid and proteinaceous substances in culture supernatant	Gopal *et al.*, 2001
Caco-2 cells	*L. rhamnosus* GG	Unknown	*Salmonella* Typhimurium	Lactic acid or a substance active at low pH[1,2]	Hudault *et al.*, 1997
Caco-2 cells	*L. casei* subsp. *rhamnosus* Lcr35	Unknown	*Escherichia coli* (EPEC, ETEC), *Klebsiella pneumoniae*	Steric hindrance, increased expression of mucins, production of biosurfactants, production of bacteriocin-like substance	Forestier *et al.*, 2001
Caco-2 cells	*L. reuteri* JCM 1081, *L. crispatus* JCM 8779	Unknown	*Escherichia coli* (ETEC), *Salmonella* Typhimurium, *Enterococcus faecalis*	Exclusion and antimicrobial activity	Todoriki *et al.*, 2001
Caco-2 cells	*L. gasseri* K7	Unknown	*Escherichia coli*	Competition and exclusion	Bogovič Matijašić *et al.*, 2006
Caco-2/TC7 cells	*L. acidophilus* IBB 801, *L. amylovorus* DCE 471, *L. casei* Shirota, *L. rhamnosus* GG	Unknown except Shirota (see below) and GG (see below)	*Salmonella enterica* serovar Typhimurium	Lactic acid	Makras *et al.*, 2006
Caco-2/TC7 cells	*L. johnsonii* La1, *L. plantarum* ACA-DC 287	Unknown except La1(see above)	*Salmonella enterica* serovar Typhimurium	Lactic acid and inhibitory substance(s)	Makras *et al.*, 2006

Caco-2 cells and intestine-407 cells	*L. casei* DN-114	Unknown	*Escherichia coli*	Steric hindrance and antimicribial compounds	Ingrassia *et al.*, 2005
Caco-2 cells and human intestinal mucus	*L. rhamnosus* GG, *L. casei* Shirota	Hydrophobic interaction (GG), at least two types adhesins (Shirota)	*Escherichia coli, Salmonella* Typhimurium*, Salmonella* enteritidis	Steric hindrance	Lee and Puong 2002; Lee *et al.*, 2003.
HT-29/cl.19A cells and Caco-2 cells	*Streptococcus thermophilus* ATCC 19258, *L. acidophilus* ATCC 4356	Unknown	*Escherichia coli* (EIEC)	Interaction with epithelial cells	Resta-Lenert and Barrett 2003
HT-29 cells	*L. plantarum* 299v and *L. rhamnosus* GG	Unknown	*Escherichia coli* (EPEC)	Induction of intestinal mucin gene expression	Mack *et al.*, 1999
C2Bbe1 cells	*L. rhamnosus* ATCC 53103 (=GG)	See above	*Escherichia coli* (EHEC)	Inhibition of internalization of EHEC	Hirano *et al.*, 2003
HeLa cells	*L. crispatus* ZJ001 and its S-layer protein	S-layer protein	*Escherichia coli* (EHEC), *Salmonella* Typhimurium	Competitive exclusion	Chen *et al.*, 2007
Kato III, MKN45 and murine gastric epithelial cells	*L. salivarius* WB1004	Unknown	*Helicobacter pylori*	Steric hindrance and/or lactic acid	Kabir *et al.*, 1997
T84 cells	*L. helveticus* R0052, *L. rhamnosus* R0011	Unknown	*Escherichia coli* (EHEC, EPEC)	Attenuation of pathogen-induced epithelial injury	Sherman *et al.*, 2005.
HEp-2 cells and T84 cells	*L. helveticus* R0052 and its S-layer protein	Unknown	*Escherichia coli* (EHEC)	Attenuation of pathogen-induced epithelial injury	Johnson-Henry *et al.*, 2007
Human intestinal mucus	*L. rhamnosus* GG, *Lactococcus lactis* subsp. *lactis*, *Propionobacterium freudenreichii* subsp. *shermanii*	Unknown exept GG (see above)	*Staphylococcus aureus*	Production of antimicrobial substances	Vesterlund *et al.*, 2006

Table 5.2 *continued*

Adhesion/invasion target	*Lactobacillus* strain	Adhesin	Pathogen	Presumed or established inhibitory mechanism	Reference
Human uroepithelial cells	*L. rhamnosus* GR-1	Lipoteichoic acid	*Escherichia coli*, *Klebsiella pneumoniae, Pseudomonas aeruginosa*	Steric hindrance of receptor sites	Chan *et al.*, 1985
Human uroepithelial cells	11 strains including *L. rhamnosus* GR-1	Unknown except GR-1 (see above)	*Escherichia coli, Klebsiella pneumoniae, Pseudomonas* spp.	Competitive exclusion	Reid *et al.*, 1987
Human uroepithelial cells	*L. reuteri* RC-14	29 kDa protein	*Enterococcus faecalis*	Biosurfactant	Heinemann *et al.*, 2000
Human vaginal epithelial cells	15 *Lactobacillus* strains including *L. crispatus*	Unknown	*Pseudomonas aeruginosa*, *Klebsiella pneumoniae*	Inhibition of adherence	Osset *et al.*, 2001
Immobilized laminin and basement membrane preparation	*L. crispatus* JCM 5810 and S-layer protein	S-layer protein (CbsA)	*Escherichia coli* (ETEC, EPEC, EIEC)	Exclusion	Horie *et al.*, 2002
Glycolipids	*L. reuteri* TM105	47 kDa protein	*Helicobacter pylori*	Competition of receptor	Mukai *et al.*, 2002
Tissue sections of chicken intestine	*L. crispatus* ST1	Unknown	*Escherichia coli* (APEC) O78 strain 789	Exclusion	Edelman *et al.*, 2003

[1]Lehto and Salminen (1997) showed that inhibition of *Salmonella* Typhimurium adhesion to Caco-2 cells with spent culture supernatant was most likely a pH effect.
[2]De Keersmaecker *et al*. (2006) concluded that strong antimicrobial activity of *L. rhamnosus* GG against *Salmonella* is mediated by lactic acid.

probiotic have not been documented – or on less specific steric inhibition. Other mechanisms, which may be relevant *in vivo*, include modulation of immune responses, increased production of intestinal mucins, and production of antimicrobial substances such as H_2O_2, lactic acid, bacteriocins, and bacteriocin-like substances (Reid and Burton 2002; Servin, 2004; Table 5.2).

The role of specific lactobacillar adhesion proteins in pathogen exclusion has been reported in a few cases. Biosurfactants from *L. reuteri* contain a 29-kDa protein that inhibits the adhesion of *E. faecalis* to human uroepithelial cells (Heinemann *et al.*, 2000). Both the *L. crispatus* JCM 5810 cells and the isolated CbsA protein inhibited the adhesion of enterotoxigenic *E. coli* to immobilized laminin and basement membrane preparation (Horie *et al.*, 2002). Similarly, the cells and the S-layer proteins of *L. crispatus* ZJ001 inhibited the adhesion of *E. coli* O157:H7 and *Salmonella* Typhimurium to HeLa cells (Chen *et al.*, 2007). Pretreatment of HEp-2 cells with S-layer protein extracts from *L. helveticus* R0052 inhibited the adhesion of enterohaemorrhagic *E. coli* and attaching–effacing lesions caused by the pathogen (Johnson-Henry *et al.*, 2007). The latter report suggested that the lactobacillar adhesion protein was not only interfering with pathogen adhesion but was also working as an effector molecule to epithelial cells or modulator of pathogen-epithelial cell cross-talk.

Lactobacillus strains have been successfully used to prevent gastrointestinal and urinary tract infections in animal and human studies (Servin, 2004; O'May and Macfarlane, 2005; Reid and Bruce, 2006). The mechanisms behind the effects have been elucidated in few studies, these include enhancement of immune responses as well as antibacterial activity (Gill, 2003). However, it is probable that the effects result from several cumulative factors, which is exemplified by the clinical study reporting that a mixture of *L. rhamnosus* GR-1 and *L. reuteri* RC-14 effectively prevented urogenital infection (Reid and Bruce, 2006). Both GR-1 and RC-14 adhere to uroepithelial cells, and the strain GR-1 produces antimicrobial substances, down-regulates host inflammatory processes and up-regulates host defence factors. The RC-14 strain produces hydrogen peroxide and biosurfactants, up-regulates host mucin production and down-regulates virulence factor expression in pathogens. Another example is the strong protective effect against gastrointestinal infections in children given by *L. reuteri* ATCC 55730 (Weizman *et al.*, 2005). This bacterium has in another clinical study been shown to adhere to and colonise the mucosal surface and also modulate the immune response (Valeur *et al.*, 2004). The production of the antimicrobial substance reuterin may also be of importance. These clinical effects are likely to be potentiated by lactobacillar adhesion.

References

Adlerberth, I., Ahrné, S., Johansson, M.L., Molin, G., Hanson, L.A., and Wold, A.E. (1996). A mannose-specific adherence mechanism in *Lactobacillus plantarum* conferring binding to the human colonic cell line HT-29. Appl. Environ. Microbiol. 62, 2244–51.

Aleljung, P., Paulsson, M., Emödy, L., Andersson, M., Naidu, A.S., and Wadström, T. (1991). Collagen binding by lactobacilli. Curr. Microbiol. 23, 33–38.

Antikainen, J., Anton, L., Sillanpää, J., and Korhonen, T.K. (2002). Domains in the S-layer protein CbsA of *Lactobacillus crispatus* involved in adherence to collagens, laminin and lipoteichoic acids and in self-assembly. Mol. Microbiol. 46, 381–394.

Antikainen, J., Kuparinen, V., Lähteenmäki, K., and Korhonen, T.K. (2007a). pH-dependent association of enolase and GAPDH of *Lactobacillus crispatus* with the cell wall and lipoteichoic acids. J. Bacteriol. *189*, 4539–4543.

Antikainen, J., Kuparinen, V., Lähteenmäki, K., Korhonen, T.K. (2007b) Enolases from Gram-positive bacterial pathogens and commensal lactobacilli share functional similarity in virulence-associated traits. FEMS Immunol. Med. Microbiol. *51*, 526–534.

Bergmann, S., Rohde, M., Preissner, K.T., and Hammerschmidt, S. (2005). The nine residue plasminogen-binding motif of the pneumococcal enolase is the major cofactor of plasmin-mediated degradation of extracellular matrix, dissolution of fibrin and transmigration. Thromb. Haemost. *94*, 304–311.

Bergonzelli, G.E., Granato, D., Pridmore, R.D., Marvin-Guy, L.F., Donnicola, D., and Corthésy-Theulaz, I.E. (2006). GroEL of *Lactobacillus johnsonii* La1 (NCC 533) is cell surface associated: potential role in interactions with the host and the gastric pathogen *Helicobacter pylori*. Infect. Immun. *74*, 425–434.

Bernet, M.F., Brassart, D., Neeser, J.R., and Servin, A.L. (1993). Adhesion of human bifidobacterial strains to cultured human intestinal epithelial cells and inhibition of enteropathogen-cell interactions. Appl. Environ. Microbiol. 59, 4121–4128.

Bernet, M.F., Brassart, D., Neeser, J.R., and Servin, A.L. (1994). *Lactobacillus acidophilus* LA 1 binds to cultured human intestinal cell lines and inhibits cell attachment and cell invasion by enterovirulent bacteria. Gut 35, 483–489.

Bernet-Camard, M.F., Liévin, V., Brassart, D., Neeser, J.R., Servin, S.L., and Hudault, S. (1997). The human *Lactobacillus acidophilus* strain LA1 secretes a nonbacteriocin antimicrobial substance(s) active in vitro and in vivo. Appl. Environ. Microbiol. *63*, 2747–2753.

Blum, S., Reniero, R., Schiffrin, E.J., Crittenden, R., Mattila-Sandholm, T., Ouwehand, A.C., Salminen, S., von Wright, A., Saarela, M., and Saxelin, M. (1999). Adhesion studies for probiotics: need for validation and refinement. Trends Food Sci. Technol. *10*, 405–410.

Boekhorst, J., Helmer, Q., Kleerebezem, M., and Siezen, R.J. (2006). Comparative analysis of proteins with a mucus-binding domain found exclusively in lactic acid bacteria. Microbiology *152*, 273–280.

Bogovič Matijašić, B., Narat, M., Peternel, M.Z., and Rogelj, I. (2006). Ability of *Lactobacillus gasseri* K7 to inhibit *Escherichia coli* adhesion in vitro on Caco-2 cells and ex vivo on pigs' jejunal tissue. Int. J. Food Microbiol. *107*, 92–96.

Boot, H.J., and Pouwels, P.H. (1996). Expression, secretion and antigenic variation of bacterial S-layer proteins. Mol. Microbiol. *21*, 1117–1123.

Boot, H.J., Kolen, C.P., van Noort, J.M., and Pouwels, P.H. (1993). S-layer protein of *Lactobacillus acidophilus* ATCC 4356: purification, expression in *Escherichia coli*, and nucleotide sequence of the corresponding gene. J. Bacteriol. *175*, 6089–6096.

Boot, H.J., Kolen, C.P., Andreadaki, F.J., Leer, R.J., and Pouwels, P.H. (1996a). The *Lactobacillus acidophilus* S-layer protein gene expression site comprises two consensus promoter sequences, one of which directs transcription of stable mRNA. J. Bacteriol. *178*, 5388–5394.

Boot, H.J., Kolen, C.P., Pot, B., Kersters, K., and Pouwels, P.H. (1996b). The presence of two S-layer-protein-encoding genes is conserved among species related to *Lactobacillus acidophilus*. Microbiology *142*, 2375–2384.

Brassard, J., Gottschalk, M., and Quessy, S. (2004). Cloning and purification of the *Streptococcus suis* serotype 2 glyceraldehyde-3-phosphate dehydrogenase and its involvement as an adhesin. Vet. Microbiol. *102*, 87–94.

Buck, B.L., Altermann, E., Svingerud, T., and Klaenhammer, T.R. (2005). Functional analysis of putative adhesion factors in *Lactobacillus acidophilus* NCFM. Appl. Environ. Microbiol. *71*, 8344–8351.

Båth, K., Roos, S., Wall, T., and Jonsson, H. (2005). The cell surface of *Lactobacillus reuteri* ATCC 55730 highlighted by identification of 126 extracellular proteins from the genome sequence. FEMS Microbiol. Lett. *253*, 75–82.

Callegari, M.L., Riboli, B., Sanders, J.W., Cocconcelli, P.S., Kok, J., Venema, G., and Morelli, L. (1998). The S-layer gene of *Lactobacillus helveticus* CNRZ 892: cloning, sequence and heterologous expression. Microbiology *144*, 719–726.

Carneiro, C.R., Postol, E., Nomizo, R., Reis, L.F., and Brentani, R.R. (2004). Identification of enolase as a laminin-binding protein on the surface of *Staphylococcus aureus*. Microbes Infect. *6*, 604–608.

Chan R.C., Reid G., and Irvin R.T. (1985). Competitive exclusion of uropathogens from human uroepithelial cells by *Lactobacillus* whole cells and cell wall fragments. Infect. Immun. *47*, 84–89.

Chauvière, G., Coconnier, M.H., Kerneis S., Darfeuille-Michaud, A., Joly, B., and Servin, A.L. (1992). Competitive exclusion of diarrheagenic *Escherichia coli* (ETEC) from human enterocyte-like Caco-2 cells by heat-killed *Lactobacillus*. FEMS Microbiol. Lett. *92*, 213–218.

Chen, X., Xu, J., Shuai, J., Chen, J., Zhang, Z., and Fang, W. (2007). The S-layer proteins of *Lactobacillus crispatus* ZJ001 is responsible for competitive exclusion against *Escherichia coli* O157:H7 and *Salmonella typhimurium*. Int. J. Food Microbiol. *115*, 307–312.

Chhatwal, G.S. (2002). Anchorless adhesins and invasins of Gram-positive bacteria: a new class of virulence factors. Trends Microbiol. *10*, 205–208.

Christie, J., McNab, R., and Jenkinson, H.F. (2002). Expression of fibronectin-binding protein FbpA modulates adhesion in *Streptococcus gordonii*. Microbiology *148*, 1615–25.

Coconnier, M.H., Bernet, M.F., Chauvière, G., and Servin A.L. (1993a). Adhering heat-killed human *Lactobacillus acidophilus* strain LB, inhibits the process of pathogenicity of diarrhoeagenic bacteria in cultured human intestinal cells. J. Diarrhoeal Dis. Res. *11*, 235–242.

Coconnier, M.H., Bernet, M.F., Kernéis, S., Chauvière, G., Fourniat, J., and Servin A.L. (1993b). Inhibition of adhesion of enteroinvasive pathogens to human intestinal Caco-2 cells by *Lactobacillus acidophilus* strain LB decreases bacterial invasion. FEMS Microbiol. Lett. *110*, 299–306.

Colloca, M.E., Ahumada, M.C., Lopez, M.E., and Nader-Macias, M.E. (2000). Surface properties of lactobacilli isolated from healthy subjects. Oral Dis. *6*, 227–233.

De Keersmaecker, S.C., Verhoeven, T.L., Desair, J., Marchal, K., Vanderleyden, J., and Nagy, I. (2006). Strong antimicrobial activity of *Lactobacillus rhamnosus* GG against *Salmonella typhimurium* is due to accumulation of lactic acid. FEMS Microbiol. Lett. *259*, 89–96.

Edelman, S., Leskelä, S., Ron, E., Apajalahti, J., and Korhonen, T.K. (2003). In vitro adhesion of an avian pathogenic *Escherichia coli* O78 strain to surfaces of the chicken intestinal tract and to ileal mucus.Vet. Microbiol. *91*, 41–56.

Forestier, C., De Champs, C., Vatoux, C., and Joly, B. (2001). Probiotic activities of *Lactobacillus casei rhamnosus*: in vitro adherence to intestinal cells and antimicrobial properties. Res. Microbiol. *152*, 167–173.

Frece, J., Kos, B., Svetec, I.K., Zgaga, Z., Mrsa, V., and Suskovic, J. (2005). Importance of S-layer proteins in probiotic activity of *Lactobacillus acidophilus* M92. J. Appl. Microbiol. *98*, 285–292.

Garrote, G.L., Delfederico, L., Bibiloni, R., Abraham, A.G., Pérez, P.F., Semorile, L., and De Antoni, G.L. (2004). Lactobacilli isolated from kefir grains:

evidence of the presence of S-layer proteins. J. Dairy Res. *71*, 222–230.

Gatti, M., Rossetti, L., Fornasari, M.E., Lazzi, C., Giraffa, G., and Neviani, E. (2005). Heterogeneity of putative surface layer proteins in *Lactobacillus helveticus*. Appl. Environ. Microbiol. *71*, 7582–7588.

Ge, J., Catt, D.M., and Gregory, R.L. (2004). *Streptococcus mutans* surface alpha-enolase binds salivary mucin MG2 and human plasminogen. Infect. Immun. *72*, 6748–6752.

Gill, H.S. (2003). Probiotics to enhance anti-infective defences in the gastrointestinal tract. Best Pract. Res. Clin. Gastroenterol. *17*, 755–773.

Gill, H.S., and Guarner, F. (2006). Probiotics and human health: a clinical perspective. Postgrad. Med. J. *80*, 516–526.

Gopal, P.K., Prasad, J., Smart, J., and Gill, H.S. (2001). In vitro adherence properties of *Lactobacillus rhamnosus* DR20 and *Bifidobacterium lactis* DR10 strains and their antagonistic activity against an enterotoxigenic *Escherichia coli*. Int. J. Food Microbiol. *67*, 207–216.

Granato, D., Perotti, F., Masserey, I., Rouvet, M., Glliard, M., Servin, A., and Brassart, D. (1999). Cell surface-associated lipoteichoic acid acts as an adhesion factor for attachment of *Lactobacillus johnsonii* La1 to human enterocyte-like Caco-2 cells. Appl. Environ. Microbiol. 65, 1071–1077.

Granato, D., Bergonzelli, G.E., Pridmore, R.D., Marvin, L., Rouvet, M., and Corthésy-Theulaz, I.E. (2004). Cell surface-associated elongation factor Tu mediates the attachment of *Lactobacillus johnsonii* NCC533 (La1) to human intestinal cells and mucins. Infect. Immun. *72*, 2160–2169.

Greene, J.D., and Klaenhammer, T.R. (1994). Factors involved in adherence of lactobacilli to human Caco-2 cells. Appl. Environ. Microbiol. *60*, 4487–4494.

Hagen, K.E., Guan, L.L., Tannock, G.W., Korver, D.R., and Allison, G.E. (2005). Detection, characterization, and in vitro and in vivo expression of genes encoding S-proteins in *Lactobacillus gallinarum* strains isolated from chicken crops. Appl. Environ. Microbiol. *71*, 6633–6643.

Harty, D.W., Oakey, H.J., Patrikakis, M., Hume, E.B., and Knox, K.W. (1994). Pathogenic potential of lactobacilli. Int. J. Food Microbiol. *24*, 179–189.

Heinemann, C., van Hylckama Vlieg, J.E., Janssen, D.B., Busscher, H.J., van der Mei, H.C., and Reid, G. (2000). Purification and characterization of a surface-binding protein from *Lactobacillus fermentum* RC-14 that inhibits adhesion of *Enterococcus faecalis* 1131. FEMS Microbiol. Lett. *190*, 177–180.

Hirano, J., Yoshida, T., Sugiyama, T., Koide, N., Mori, I., and Yokochi, T. (2003). The effect of *Lactobacillus rhamnosus* on enterohemorrhagic *Escherichia coli* infection of human intestinal cells in vitro. Microbiol. Immunol. *47*, 405–409.

Horie, M., Ishiyama, A., Fijihira-Ueki, Y., Sillanpää, J., Korhonen, T.K., and Toba, T. (2002). Inhibition of the adhesion of Escherichia coli strains to basement membrane by *Lactobacillus crispatus* expressing an S-layer. J. Appl. Microbiol. 92, 396–403.

Hudault, S., Liévin, V., Bernet-Camard M.F., and Servin A.L. (1997). Antagonistic activity exerted in vitro by *Lactobacillus casei* (Strain GG) against *Salmonella typhimurium* C5 infection. Appl. Environ. Microbiol. *63*, 513–518.

Hurmalainen, V., Edelman, S., Antikainen, J., Baumann, M., Lähteenmäki, K., and Korhonen, T.K. (2007). Extracellular proteins of *Lactobacillus crispatus* enhance activation of human plasminogen. Microbiology *153*, 1112–1122.

Hynönen, U., Westerlund-Wikström, B., Palva, A., and Korhonen, T. (2002). Identification by flagellum display of an epithelial cell- and fibronectin-binding function in the SlpA surface protein of *Lactobacillus brevis*. J. Bacteriol. *184*, 3360–3367.

Ingrassia, I., Leplingard, A., and Darfeuille-Michaud, A. (2005). *Lactobacillus casei* DN-114 001 inhibits the ability of adherent-invasive *Escherichia coli* isolated from Crohn's disease patients to adhere to and to invade intestinal epithelial cells. Appl. Environ. Microbiol. *71*, 2880–2887.

Jakava-Viljanen, M., Åvall-Jääskeläinen, S., Messner, P., Sleytr, U.B., and Palva, A. (2002). Isolation of three new surface layer protein genes (*slp*) from *Lactobacillus brevis* ATCC 14869 and characterization of the change in their expression under aerated and anaerobic conditions. J. Bacteriol. *184*, 6786–6795.

Jin, H., Song, Y.P., Boel, G., Kochar, J., and Pancholi, V. (2005). Group A streptococcal surface GAPDH, SDH, recognizes uPAR/CD87 as its receptor on the human pharyngeal cell and mediates bacterial adherence to host cells. J. Mol. Biol. *350*, 27–41.

Johnson-Henry, K.C., Hagen, K.E., Gordonpour, M., Tompkins, T.A., and Sherman, P.M. (2007). Surface-layer protein extracts from *Lactobacillus helveticus* inhibit enterohaemorrhagic *Escherichia coli* O157:H7 adhesion to epithelial cells. Cell. Microbiol. *9*, 356–367.

Kabir, A.M., Aiba, Y., Takagi, A., Kamiya, S., Miwa, T., Koga, Y. (1997). Prevention of Helicobacter *pylori* infection by lactobacilli in a gnotobiotic murine model. Gut. *41*, 49–55.

Kahala, M., and Palva, A. (1999). The expression signals of the *Lactobacillus brevis* slpA gene direct efficient heterologous protein production in lactic acid bacteria. Appl. Microbiol. Biotechnol. *51*, 71–78.

Kahala, M., Savijoki, K., and Palva, A. (1997). In vivo expression of the *Lactobacillus brevis* S-layer gene. J. Bacteriol. *179*, 284–286.

Kapczynski, D.R., Meinersmann, R.J., and Lee, M.D. (2000). Adherence of *Lactobacillus* to intestinal 407 cells in culture correlates with fibronectin binding. Curr. Microbiol. *41*, 136–141.

Lee, Y.K. and Puong, K.Y. (2002). Competition for adhesion between probiotics and human gastrointestinal pathogens in the presence of carbohydrate. Br. J. Nutr. *88*, S101–S108.

Lee, Y.K., Puong, K.Y., Ouwehand, A.C., and Salminen, S. (2003). Displacement of bacterial pathogens from mucus and Caco-2 cell surface by lactobacilli. J. Med. Microbiol. 52, 925–930.

Lee, J.S., Poo, H., Han, D.P., Hong, S.P., Kim, K., Cho, M.W., Kim, E., Sung, M.H., Kim, C.J. (2006). Mucosal immunization with surface-displayed severe acute respiratory syndrome coronavirus spike protein on *Lactobacillus casei* induces neutralizing antibodies in mice. J. Virol. *80*, 4079–4087.

Lehto, E.M., and Salminen, S. (1997). Inhibition of *Salmonella typhimurium* adhesion to Caco-2 cell cultures by *Lactobacillus* strain GG spent culture supernate: only a pH effect? FEMS Immunol. Med. Microbiol. *18*, 125–132.

Lijnen, H.R., and Collen, D. (1995). Mechanisms of physiological fibrinolysis. Bailliere's Clin. Haematol. *8*, 277–290.

Liu, X., Lagenaur, L.A., Simpson, D.A., Essenmacher, K.P., Frazier-Parker, C.L., Liu, Y., Tsai, D., Rao, S.S., Hamer, D.H., Parks, T.P., Lee, P.P., Xu, Q. (2006). Engineered vaginal *Lactobacillus* strain for mucosal delivery of the human immunodeficiency virus inhibitor cyanovirin-N. Antimicrob. Agents Chemother. *50* 3250–3259.

Lorca, G., Torino, M.I., Font de Valdez, G., and Ljungh, A.A. (2002). Lactobacilli express cell surface proteins which mediate binding of immobilized collagen and fibronectin. FEMS Microbiol. Lett. *206*, 31–37.

Lähteenmäki, K., Edelman, S., and Korhonen, T.K. (2005). Bacterial metastasis: the host plasminogen system in bacterial invasion. Trends Microbiol. *13*, 79–85.

Mack, D.R., Michail, S., Wei, S., McDougall, L., and Hollingsworth, M.A. (1999). Probiotics inhibit enteropathogenic *E. coli* adherence in vitro by inducing intestinal mucin gene expression. Am. J. Physiol. *276*, G941–G950.

Makras, L., Triantafyllou, V., Fayol-Messaoudi, D., Adriany, T., Zoumpopoulou, G., Tsakaidou, E., Servin, A., and De Vuyst, L. (2006). Kinetic analysis of the antibacterial activity of probiotic lactobacilli towards *Salmonella enterica* serovar Typhimurium reveals a role for lactic acid and other inhibitory compounds. Res. Microbiol. *157*, 241–247.

Marraffini L.A., Dedent A.C., and Schneewind O. (2006). Sortases and the art of anchoring proteins to the envelopes of gram-positive bacteria. Microbiol. Mol. Biol. Rev. *70*, 192–221.

Martínez, B., Sillanpää, J., Smit, E., Korhonen, T.K., and Pouwels, P.H. (2000). Expression of cbsA encoding the collagen-binding S-protein of *Lactobacillus crispatus* JCM5810 in *Lactobacillus casei* ATCC 393(T). J. Bacteriol. *182*, 6857–6861.

Masuda, K., and Kawata, T. (1980). Reassembly of the regularly arranged subunits in the cell wall of *Lactobacillus brevis* and their reattachment to cell walls. Microbiol. Immunol. *24*, 299–308.

Masuda, K., and Kawata, T. (1981). Characterization of a regular array in the wall of *Lactobacillus buchneri* and its reattachment to the other wall components. J. Gen. Microbiol. *124*, 81–90.

Merk, K., Borelli, C., and Korting, H.C. (2005). Lactobacilli – bacteria-host interactions with special regard to the urogenital tract. Int. J. Med. Microbiol. *295*, 9–18.

Mignatti, P., and Rifkin, D.B. (1993). Biology and biochemistry of proteinases in tumor invasion. Physiol. Rev. *73*, 161–195.

Miyoshi, Y., Okada, S., Uchimura, T., and Satoh, E. (2006). A mucus adhesion promoting protein, MapA, mediates the adhesion of *Lactobacillus reuteri* to Caco-2 human intestinal epithelial cells. Biosci. Biotechnol. Biochem. *70*, 1622–1628.

Mukai, T., Asakawa, T., Sato, E., Mori, K., Matsumoto, M., and Ohori, H. (2002). Inhibition of binding of *Helicobacter pylori* to the glycolipid receptors by probiotic *Lactobacillus reuteri*. FEMS Immunol. Med. Microbiol. 32, 105–110.

Myöhänen, H., and Vaheri, A. (2004). Regulation and interactions in the activation of cell-associated plasminogen. Cell Mol. Life Sci. *61*, 2840–2858.

Neeser, J.R., Granato, D., Rouvet, M., Servin, A., Teneberg, S., Karlsson, K.A. (2000). *Lactobacillus johnsonii* La1 shares carbohydrate-binding specificities with several enteropathogenic bacteria. Glycobiology *10*, 1193–1199.

Ocaña, V.S., Bru, E., De Ruiz Holgado, A.A., and Nader-Macias, M.E. (1999). Surface characteristics of lactobacilli isolated from human vagina. J. Gen. Appl. Microbiol. *45*, 203–212.

O'May, G.A. and Macfarlane, G.T. (2005). Health claims associated with probiotics. In Probiotic Dairy Products, A.Y. Tamime, ed. (Oxford: Blackwell Publishing), pp. 138–166.

Osset, J., Bartolomé, R.M., García, E., and Andreu, A. (2001). Assessment of the capacity of *Lactobacillus* to inhibit the growth of uropathogens and block their adhesion to vaginal epithelial cells. J. Infect. Dis. *183*, 485–491.

Pancholi, V. (2001). Multifunctional alpha-enolase: its role in diseases. Cell Mol. Life Sci. *58*, 902–920.

Pancholi, V., and Chhatwal, G.S. (2003). Housekeeping enzymes as virulence factors for pathogens. Int. J. Med. Microbiol. *293*, 391–401.

Pancholi, V., and Fischetti, V.A. (1992). A major surface protein on group A streptococci is a glyceraldehyde-3-phosphate-dehydrogenase with multiple binding activity. J. Exp. Med. *176*, 415–426.

Pankov, R., and Yamada, K.M. (2002). Fibronectin at a glance. J. Cell. Sci. *115*, 3861–3863.

Paterson, G.K. and Mitchell, T.J. 2004. The biology of Gram-positive sortase enzymes. Trends Microbiol. *12*, 89–95.

Peña, J.A., and Versalovic, J. (2003). *Lactobacillus rhamnosus* GG decreases TNF-alpha production in lipopolysaccharide-activated murine macrophages by a contact-independent mechanism. Cell. Microbiol. *5*, 277–285.

Plow, E.F., Ploplis, V.A., Carmeliet, P., and Collen, D. (1999). Plasminogen and cell migration in vivo. Fibrinolysis and Proteolysis *13*, 49–53.

Pretzer, G., Snel, J., Molenaar, D., Wiersma, A., Bron, P.A., Lambert, J., de Vos, W.M., van der Meer, R., Smits, M.A., and Kleerebezem, M. (2005). Biodiversity-based identification and functional characterization of the mannose-specific adhesin of *Lactobacillus plantarum*. J. Bacteriol. *187*, 6128–6136.

Reid, G. and Bruce, A.W. (2006). Probiotics to prevent urinary tract infections: the rationale and evidence. World J. Urol. *24*, 28–32.

Reid, G. and Burton, J. (2002). Use of *Lactobacillus* to prevent infection by pathogenic bacteria. Microbes Infect. *4*, 319–324.

Reid, G., Cook, R.L., and Bruce, A.W. (1987). Examination of strains of lactobacilli for properties which may influence bacterial interference in the urinary tract. J. Urol. *138*, 330–335.

Resta-Lenert, S. and Barrett, K.E. (2003). Live probiotics protect intestinal epithelial cells from the effects of infection with enteroinvasive *Escherichia coli* (EIEC). Gut 52, 988–997.

Rojas, M., Ascencio, F., and Conway, P.L. (2002). Purification and characterization of a surface protein from *Lactobacillus fermentum* 104R that binds to porcine small intestinal mucus and gastric mucin. Appl. Environ. Microbiol. *68*, 2330–2336.

Roos, S., and Jonsson, H. (2002). A high-molecular-mass cell-surface protein from *Lactobacillus reuteri* 1063 adheres to mucus components. Microbiology *148*, 433–442.

Roos, S., Aleljung, P., Robert, N., Lee, B., Wadström, T., Lindberg, M., and Jonsson, H. (1996). A collagen binding protein from *Lactobacillus reuteri* is part of an ABC transporter system? FEMS Microbiol. Lett. *144*, 33–38.

Rossmann, M.G. (1989). The canyon hypothesis. Viral Immunol. *2*, 143–161.

Sára, M., and Sleytr, U.B. (2000). S-Layer proteins. J. Bacteriol. *182*, 859–868.

Sarén, A., Virkola, R., Hacker, J., and Korhonen, T.K. (1999). The cellular form of human fibronectin as an adhesion target for the S fimbriae of meningitis-associated *Escherichia coli*. Infect. Immun. *67*, 2671–2676.

Savijoki, K., Kahala, M., and Palva, A. (1997). High level heterologous protein production in *Lactococcus* and *Lactobacillus* using a new secretion system based on the *Lactobacillus brevis* S-layer signals. Gene *186*, 255–262.

Schneitz, C., Nuotio, L., and Lounatmaa, K. (1993). Adhesion of *Lactobacillus acidophilus* to avian intestinal epithelial cells mediated by the crystalline bacterial cell surface layer (S-layer). J. Appl. Bacteriol. *74*, 290–294.

Schwarz-Linek, U., Höök, M., and Potts, J.R. (2004). The molecular basis of fibronectin-mediated bacterial adherence to host cells. Mol. Microbiol. 52, 631–641.

Seifert, K.N., McArthur, W.P., Bleiweis, A.S., and Brady, L.J. (2003). Characterization of group B streptococcal glyceraldehyde-3-phosphate dehydrogenase: surface localization, enzymatic activity, and protein-protein interactions. Can. J. Microbiol. *49*, 350–356.

Servin, A.L. (2004). Antagonistic activities of lactobacilli and bifidobacteria against microbial pathogens. FEMS Microbiol. Rev. *28*, 405–440.

Sherman, P.M., Johnson-Henry, K.C., Yeung, H.P., Ngo, P.S.C., Goulet, J., and Tompkins, T.A. (2005). Probiotics reduce enterohemorrhagic *Escherichia coli* O157:H7- and enteropathogenic *E. coli* O127:H6-induced changes in polarized T84 epithelial cell monolayers by reducing bacterial adhesion and cytoskeletal rearrangements. Infect. Immun. *73*, 5183–5188.

Sillanpää, J., Martínez, B., Antikainen, J., Toba, T., Kalkkinen, N., Tankka, S., Lounatmaa, K., Keränen, J., Höök, M., Westerlund-Wikström, B., Pouwels, P.H., and Korhonen, T.K. (2000). Characterization of the collagen-binding S-layer protein CbsA of *Lactobacillus crispatus*. J. Bacteriol. *182*, 6440–6450.

Smit, E., and Pouwels, P.H. (2002). One repeat of the cell wall binding domain is sufficient for anchoring the *Lactobacillus acidophilus* surface layer protein. J. Bacteriol. *184*, 4617–4619.

Smit, E., Oling, F., Demel, R., Martinez, B., and Pouwels, P.H. (2001). The S-layer protein of *Lactobacillus acidophilus* ATCC 4356: identification and characterisation of domains responsible for S-protein assembly and cell wall binding. J. Mol. Biol. *305*, 245–257.

Smit, E., Jager, D., Martinez, B., Tielen, F.J., and Pouwels, P.H. (2002). Structural and functional analysis of the S-layer protein crystallisation domain of *Lactobacillus acidophilus* ATCC 4356: evidence for protein-protein interaction of two subdomains. J. Mol. Biol. *324*, 953–964.

Spellerberg, B., Rozdzinski, E., Martin, S., Weber-Heynemann, J., Schnitzler, N., Lutticken, R., and Podbielski, A. (1999). Lmb, a protein with similarities to the LraI adhesin family, mediates attachment of *Streptococcus agalactiae* to human laminin. Infect Immun. *67*, 871–8.

Steidler, L. (2002). In situ delivery of cytokines by genetically engineered *Lactococcus lactis*. Antonie Van Leeuwenhoek. *82* 323–331.

Styriak, I., Zatkovic, B., and Marsalkova, S. (2001). Binding of extracellular matrix proteins by lactobacilli. Folia Microbiol. (Praha) *46*, 83–85.

Sun, H., Ringdahl, U., Homeister, J.W., Fay, W.P., Engleberg, N.C., Yang, A.Y., Rozek, L.S., Wang, X., Sjöbring, U., and Ginsburg, D. (2004). Plasminogen is a critical host pathogenicity factor for group A streptococcal infection. Science *305*, 1283–1286.

Tao, Y., Drabik, K.A., Waypa, T.S., Musch, M.W., Alverdy, J.C., Schneewind, O., Chang, E.B., and Petrof, E.O. (2006). Soluble factors from *Lactobacillus* GG activate MAPKs and induce cytoprotective heat shock proteins in intestinal epithelial cells. Am. J. Physiol. Cell. Physiol. *290*, C1018–30.

Toba, T., Virkola, R., Westerlund, B., Björkman, Y., Sillanpää, J., Vartio, T., Kalkkinen, N., and Korhonen, T.K. (1995). A Collagen-Binding S-Layer Protein in *Lactobacillus crispatus*. Appl. Environ. Microbiol. *61*, 2467–2471.

Todoriki, K., Mukai, T., Sato, S. and Toba, T. (2001). Inhibition of adhesion of food-borne pathogens to Caco-2 cells by *Lactobacillus* strains. J. Appl. Microbiol. *91*, 154–159.

Tuomola, E.M., Ouwehand, A.C., and Salminen, S.J. (2000). Chemical, physical and enzymatic pre-treatments of probiotic lactobacilli alter their adhesion to human intestinal mucus glycoproteins. Int. J. Food Microbiol. *60*, 75–81.

Turner, M.S., Woodberry, T., Hafner, L.M., and Giffard, P.M. (1999). The *bspA* locus of *Lactobacillus fermentum* BR11 encodes an L-cystine uptake system. J. Bacteriol. *181*, 2192–2198.

Turner, M.S., Hafner, L.M., Walsh, T., and Giffard, P.M. (2003). Peptide surface display and secretion using two LPXTG-containing surface proteins from *Lactobacillus fermentum* BR11. Appl. Environ. Microbiol. *69*, 5855–63.

Vaarala, O. (2003). Immunological effects of probiotics with special reference to lactobacilli. Clin. Exp. Allergy 33, 1634–1640.

Valeur, N., Engel, P., Carbajal, N., Connolly, E., and Ladefoged, K. (2004). Colonization and immunomodulation by *Lactobacillus reuteri* ATCC 55730 in the human gastrointestinal tract. Appl. Environ. Microbiol. *70*, 1176–81.

van Pijkeren, J.P., Canchaya, C., Ryan, K.A., Li, Y., Claesson, M.J., Sheil, B., Steidler, L., O'Mahony, L., Fitzgerald, G.F., van Sinderen, D., and O'Toole, P.W. (2006). Comparative and functional analysis of sortase-dependent proteins in the predicted secretome of *Lactobacillus salivarius* UCC118. Appl. Environ. Microbiol. 72, 4143–4153.

Ventura, M., Jankovic, I., Walker, D.C., Pridmore, R.D., and Zink, R. (2002). Identification and characterization of novel surface proteins in *Lactobacillus johnsonii* and *Lactobacillus gasseri*. Appl. Environ. Microbiol. *68*, 6172–6181.

Verbelen, C., Antikainen, J., Korhonen, T.K., and Dufrêne, Y.F. (2007). Exploring the molecular forces within and between CbsA S-layer proteins using single molecule force spectroscopy. Ultramicroscopy *107*, 10–11.

Vesterlund, S., Karp, M., Salminen, S., and Ouwehand, A.C. (2006). *Staphylococcus aureus* adheres to human intestinal mucus but can be displaced by certain lactic acid bacteria. Microbiology *152*, 1819–1826.

Vidgren, G., Palva, I., Pakkanen, R., Lounatmaa, K., and Palva, A. (1992). S-layer protein gene of *Lactobacillus brevis*: cloning by polymerase chain reaction and determination of the nucleotide sequence. J. Bacteriol. *174*, 7419–7427.

Walter, J., Chagnaud, P., Tannock, G.W., Loach, D.M., Dal Bello, F., Jenkinson, H.F., Hammes, W.P., and Hertel, C. (2005). A high-molecular-mass surface protein (Lsp) and methionine sulfoxide reductase B (MsrB) contribute to the ecological performance of *Lactobacillus reuteri* in the murine gut. Appl. Environ. Microbiol. *71*, 979–986.

Weizman, Z., Asli, G., and Alsheikh, A. (2005). Effect of a probiotic infant formula on infections in child care centers: comparison of two probiotic agents. Pediatrics *115*, 5–9.

Westerlund, B., and Korhonen, T.K. (1993). Bacterial proteins binding to the mammalian extracellular matrix. Mol. Microbiol. *9*, 687–694.

Åvall-Jääskeläinen, S., Kylä-Nikkila, K., Kahala, M., Miikkulainen-Lahti, T., and Palva, A. (2002). Surface display of foreign epitopes on the *Lactobacillus brevis* S-layer. Appl. Environ. Microbiol. 68, 5943–5951.

Åvall-Jääskeläinen, S., and Palva, A. (2005). *Lactobacillus* surface layers and their applications. FEMS Microbiol. Rev. *29*, 511–529.

Lactobacillus Stress Responses

Graciela L. Lorca and Graciela Font de Valdez

Abstract

Within the lactic acid bacteria group, the genus *Lactobacillus* is widely used for food fermentation and preservation, and as food additives because its probiotic properties. Lactobacilli are usually exposed to harsh stress conditions such as starter handling and storage (freeze-drying or freezing), during food processing (heat, cold, high concentration of NaCl, and high hydrostatic pressure) and in their passage through the gastrointestinal tract (acidity and bile salts). The optimal performance of these strains depends on the stabilization of their survival potential and their metabolic activity. The purpose of this chapter is to summarize the current knowledge of the survival strategies developed by *Lactobacillus* in order to survive under stress conditions.

Introduction

Lactobacilli species have been used for centuries in food preservation, as starters for dairy products, fermented vegetables, fish and sausages (Mäyrä-Mäkinen and Bigret, 1998; Ross *et al.*, 2002), as well as probiotics because of their potential therapeutic and prophylactic attributes (Ouwehand *et al.*, 2002; O'Mahony *et al.*, 2005). The general utility of *Lactobacillus* species is related to their GRAS (Generally Recognized as Safe) status. Therefore, these bacteria must resist the adverse conditions encountered in the industrial processes, such as the starter handling and storage (high cell density fermentations, freeze-drying, freezing) or during the passage through the gastrointestinal tract (low pH, bile salts) in order to accomplish their task.

The ability to quickly respond to a given stress condition is essential for survival. Transcriptome analyses in combination with high-resolution two-dimensional polyacrylamide gel electrophoresis (2D-PAGE) and mass spectrometry have been extensively applied for the description of stress responses of *Escherichia coli* and *Bacillus subtilis* (Schweder and Hecker, 2004). Nowadays, these new technologies have become available for lactobacilli shedding some light on the mechanisms involved in the resistance to stress conditions in this genus. However, the genetic modification of lactobacilli still represents a big challenge. These facts are reflected in the few mutant strains constructed and analysed (Table 6.1) in relation to the stress response.

Like in other genus, *Lactobacillus* develops an adaptive response when growing under moderate stress conditions that might, in some cases, cross-protect them to other stress. These survival strategies may include changes in the lipid and protein profiles of the cell membrane (Couto *et al.*, 1996), the induction of stress proteins (Olson 1993; Hecker *et al.*, 1996), exopolysaccharide production (Torino *et al.*, 2001) or shifts in the fermentation pattern towards the formation of neutral end products (Magni *et al.*, 1999).

In general, the physiological status of the cell and other environmental factors such as growth media (rich or poor), pH, salt content and water activity will affect the mechanism of resistance to stress. The resistance systems can be divided in three classes.

Table 6.1 Summary of the phenotypic effects obtained by mutation of stress-related genes

Microorganism	Gene disrupted	Putative function	Phenotypic effect	Reference
L. acidophilus	*lisRK*	2CS[1]	Decreased acid tolerance	Azcarate-Peril *et al.*, 2005
L. acidophilus	*treB-treC*	Trehalose phosphotransferase transport system and trehalose-6-phosphate hydrolase	Reduced cryotolerance	Duong *et al.*, 2006
L. acidophilus	*bshA*	Bile salt hydrolase A	Reduced ability to hydrolyse TCDCA[2] and GCDCA[3] bile salts	McAuliffe *et al.*, 2005
L. acidophilus	*bshB*	Bile salt hydrolase B	Inability to hydrolyse any bile salt conjugated to taurine	McAuliffe *et al.*, 2005
L. acidophilus	La57	Glutamate/GABA antiporter	Decreased acid tolerance	Azcarate-Peril *et al.*, 2004
L. acidophilus	La867	Transcriptional regulator	Decreased acid tolerance	Azcarate-Peril *et al.*, 2004
L. acidophilus	La995	Amino acid permease	Decreased acid tolerance. Decreased ethanol tolerance	Azcarate-Peril *et al.*, 2004
L. acidophilus	La996	Ornithine decarboxylase	Decreased acid tolerance. Increased bile tolerance	Azcarate-Peril *et al.*, 2004
L. amylovorus	*bsh*	Bile salt hydrolase	Decreased specific growth rate in presence of bile but higher overall tolerance to bile toxicity	Grill *et al.*, 2000
L. casei	*cspA*	RNA chaperon	Reduced growth at normal and low temperature	Sauvageot *et al.*, 2006
L. helveticus	*htrA*	Serine-protease	Reduced heat tolerance	Smeds *et al.*, 1998
L. plantarum	*bsh*	Bile salt hydrolase	Increased resistance to glycocholic acid	De Boever and Verstraete, 1999
L. plantarum	*ccpA*	Catabolite control protein A	Decreased heat tolerance	Castaldo *et al.*, 2006
L. sakei	*pfk*	Pyruvate formate kinase	Decreased survival at 30°C, 4°C and 4% NaCl	Marceau *et al.*, 2004
L. sakei	*rrp-1*	2CS- Hypothetical response regulator protein	Decreased acid tolerance	Morel-Deville *et al.*, 1998
L. sakei	*rrp-2*	2CS- putative alkaline phosphatase synthesis transcriptional regulator PhoP	Temperature-sensitive. Increased resistance to oxidative stress	Morel-Deville *et al.*, 1998
L. sakei	*rrp-31*	2CS-putative transcriptional activator VanR	Poor growth and viability under normal laboratory conditions and increased susceptibility to heat, acid pH, oxygen and hydrogen peroxide. Increased resistance to vancomycin and teicoplanin	Morel-Deville *et al.*, 1998
L. sakei	*rrp-48*	2CS- putative alkaline phosphatase synthesis transcriptional regulator PhoP	Increased resistance to oxidative stress. Decreased acid tolerance.	Morel-Deville *et al.*, 1998

[1]2CS: Two-component system
[2]TCDCA: taurochenodeoxycholic acid.
[3]GCDCA: glycochenodeoxycholic acid.

1 Specific, induced by a sublethal dose of stress; the adaptive response usually is associated with the log-phase of growth and involves the induction of specific groups of genes or regulon designed to cope with a specific stress condition.
2 General systems, where the adaptation to one stress condition can render cells resistant to other stress condition.
3 Stationary-phase-associated stress response; it involves the induction of numerous regulons designed to overcome several stress conditions. Unlike the adaptive response, the stationary-phase-associated response does not require any pre-exposure to stress to develop the response.

The focus of this chapter is to gather the information available on the current efforts towards understanding of the physiological answers of lactobacilli in response to acid, heat, cold, osmotic, bile and high hydrostatic pressure stress and how these responses are regulated at the molecular level.

The acid stress response

Lactic acid bacteria (LAB) growth is characterized by the production of lactic acid. It is formed as the end-product of glycolysis when pyruvate is reduced to lactate and secreted into the culture medium. The toxicity of lactate results in reduced end products formation and affects the viability of the cells by decreasing the internal pH (pHi). When low pHi values are attained, LAB reached the stationary phase of growth even if nutrients are still available (Hutkins and Nannen, 1993). The inhibitory effect of organic acids is mainly caused by the undissociated form of the molecule, which diffuses across the cell membrane towards the more alkaline cytosol (Hutkins and Nannen, 1993; Axe and Bailey, 1995). Several theories have been postulated to explain the toxic effect of organic acids including the dissipation of the membrane potential, acidification of the cytosol and intracellular anion accumulation (Kashket, 1987; Axe and Bailey, 1995; Pieterse *et al.*, 2005). The inducible survival mechanisms, which protect the cells from acid death, are referred as the acid tolerance response (ATR) (Foster and Hall, 1991).

In *Lactobacillus* the ATR to lactic acid has been studied in *L. acidophilus* (Lorca *et al.*, 1998, 2001a, 2002b, 2003, Azcarate-Peril *et al.*, 2004), *L. helveticus* (Cappa *et al.*, 2005), *L. bulgaricus* (Penaud *et al.*, 2006), *L. delbruekii* subsp. *bulgaricus* (Lim *et al.*, 2001), *L. collinoides* (Laplace *et al.*, 1999), *L. sanfranciscensis* (De Angelis *et al.*, 2001) and *L. plantarum* (Pieterse *et al.*, 2005). The acid tolerance increases at least in two different physiological states. In *L. acidophilus* two acid tolerance responses (ATRs) can be induced in exponential phase upon exposure to sublethal pH (3.8–6.0) (Lorca *et al.*, 1998): (i) a homeostatic response related to an increase in ATPase activity (independent of *de novo* protein synthesis) induced by adaptation at pH 4.2 for 15 min (Lorca and Font de Valdez, 2001a), and (ii) a protein-dependent ATR induced at pH 5.0 for 60 min (Lorca *et al.*, 2002). Interestingly, the survival of stationary-phase cells of *L. acidophilus* CRL 639 to acid stress is a function of the low pH attained by the cultures (Lorca and Font the Valdez, 2001b). The study of protein profiles showed that seven proteins were expressed as result of the stationary phase itself, while nine proteins were exclusively induced as a result of the drop in culture pH during free fermentation runs. Experiments performed with cultures grown at different controlled pH values (pH 4.5, 5.0, 5.5 or 6.0) using lactic acid or hydrochloric acid to adjust the pH of the media determined that the stationary phase response was linked to the concentration of undissociated lactic acid. When using lactic acid to adjust the pH of the medium, five log-cycles decrease in survival was observed after challenge at pH 3.0 for cultures grown at pH 5.5. On the contrary, those grown at pH 4.5 or 5.0 proved to be very resistant. With hydrochloric acid, the resulting cultures were very sensitive to acid shock regardless of the pH. Thus, the ATR induced in *L. acidophilus* would be a combined effect of the pH and the concentration of lactic acid in the medium. An initial concentration of undissociated lactic acid in the growth medium as low as 0.6 mM would be enough to trigger the ATR in *L. acidophilus* CRL 639 (Lorca *et al.*, unpublished results).

The transcription profiles of *L. plantarum* WCFS1 grown in steady state cultures that varied in lactate/lactic acid concentration, pH, osmolarity and absolute and relative growth rate, were compared by microarray analysis (Pieterse *et al.*, 2005). Interestingly, several cell surface protein-encoding genes were induced. These could be associated with the appearance of a rough surface on the cells when cultured at low pH as shown by scanning electron microscopy (Pieterse *et al.*, 2005). Other genes that showed a growth rate-independent lactic acid response were: the ClpE (ATP-binding subunit of the Clp protease), exinuclease (subunit C), catalase and Dpr-like protein (Pieterse *et al.*, 2005). The overexpression of these genes has been previously associated with a general stress response (Hartke *et al.*, 1996; Price *et al.*, 2001; Quivey *et al.*, 1995; Yamamoto *et al.*, 2000).

Mechanisms of resistance

Several mechanisms have been implicated in acid adaptation in *Lactobacillus*. Some of them have a general effect (i.e. changes in cell envelope, DNA repair) but others have an active role in the control of the internal pH and proton motive force. These mechanisms include proton pumps and enzymes involved in the production of alkaline compounds (Fig. 6.1).

Proton pumps

Among the several mechanisms that regulate the pHi homeostasis is the F_1F_0-ATPase. This multimeric enzyme complex can either synthesize ATP using protons or it can expulse protons out of the cell at the expense of ATP hydrolysis. In *Lactobacillus* the sole function of the F_1F_0-ATPase is the extrusion of protons (H^+) and consequently the establishment of pH homeostasis. The presence of F_1F_0-ATPase is ubiquitous in LAB (Lorca *et al.*, 2007). The early studies on this system were performed in *Enterococcus hirae*, in which ATPase activity is controlled primary at the level of pH-dependent subunit assembly (Kobayashi *et al.*, 1984, 1986; Arikaido *et al.*, 1999). A decrease in the cytoplasmic pH results in increased ATPase activity (Arikaido *et al.*, 1999). Similarly, in *L. acidophilus* CRL639 a homeostatic response related to an increase in ATPase activity is induced by adaptation at pH 4.2 for 15 min (Lorca and Font de Valdez, 2001a). Regulation of transcription or translation does not seem to play an important role. However, Kullen and Klaenhammer (1999) described the regulation of the ATPase in *L. acidophilus* at the level of transcription. *L. helveticus* strains with decreased H^+-ATPase activity showed impaired growth at low pH (Yamamoto *et al.*, 1996). In *L. plantarum*, on the contrary, no differential expression of the F_1F_0-ATPase, was observed on cultures grown at pH 4.8 (Pieterse *et al.*, 2005). Similar results were obtained with *L. bulgaricus* (Penaud *et al.*, 2006).

In addition to the F_1F_0-ATPase, P-type ATPases such us K^+-ATPases can contribute to pH homeostasis through the exchange of K^+ for H^+,K^+-ATPases have been related to pH homeostasis in *Streptococcus mutants* (Dashper and Reynolds, 1992), *Lactoccocus lactis* (Kashket and Baker, 1977) and *Ent. hirae* (Kobayashi, 1982; Bakker and Harold, 1980). The analysis of the transport capabilities of eleven gram-positive bacteria showed that all of them have between 4 and 13 P-type ATPases per organism. These could be classified according to the substrates most likely transported, based on sequence similarity to functionally characterized systems (Lorca *et al.*, 2007). The K^+-ATPase encoded in *L. bulgaricus* did not show changes in expression during acid adaptation however, three heavy metal transporting CPX-type ATPases were induced (Penaud *et al.*, 2006). Protein homology and the presence of a conserved regulatory motif in the promoter regions strongly suggest that they might be involved in copper homeostasis as previously described in *Ent. hirae* (Penaud *et al.*, 2006; Solioz and Odermatt, 1995).

Amino acid decarboxylases function as an acid tolerance mechanism increasing the internal pH by consuming hydrogen ions as part of the decarboxylating reaction. The ability of microorganisms to decarboxylate amino acids is highly variable depending not only on the species, but also on the strain and environmental conditions. Among the known decarboxylases (lysine, arginine and glutamate decarboxylases) only the glutamate decarboxylase (GAD) has been associated with pH control in *Lact. lactis* and *L. acidophilus* (Sanders *et al.*, 1998; Azcarate-Peril *et al.*, 2004). In *L. brevis* biochemical studies showed that the glutamate decarboxylase has an optimum pH of 4.2 (Ueno *et al.*, 1997). In *Lactobacillus* sp. strain E1 the decarboxylation

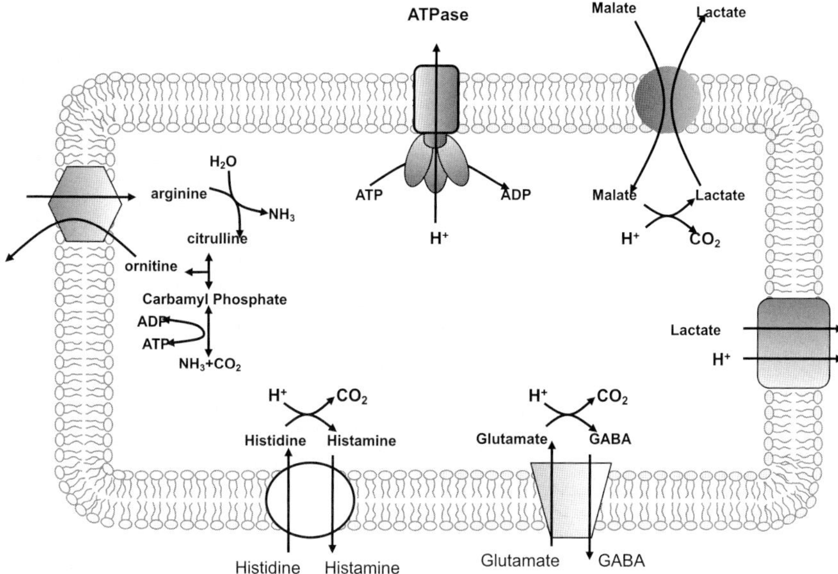

Figure 6.1 Summary of the membrane-associated mechanisms involved in acid stress tolerance in lactobacilli.

process from glutamate to γ-aminobutyrate can be coupled with ATP production via F_1F_0-ATPase dependent on the proton motive force (Higuchi *et al.*, 1997). The significance of the GAD system is linked to the abundance of glutamate in foods (Cotter and Hill, 2003).

Decarboxylation of carboxylic acids also leads to internal consumption of protons. Studies of the energetics of the L-malic acid metabolism (Konings, 2002; Cox and Henick-Kling, 1989; Olsen *et al.*, 1991; Salema *et al.*, 1994) have demonstrated that malolactic fermentation is an energy-producing pathway. It is based on the electrogenic uptake of L-malate, its intracellular decarboxylation by the malolactic enzyme, and the efflux of the monocarboxylic product, L-lactic acid. The L-lactate is excreted via a lactate/malate antiporter as described in *Lact. lactis* (Poolman *et al.*, 1991) or by an electrogenic uniport as reported in *Oenococcus oeni* and *L. plantarum* (Salema *et al.*, 1994; Olsen *et al.*, 1991). In either case, the exchange pathway results in the generation of a proton motive force sufficient to drive ATP synthesis via the membrane-bound F_1F_0-ATPase (Poolman *et al.*, 1991) and the proton consumption during the decarboxylation of L-malate participates in the regulation of the pH homeostasis (Salema *et al.*, 1996). Additionally, due to the pK_a difference of the carboxylic groups of malate and lactate, the external replacement of malate by lactate results in alkalinization of the medium (Cotter and Hill, 2003).

Shifts from homolatic to heterolactic fermentation at low pH has been reported for certain homofermentative lactobacilli (Rhee and Pack, 1980; Borch *et al.*, 1991; Torino *et al.*, 2001) and lactococci (Garrigues *et al.*, 1997), especially for species that are able to utilize citrate as carbon source (Cogan *et al.*, 1981). The uptake of citrate coupled to efflux of lactate produced during glycolysis is a regulatory mechanism for maintaining the pH homeostasis through the consumption of protons in the decarboxylation of oxalacetate (Magni *et al.*, 1999; Martin *et al.*, 2004).

Production of basic compounds. The arginine deiminase (ADI) pathway catalyses the conversion of arginine to ornithine, ammonia, and carbon dioxide and concomitantly generates 1 mol of ATP per mol of arginine consumed. The ADI pathway is widely spread in LABs (Cunin *et al.*, 1986; Nehme *et al.*, 2005; Zuniga *et al.*, 1998; Tonon *et al.*, 2001; Dong *et al.*, 2002; Barcelona-Andres *et al.*, 2002) and includes three enzymes: arginine deiminase (ADI, *arcA*), ornithine transcarbamoylase (OTC, *arcB*), and carbamate kinase (CK, *arcC*). A fourth gene that encodes a transport protein catalysing an electroneutral exchange between arginine and ornithine has been identified (*arcD*). Additionally, a putative transaminase-encoding gene (*arcT*) and a transcription factor belonging to the Crp-Fnr family (*arcR*) are associated with the pathway. The organization and genes included in the different transcriptional units is particularly complex in LAB if compared to other organisms (Zuniga *et al.*, 2002; Fernandez and Zuniga, 2006). In *L. sakei* and *L. reuteri* ADI is associated with a higher survival after preincubation at a sublethal pH or during the stationary phase (Champomier Verges *et al.*, 1999; Rollan *et al.*, 2003). The protection of ADI pathway against damage caused by acidic environments depends on the pH rise associated with the ammonia production (Marquis *et al.*, 1987) at pH values below the minimal for growth or glycolysis.

The complete sequencing of many LABs has allowed the identification of several additional clusters and transport systems that might be involved in stress tolerance (Makarova *et al.*, 2006; Lorca *et al.*, 2007). However, it still remains to be investigated.

Protection and repair of macromolecules

The proteomic analysis of the ATR in several lactobacilli has revealed that the heat-shock associated chaperons were always up-regulated. The DnaK–DnaJ–GrpE chaperone team maintains nascent or pre-existing proteins in un-folded states, while the GroES–GroEL complex interact with partially folded polypeptides and assist in additional folding (Hartl *et al.*, 1992). Western blot experiments showed that GroES, DnaJ and GrpE were induced during acid adaptation in *L. acidophilus* (Lorca *et al.*, 2002) and DnaK, GroES and GroEL were induced in *L. delbruekii* subsp. *bulgaricus* (Lim *et al.*, 2000; Silva *et al.*, 2005).

The bacteria have developed different ways to defend their genomes against DNA-damaging agents. The consequences of intracellular acidification include loss of purines and pyrimidines from DNA at a greater rate than at alkaline pH (Lindahl and Nyberg, 1972). One of the most important repair systems is nucleotide excision repair (Lin and Sancar, 1992). It is carried out by Uvr endonucleases and repair DNA by removal of a 12- to 13-base-long oligonucleotide containing the lesion (Sancar and Rupp, 1983). An *uvrA* mutant of *Strep. mutans* was more sensitive than the wild type to growth at pH 5.0, and acid adapted mutants were unable to survive exposure to pH 3.0 (Hanna *et al.*, 2001). In *L. helveticus uvrA* was induced in response to acid and oxidative stress (Cappa *et al.*, 2005).

Heat shock response

A quick increase in temperature is a serious threat to the integrity of almost all cellular macromolecules. Protein misfolding and subsequent aggregation is accelerated under heat stress (Gross, 1996). Additionally, destabilization of ribosomes and RNA, and alterations of the membrane fluidity were also described (Narberhaus *et al.*, 2006; Mansilla *et al.*, 2004). Responses to high-temperature shock have been, and continue to be, extensively studied (for a review see Yura *et al.*, 2000; Ventura *et al.*, 2006). In several lactobacilli, it has been reported that the pre-incubation of cells at a sub-lethal temperature (~10°C above the normal growth temperature) increases their tolerance factor [calculated as the ratio: survival of adapted cells (%)/survival of control cells (%)] from 5 to 1000 (De Angelis and Gobbetti, 2004). The heat stress response has been studied by analyzing it's the effect on growth, heat tolerance and protein synthesis.

Heat shock proteins (HSPs) are a group of ubiquitous intracellular molecules that function as molecular chaperones and proteases in numerous processes that either rescues misfolded proteins or promotes their degradation (Gross, 1996; Yura *et al.*, 2000). HSPs have been classified according to their molecular mass into families designated Hsp100 (ClpB, ClpC, ClpX), ClpP, Hsp90 (HptG), Hsp70 (DnaK), DnaJ, GrpE, Hsp60 (GroEL), Charonin (Lon), Hsp33 and small HSPs (Hsp16, Hsp18) (Ventura *et al.*, 2006). The cellular concentration of heat-shock proteins is regulated primarily at the level of transcription and involves positive control by alternative sigma factors or negative control by repressor proteins (see next section).

Among the HSPs frequently identified by proteomic studies in lactobacilli are the chaperons DnaK and GroEL complexes and the Hsp100, ClpP protein and Clp holoenzyme. Both GroEL and DnaK recognize hydrophobic surfaces of unfolded proteins and they require specific co-chaperons (GroES and DnaJ-GrpE, respectively) and ATP in order to complete their co- and post-translational protein folding activities. The Hsp100 chaperon family includes ATP-hydrolysing proteins that promotes changes in the folding and assembly of other proteins. The ClpP multimeric complex is a serine protease that degrades short peptides. During heat shock the ClpP protease associates with different Clp-ATPases (Hsp100 family) that will determine the specificity of the peptides to be degraded. Four Clp-ATPases, *clpB*, *clpC*, *clpE* and *clpX* have been described in *Lact. lactis* (Ingmer *et al.*, 1999; Skinner and Trempy, 2001). ClpP mutants were more sensitive to heat shock and puromycin, an aminoacyl-tRNA analogue which leads to truncated and misfolded proteins (Frees and Ingmer, 1999). An increase in the expression of the *clpE* was reported in *L. plantarum* in response to acid stress (Pieterse *et al.*, 2005) while in *L. sanfranciscensis* the Clp protease seems to be induced preferentially under high hydrostatic pressure (Hormann *et al.*, 2006).

Induction of HSP proteins has been reported in *L. acidophilus*, *L. casei*, *L. helveticus* (Broadbent *et al.*, 1997), *L. collinoides* (Laplace *et al.*, 1999), *L. delbruekii* subsp. *bulgaricus* (Gouesbert *et al.*, 2002), *L. johnsonii* (Zink *et al.*, 2000), *L. plantarum* (De Angelis *et al.*, 2004), *L. rhamnosus* (Prasad *et al.*, 2003), *L. paracasei* and *L. salivarius* (Gardiner *et al.*, 2000). Proteomic approaches have shown that the specific number of HSPs induced fluctuates from one microorganism to another (i.e. from 15 in *L. casei* to 36 in *L. collinoides*) (Broadbent *et al.*, 1997; Laplace *et al.*, 1999). In *L. plantarum* the proteins induced as result of heat adaptation were separated by two-dimensional gel electrophoresis and further identified by N-terminal sequencing. These proteins showed homology to DnaK, GroEL, trigger factor, ribosomal proteins L1, L11, L31, and S6, DNA-binding protein II HlbA, and CspC. This study showed that the heat resistance of *L. plantarum* is a complex process involving proteins with various roles in cell physiology, including chaperone activity, ribosome stability, stringent response mediation, temperature sensing, and control of ribosomal function (De Angelis *et al.*, 2004).

HtrA, also known as DegP, is a heat shock-induced serine protease that is active in the periplasm of *E. coli* (for a review see Pallen and Wren, 1997). Homologues of HtrA have been described in a wide range of bacteria and in eukaryotes. Degradation of abnormal proteins in the periplasm has been suggested to be the main physiological role of HtrA. The expression of the *L. helveticus htrA* gene in a variety of stress conditions was analysed at the transcriptional level (Smeds *et al.*, 1998). The strongest induction (~ eightfold increase) was found in growing *L. helveticus* cells exposed to 4% (wt/vol) NaCl. Enhanced *htrA* mRNA expression was also seen in after exposure to puromycin, ethanol, or heat. Interestingly the presence of an intact *htrA* gene facilitated growth under heat stress but not under salt stress (Smeds *et al.*, 1998).

Cold shock adaptation

Freezing and freeze-drying are commonly used for the preservation and storage of microorganisms as well as for the production of starter cultures for the food industry. The optimal performance of the strains depends on the stabilization of their survival potential and metabolic activity.

In response to temperature downshift, a number of changes occur in cellular physiology such as decrease in membrane fluidity, stabilization of secondary structures of nucleic acids leading to reduced efficiency of mRNA translation and transcription, inefficient folding of some proteins, and hindered ribosome function (for a review see Phadtare, 2004). Although cold-induced activation of specific promoters has been implicated in up-regulating some cold-shock genes, post-transcriptional mechanisms play a major role in cold adaptation. The Cold-shock response and adaptation has been extensively studied in *E. coli* and *B. subtilis* (Phadtare, 2004).

Small acidic cold shock proteins (CSPs) are highly induced in response to cold shock. These proteins are found in a wide range of bacteria, and multiple paralogues are often present. In *E. coli*, for example, nine CSPs (CspA to CspI) are known (Yamanaka *et al.*, 1998); of these, CspA,

CspB, CspG, and CspI are cold inducible, with CspA (an RNA chaperone) being the most strongly induced (Jiang *et al.*, 1997). CspA binds to RNA in a cooperative manner and prevents the formation of stable secondary structures. Since RNA duplexes are more stable at low temperatures, an increased level of CSPs may be essential for protein synthesis to proceed under these conditions (Graumann and Marahiel, 1998). A drastic stabilization of the cold-shock transcripts as well as their preferential translation has been observed (Gualerzi *et al.*, 2003). This preferential translation at low temperature is due to *cis* elements present in the 5′ untranslated region of at least some cold-shock mRNAs (Yamanaka *et al.*, 1999).

In contrast to *E. coli* (Jones *et al.*, 1987), LABs can rapidly adapt to a temperature downshift (van de Guchte *et al.*, 2002). *L. acidophilus* and *L. plantarum* continue to grow at a reduced rate after a temperature decrease of 15°C (Lorca and Font de Valdez, 1998, 1999; Bâati *et al.*, 2001; Derzelle *et al.*, 2003). In lactobacilli, in general, the cold adaptation response has been studied in relation with its effect on the cryotolerance. Exponential-phase cultures of *L. acidophilus* grown at suboptimal temperature (25°C) were more resistant to freezing, acid, oxidative, ethanol, osmotic and heat stress than the same cells grown at 37°C (Lorca and Font de Valdez, 1999). Similarly, Bâati *et al.* (2001) reported that in *L. acidophilus*, the adaptation at low temperature or slow freezing of cell suspensions leads to the development of cryotolerance. These results agree with those obtained by Kim and Dunn (1997), who had demonstrated that when cultures were cold shocked at 10°C for 2 h prior to freezing, the viability of *Lact. lactis* and *Pediococcus pentosaceus* was significantly improved.

As part of the cold adaptative response, CSP or its corresponding genes have been identified in several LABs (Chapot-Chartier *et al.*, 1997; Woulters *et al.*, 2000) including *L. delbruekii* subsp. *bulgaricus* (Serror *et al.*, 2003), *L. plantarum* (Derrè *et al.*, 1999; Mayo *et al.*, 2000; Derzelle *et al.*, 2000, 2002) and *L. casei* (Sauvageot *et al.*, 2006). The number of CSP genes may vary between subspecies and strains (Chapot-Chartier *et al.*, 1997). A family of five *csp* genes (*cspA* to *cspE*) is present in *Lact. lactis*. Their expression is strongly induced upon cold shock, except for CspE, which is already abundant at optimal temperature (Wouters *et al.*, 1998, 1999). Three *csp* genes (*cspL*, *cspC*, and *cspP*) have been identified in *L. plantarum* (Mayo *et al.*, 1997; Derzelle *et al.*, 2000, 2002) being *cspL* mRNA the highest cold-shock induced of the three *csp* transcripts (Derzelle *et al.*, 2000). Derzelle *et al.* (2003) investigated the effect of overproducing each of the three cold shock proteins. CspL overproduction alleviated the reduction in growth rate triggered by exposing exponentially growing cells to cold shock (8°C). These suggest that CspL is involved in cold adaptation while overproduction of CspP led to an enhanced freezing survival. The strain overproducing CspC resumed growth more rapidly when stationary-phase cultures were diluted into fresh medium, indicating a role in the adaptation and recovery of nutritionally deprived cells. Therefore, CSPs in *L. plantarum* may have a role in fermentation carried out either at cold (CspL) or optimal temperature (CspC). The *L. delbrueckii* ssp. *bulgaricus* CspA and CspB have between 81% and 77% identity with proteins CspL and CspP of *L. plantarum*. Northern blot analyses showed that *cspA* is transcribed as a single gene and that its transcription increased after a temperature downshift from 42 to 25°C while *cspB* is part of an operon transcribed at constant level irrespective of the temperature. A *cspA* mutant strain of *L. casei* showed reduced growth rate compared with the wild type at both optimal and low temperatures, demonstrating that CspA plays an important role in the physiology of this bacterium (Sauvageot *et al.*, 2006).

Control of membrane fluidity is another response to cold shock. The mechanism of regulation, in general, occurs via the incorporation of unsaturated fatty acids as the temperature decreases. The result is a decrease in the solid-to-fluid transition temperature and thus, an increase in membrane fluidity (for a review see Mansilla *et al.*, 2004). Consequently, the ratio between unsaturated and saturated fatty acids is inversely correlated with the growth temperature, as shown in *L. plantarum* (Lonvaud-Funel and Desens, 1990) and in *L. fermentum* (Suutari and Laakso, 1992). *L. acidophilus* grown at

25°C showed a twofold increase in glycolipids in relation to phospholipids, a twofold increase in the $C_{16:0}$ and a fourfold increase in the $C_{18:2}$ fatty acids. In contrast, the C_{19}-cyc [11,12 methyleneoctadecanoic acid (lactobacillic acid] and the 10-hydroxy acid ($C_{18:0-10}$ OH) species showed a noticeable decrease (Fernandez Murga *et al.*, 1999). The $C_{18:1}$ fatty acid concentration increased in response to low temperature in *L. plantarum* (Russell *et al.*, 1995). A high concentration of $C_{19:0}$ cyc favoured the cryotolerance of *L. bulgaricus*, *L. helveticus*, and *L. acidophilus* (Gomez Zavaglia *et al.*, 2000; Wang *et al.*, 2005).

The mechanisms by which freezing causes cell damages include the efflux of water from the cell (Mazur, 1970), mechanical stress to cellular components, and rupture of cell membranes due to ice crystal formation (Souzu, 1989; Souzu *et al.*, 1989). The addition of cryoprotectants to cell suspensions prior to low-temperature exposure reduced these deleterious effects. The efflux of water from the cell resulted in osmotic shock as result of the higher concentration of the internal solutes. In LABs the accumulation of compatible solutes such as glycine betaine, proline and carnitine play an essential role in osmoprotection and cold adaptation (De Angelis and Gobbetti, 2004). Font de Valdez *et al.* (1983) reported that bovine albumin, glycogen, dextran, polyethylene glycol (PEG) 1000, PEG 4000, PEG 6000, glycerol, beta-glycerophosphate, sodium glutamate, asparagine, or cysteine exhibited marked variations in protecting the LABs against freeze-drying. Glycerol provided effective protection for *L. leichmannii* (90% survival), while *L. bulgaricus* reached a viability of 78% with 0.04 M cysteine (Font de Valdez *et al.*, 1983). In *L. delbrueckii* subsp. *bulgaricus* the tolerance to freezing was induced by pretreatment with dimethyl sulphoxide, glycerol, lactose, sucrose, and trehalose (Panoff *et al.*, 2000; Gomez Zavaglia *et al.*, 2003). The protective action of trehalose and sucrose has been ascribed to its ability to replace water in proteins and membrane structures (Leslie *et al.*, 1995). In *L. acidophilus* trehalose contributes to the survival and long-term stability of the organism after freeze-drying (Conrad *et al.*, 2000). Duong *et al.* (2006) investigated the genetic and biochemical basis for utilization of trehalose in *L. acidophilus*. By disrupting the trehalose transporter, the authors showed that trehalose must be internalized to confer cryoprotection.

Osmotic shock response

LABs are usually exposed to changes in the solute concentrations in their natural habitats or during food processing. The consequences in increase in the osmolarity and concomitant movement of water from the cell to the outside are the loss of cell turgor pressure, changes in the intracellular solute concentration and modifications in the cell volume. Microorganisms respond to a decrease in water activity in the medium by accumulating compatible solutes. They are not inhibitory to cellular processes even at very high concentrations (submolar). Compatible solutes include amino acids (glutamate, glutamine, and proline), amino acid derivatives (betaines, peptides, and N-acetylated amino acids), polyols, and sugars (trehalose and sucrose) (Csonka and Hanson, 1991). Intracellular pools of compatible solutes can be raised by increased synthesis or uptake from the medium. Both proline and glycine betaine, when supplied in the medium, are accumulated by enteric bacteria and serve as osmoprotectants (Perroud and Le Rudulier, 1985; Whatmpre *et al.*, 1990). Unlike the enteric bacteria and *B. subtilis*, in which glycine betaine is synthesized, lactobacilli rely on the uptake of such compounds from the culture medium (Jewell and Kashket, 1991; Hutkins *et al.*, 1987; Glaasker *et al.*, 1996a,b).

Hutkins *et al.* (1987) reported that *L. acidophilus* IFO 3532 (reclassified as *L. casei* subsp. *rhamnosus*) accumulated betaine when exposed to high osmotic pressure (1.8 M NaCl). The uptake of betaine is activated but not induced by exposure of the cells to high osmotic pressure. Furthermore, these organisms did not accumulate carbohydrates (reducing sugars), K^+, or amino acids, except for proline. The accumulation of glycine betaine and carnitine has been shown to enhance survival of *L. plantarum* after drying (Kets and de Bont, 1994). The accumulation of quarternary ammonium compounds in *L. plantarum* is mediated via a single transport system (Glaasker *et al.*, 1998a). The QacT transport system, which belongs to the ABC transport superfamily has a high affinity for glycine betaine (apparent *Km* of 18 mM) and carnitine and a low affinity for proline (apparent K_m of 950 mM) and other analogues (Glaasker *et al.*, 1998a). Similarly, as described for the uptake of betaine in *L. acidophilus*, the QacT transport system in *L. plantarum* is activated by hyperosmotic shock through diminished inhibition by *trans* substrate as well as through a turgor-related increase in the activity. Both effects result in an increase of the V_{max} when the medium osmolality is raised (Glaasker *et al.*, 1998a). In contrast, the glycine-betaine uptake system in *Lact. lactis* (OpuA or BusA) is subject to osmoregulation at transcriptional level as well as increased transport activity (Obis *et al.*, 1999; van der Heide and Poolman, 2000).

In contrast to KCl and NaCl osmotic shock, high concentrations of sugars (lactose or sucrose) in *L. plantarum* impose only a transient osmotic stress because external and internal sugars equilibrate shortly after the exposure. Analysis of lactose (and sucrose) uptake also indicated that the corresponding transport systems were not significantly induced or activated by hyperosmotic conditions (Glaasket *et al.*, 1998b).

Like for other kinds of stress, cross-protection has been observed after pre-exposure of cultures to sublethal concentrations of NaCl. *L. acidophilus* cells that were pre-exposed to NaCl increased its survival to bile and heat stress around 1000 times (Kim *et al.*, 2001). In *L. rhamnosus* the osmotic shock increase the storage stability of dried cultures (Prasad *et al.*, 2003). In *L. salivarius* the heterologous expression of the betaine uptake system from *Listeria monocytogenes*, BetL, resulted in a significant increase in resistance to osmotic, freezing, high pressure, spray and freeze-drying (Sheehan *et al.*, 2006).

By 2D electrophoresis, Marceau *et al.* (2004) detected 21 proteins which were modified during growth of *L. sakei* at 4°C or with 4% NaCl. Out of the six proteins that were identified by N-terminal sequencing, two belong to carbon metabolic pathways (*glpD*, *pfk*), and four could be clustered as general stress proteins (*msrA*, *asp-2*, *ohr*, and *usp*). Only the survival of the *pfk* mutant was affected at 30°C, at 4°C, and with 4% NaCl (Marceau *et al.*, 2004). Microarray analysis performed in *L. plantarum* grown with NaCl 0.8 M (Pieterse *et al.*, 2005) showed a big overlap between genes induced under osmotic and lactic acid stress. In *Lact. lactis*, analysis of 2D gels allowed the identification of the general stress proteins DnaK, GroEL, and GroES, but no specifically NaCl-induced proteins could be detected (Kilstrup *et al.*, 1997). Similarly, the expression of the chaperon DnaK is induced in *L. sanfranciscensis* (Hörmann *et al.*, 2006), *L. sakei* (Schmidt *et al.*, 1999), *L. acidophilus* and *L. johnsonii* (Zink *et al.*, 2000) in response to osmotic shock.

Modifications in the membrane properties have also been observed in response to growth in a high osmolal medium. The changes appear related to an increase in the sugar content of the whole pool of lipids and in the saturated fatty acid residues (Tymczyszyn *et al.*, 2005). In *L. casei* ATCC 393 (reclassified *L. zeae*) the glycolipid AcylH3DG is only present in membranes of NaCl containing medium. H4DG increase its concentration while H2DG undergoes a significant decrease (Machado *et al.*, 2004). Changes in the structural properties of the cell wall have also been reported (Piuri *et al.*, 2005). Analysis of peptidoglycan from *L. casei* stressed cells showed that the modification was at the structural level in accordance with a decreased peptidoglycan cross-link involving penicillin-binding proteins (Piuri *et al.*, 2005). Additionally, in high osmolarity medium, the cell envelope-associated proteinase (PrtP), and peptidases (PepX, PepI) increased their activity. In the presence of peptides, derepression of the peptidases was observed at the transcriptional level. However, the twofold activation of PrtP was through an increase of the apparent Vmax of the enzyme in hyperosmotic medium (Piuri *et al.*, 2003).

Bile tolerance response

Bile is a digestive secretion that plays a major role in the emulsification and solubilization of lipids (Vlahcevic *et al.*, 1990). It is synthesized in the liver but is stored and concentrated in the gall-bladder during the fasting state. Bile is primarily composed of bile acids (12% by weight) which are found as sodium salts under physiological conditions. These biological detergents are synthesized from cholesterol where the steroid nucleus is conjugated with either glycine or taurine (Hofmann and Mysels, 1992). The deconjugation of bile salts is catalysed by bile salt hydrolase

(BSH) enzymes (EC 3.5.1.24), which hydrolyse the amide bond and liberate the glycine/taurine moiety from the steroid core. The resulting acids are termed free, unconjugated or deconjugated bile acids. The bacterial deconjugation of bile acids has been related to the serum cholesterol reduction in humans since deconjugated bile acids are excreted more rapidly than are conjugated bile acids (Chikai *et al.*, 1987; Taranto *et al.*, 1997, 1998, 2000). As a consequence, more cholesterol is needed to synthesize new bile acids which in turn might reduce the total cholesterol in the body (Driessen and de Boer, 1989).

Bile salts are part of the body's physicochemical defense system, they possess potent antimicrobial activity and have the ability to dissolve the phospholipids, cholesterol, and proteins of cell membranes, causing cellular lysis (Hofmann, 1994). Considering these facts, an important factor in selecting *Lactobacillus* strains as probiotic cultures or as a food adjunct is the bile tolerance, which enables a selected strain to survive, grow, and perform therapeutic benefits in the intestinal tract (Gilliland and Walker, 1989; Salminen and von Wright, 1993).

Many reports have been published on the deconjugation of bile acids by lactobacilli (Bateup *et al.*, 1995; Christiaens *et al.*, 1992; Dashkevicz and Feighner, 1989; De Boever and Verstraete, 1999; De Smet *et al.*, 1995; Elkins and Savage, 1998; Grill *et al.*, 2000; Lundeen and Savage, 1990, 1992; McAuliffe *et al.*, 2005; Pereira *et al.*, 2003; Tanaka *et al.*, 1999; Tannock *et al.*, 1989; Taranto and Font de Valdez, 1999) and has recently been reviewed in Begley *et al.* (2006). However, statistical analysis showed no significant correlation between bile tolerance and bile salt deconjugation (Gilliland and Speck, 1977; Usman and Hosono, 1999; Moser and Savage, 2001). Comparative studies performed in *L. acidophilus* (McAuliffe *et al.*, 2005), *L. amylovorus* (Grill *et al.*, 2000) and *L. plantarum* (De Smet *et al.*, 1995; De Boever and Verstraete, 1999) with *bsh*– variants showed that the mutant strains displayed a decrease in the growth rate in the presence of bile salts indicating that the BSH activity is related to the resistance to bile. Despite how well widespread is the BSH activity among lactobacilli, its significance is still far from understood. Since the free bile acids are more inhibitory than the conjugated bile salts themselves, it is most likely not a detoxification mechanism for these organisms. However, the free bile salts (cholic or deoxycholic) have lower solubility at low pH and precipitate *in vitro* as result of the fermentative metabolism of the lactic acid bacteria.

In Gram-positive bacteria, like *Listeria monocytogenes* (Begley *et al.*, 2002) and *Ent. faecalis* (Flahaut *et al.*, 1996) the toxicity pattern of bile acids resembles that of detergents such as SDS. However, the actual effects of bile acids on the bacterial cell, and consequently the mechanisms of tolerance/resistance, are just starting to be unravelled (Fig. 6.2). The exposure of *L. reuteri* CRL 1098 to bile salts produces changes in the cellular ultra-structure as evidenced by the presence of folds and buds in the cell membrane which shows a flat and disorganized distribution (Valdez *et al.*, 1996). Further studies performed with free bile acids confirm that the structural distortions of the cell surface were due to the action of unconjugated bile acid (Taranto *et al.*, 2006). Changes in the lipid profile of these cells also supported the fact that the cell surface may be the target of the unconjugated bile acid action. In the presence of deoxycholic acid, cells showed a 4-fold increase in the glycolipid to phospholipid ratio compared to untreated cells, mainly due to a decrease in phospholipids, the main structural component of the cell membrane (Taranto *et al.*, 2003, 2006). The relative increase in glycolipids has been reported for bifidobacteria (Gómez-Zavaglia *et al.*, 2002) and could enhance the stability of the cell membrane when confronted with environmental changes due to the ability of these compounds to undergo interlipid hydrogen bonding via glycosyl head groups (Curatolo, 1987). Additionally, deoxycholic acid produced complete permeabilization of cells and abolished glucose uptake (Taranto *et al.*, 2006). Changes in the antibiotic susceptibility of lactobacilli have also been reported. The presence of bile salts resulted in a complete loss of resistance to polymyxin B and the aminoglycosides gentamicin, kanamycin, and streptomycin (Charteris *et al.*, 2000).

The identification of a relative high number of genes encoding for cell envelope functions confirmed a major impact of bile acids on the integrity and functionality of the cell membrane

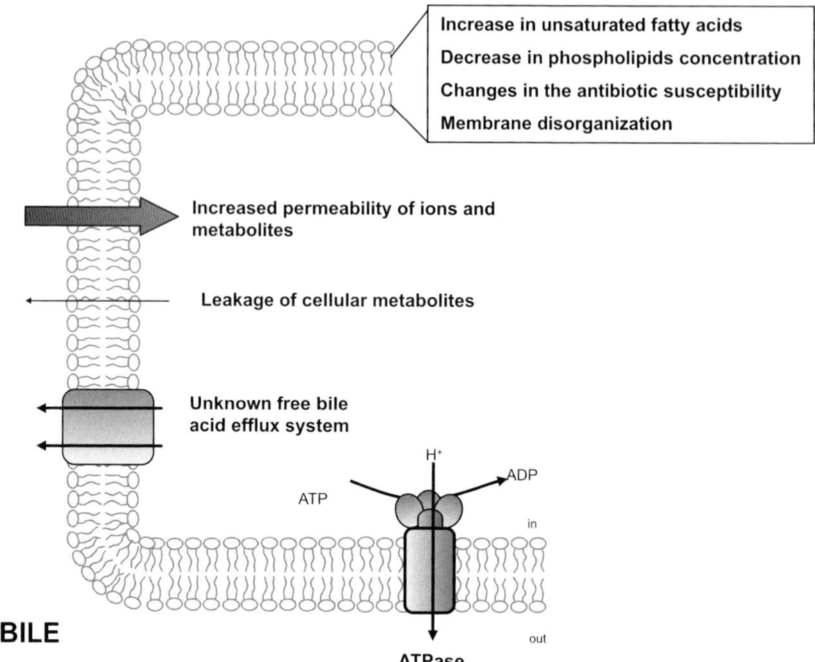

Figure 6.2 General consequences of lactobacilli exposure to free bile acids.

and cell wall (Bron *et al.*, 2004, 2006). In *L. plantarum*, Bron *et al.* (2004, 2006) showed that the expression levels of at least 12 gene clusters appeared to be regulated by bile. Seven gene clusters encoding typical stress functions were identified including the glutathione reductase and the *met-cysK* operon which have been previously linked to the response to bile stress in *Propionibacterium* (Leverrier *et al.*, 2003). The FtsH metallopeptidase (involved in general stress responses) as well as genes related to the osmotic and acid tolerance response (i.e. glutamate decarboxylase, glutamate dehydrogenase and glutamate tRNA-ligase) were also induced. In contrast, six transporters (ABC- or PTS-type) were down regulated as well as the expression of genes encoding for multiple extracellular proteins (Bron *et al.*, 2006). The only membrane-associated function of which the transcription levels are enhanced during bile stress is the F_1F_0-ATPase system. Apparently, maintenance of the proton motif force is crucial during bile stress. Similar results were obtained in *Bifidobacterium* (Sanchez *et al.*, 2006).

Kurdi *et al.* (2000) showed that cholic acid is accumulated in different *Lactobacillus* strains, apparently passively and according to the ΔpH. The quantitative relationship between the amount of cholic acid and the internal pH value suggests that the accumulation is mediated not by a protein carrier but by the well-known spontaneous distribution of hydrophobic weak acids across the bacterial membranes. The authors also proposed that cholic acid provokes membrane damage in the 2–4 mM range, leading to increased permeability of ions and metabolites. However, a complete dissipation of proton motive force and the collapse of other ion concentration gradient(s) as well as probably the leakage of cellular metabolites was observed at concentrations around the minimal inhibitory concentration, where the viability of the cell was dramatically decreased (Kurdi *et al.*, 2006).

A few species of bacteria have been described as possessing active bile acid/salt exporters, which make them resistant to these compounds. In *E. coli* multidrug resistant (MDR) transporters have been related to the extrusion of bile (Thanassi *et al.*, 1997). In contrast, an ATP-dependent multispecific organic anion transporter is involved in the increased resistance of *Lact. lactis* MG 1363 mutants to cholic acid (Yokota *et al.*, 2000). In lactobacilli the only MDR system genetically characterized is the HorA ABC transport system which is involved in the resistance of *L. brevis* to hop and several toxic compounds (Sami *et al.*, 1997; Sakamoto *et al.*, 2001). However, its putative involvement in bile tolerance has not been established. In silico analysis of the transport capabilities of several LABs showed that four of the six prokaryotic ABC drug efflux families (Drug E2, Drug RA1, Drug RA2 and MacB) are present in the genomes of *L. brevis, L. casei, L. debruekii* subsp. *bulgaricus* and *L. gasseri* (Lorca *et al.*, 2007). The role of these systems in resistance to free bile acids is as yet unknown.

High hydrostatic pressure stress

High hydrostatic pressure treatment of food systems was first reported 100 years ago by Hite (1899) and has received considerable attention since it was rediscovered few years ago (Knorr, 1999). It has been at the center of food research since it causes less undesirable chemical changes in food components than heat preservation (Knorr, 1999). However, it causes severe stress to microorganisms. A wide variety of high pressure-induced phenomenon has been reported (for a review see Hoover *et al.*, 1989), including changes in cellular morphology, biochemical reactions, genetic mechanisms, and membrane integrity. The ribosomes and membranes have been suggested as important determinants in pressure sensitivity (Earnshaw *et al.*, 1995).

In contrast to other stress conditions a specific high hydrostatic pressure stress response cannot be expected from lactobacilli since it is normally not exposed to high pressure. Some of the effects of high pressure are similar to other stresses which might explain the development of cross-protection observed in *L. sanfranciscensis* (Scheyhing *et al.*, 2004). Preexposure of exponential-phase cultures to sublethal conditions of pressure, osmotic, cold and acid stress led to an increase of barotolerance dependent on *de novo* protein synthesis (Scheyhing *et al.*, 2004). Similar results were found in *L. rhamnosus* (Ananta and Knorr, 2004). Wouters *et al.* (1998) found no morphological changes in the cytoplasmic membrane upon lethal pressure treatment of *L. plantarum* however, the membrane permeability was increased, and the efflux of protons as well as the ability to maintain a ΔpH across the membrane was impaired.

The effect of sublethal hydrostatic pressure on *L. sanfranciscensis* has been analysed from two perspectives: the proteome and the transcriptome response. Pavlovic *et al.* (2005) using a shot-gun-microarray identified ribosomes and impaired translation among primary targets for high pressure treatment. The majority of high pressure affected genes were found to encode either translation factors (EF-G, EF-TU), ribosomal proteins (S2, L6, L11), genes changing translational accuracy or molecular chaperones (GroEL, ClpL). The induction of heat as well as cold shock genes (e.g. hsp60, gyrA) may be explained as a result of high pressure affected protein synthesis (Pavlovic *et al.*, 2005). The proteome analysis of *L. sanfranciscensis* before and upon high-pressure incubation revealed the differential expression of 16 proteins (Hörmann *et al.*, 2006) that in general agree with the results reported by Pavlovic *et al.*, (2005). Nine of them were up-regulated whereas seven were repressed under these conditions. One of the most strongly increased protein spots was a 60 kDa protein with homology to ribokinases (Drews *et al.*, 2002; Hörmann *et al.*, 2006). Interestingly, this protein is also overexpressed after exposure to different stress conditions such as osmotic, cold, acid. The specifics of this induction remains unknown (Hörmann *et al.*, 2006). The highest similarity in the high-pressure proteome was found to be with cold- and NaCl-stressed cells, with 11 overlapping proteins. At the proteome level, *L. sanfranciscensis* appears to use overlapping subsets of stress-inducible proteins rather than stereotype responses. As expected, a specific pressure response might not exist in these bacteria (Hörmann *et al.*, 2006).

Regulation of the stress response

Among the stress responses, the heat shock has been widely studied in bacteria (for a review see Schumann, 2003). In *B. subtilis*, a model organism for Gram-positive bacteria, the heat shock stimulon comprises at least six classes of heat shock genes classified depending on the different type of transcriptional regulation that they undergo (Schumann, 2003; Darmon *et al.*, 2002; Helman *et al.*, 2001). Class I and class III are controlled by two transcriptional repressors (HrcA and CtsR), Class II by an alternative sigma factor (σ^B, Hecker *et al.*, 1996), Class IV by a still unknown transcriptional activator, Class V by a two-component signal transduction system, and Class VI for which the regulation remains unknown (Schumann, 2003). In lactobacilli, the information about the regulation of the stress response is still scattered. The most relevant difference between *Bacillus* and LABs, in general, is the absence of a σ^B orthologue in LABs (Spano and Massa, 2006). In lactobacilli, the stress response is negatively regulated by the one-component system proteins HrcA or CtsR and by two-component signal transduction systems.

Two-component signal transduction systems (2CSs)

2CSs basically consist of a membrane-associated histidine kinase sensor and a cytoplasmic response regulator. In presence of an extracellular stimulus, the histidine kinase is activated by autophosphorylation on a conserved histidine residue and serves as phospho-donor for a conserved aspartate residue on its cognate response regulator. The response regulator, which is in general a DNA-binding protein, on its phosphorylated form will activate the transcription of a specific set of genes (for reviews see Parkinson, 1993; Stock *et al.*, 1995; Guedon *et al.*, 2001). 2CSs are widely spread in nature (Alex and Simon, 1994) and they function in bacterial adaptation, survival and virulence. In *L. sakei* five different response regulators were amplified by PCR using degenerate oligonucleotides and their cognate histidine protein kinase identified (Morel-Deville *et al.*, 1998). Insertional inactivation showed that they are implicated in the susceptibility to the glycopeptide antibiotic vancomycin, to extreme pH, temperature and oxidative conditions (Morel-Deville *et al.*, 1998). The sequencing and analysis of several bacterial genomes has uncovered the presence of many 2CSs. *Lact. lactis* IL1403 encodes for eight response regulators and seven histidine kinases (Bolontin *et al.*, 2001). Characterization of the insertional mutants revealed phenotypic differences which ranged from acid sensitivity/resistance, altered phosphatase activity, increased salt sensitivity and H_2O_2 sensitivity (O'Connell-Motherway *et al.*, 2000). However, neither in *L. sakei* nor in *Lact. lactis* the specific target genes were identified. A more extensive analysis with global methods such as DNA microarrays is necessary to define their potential regulons in LAB. This approach has been described for *B. subtilis* (Kobayashi *et al.*, 2001), *Staphylococcus aureus* and *Strep. pneumoniae* (Clausen *et al.*, 2003; Throup *et al.*, 2000). In *L. acidophilus* whole-genome microarray analysis was performed to identify the genes responsible for the decreased acid tolerance in a *lisRK* mutant strain (Azcarate-Peril *et al.*, 2005). This system was previously described in *Listeria monocytogenes* and participates in both stress response and virulence (Cotter *et al.*, 1999). The microarray data revealed an altered expression pattern on numerous ORFs encoding genes for the major components of the proteolytic system (Azcarate-Peril *et al.*, 2005).

One-component regulatory system

In the class I of the heat-induced genes are the operon encoding for the major chaperone complexes, GroES-GroEL and GrpE-DnaK-DnaJ. In lactobacilli the induction in the expression of these genes has been linked to acid, ethanol, cold, osmotic, starvation and heat stress. In *B. subtilis* the regulation of *groES-groEL* and *grpE-dnaK-dnaJ* involves a σ^A-dependent promoter and highly conserved inverted repeat (Controlling Inverted Repeat of Chaperone Expression, CIRCE: ttagcagtc-N9-gagtgctaa) which is the binding site for the HrcA repressor (Zuber and Schumann, 1994). In *L. sakei*, a putative CIRCE sequence was found upstream of the *hrcA-grpE-dnaK* (Schmidt *et al.*, 1999). Northern hybridization analysis revealed that the transcription of the genes was induced by heat shock as well as by salt or ethanol stress. The regulatory function of the CIRCE element in *L. sakei* was corroborated using transcriptional fusions (Schmidt *et al.*, 1999). CIRCE elements have also been identified in the promoters of *groES-groEL* operon in *L. helveticus* (Broadbent *et al.*, 1998), *L. johnsonii* (Walker *et al.*, 1999) and *L. acidophilus* (Lorca *et al.*, 2002). A similar arrangement has been described for the *dnaK* and *groESL* operons in *L. plantarum* (Kleerebezem *et al.*, 2003). Castaldo *et al.* (2006) reported an additional system of regulation of the *dnaK* operon in *L. plantarum*, based on a positive control exerted by the catabolite control protein (CcpA) protein, which would interact with *cre* sequences present in the regulatory region of the *dnaK* and *groESL* operons. The absence of the CcpA protein results in a lower induction of the chaperon coding operons, with a consequent lower per cent of survival of the LM3–2 mutant strain population with respect to the wild type when challenged to heat shock (Castaldo *et al.*, 2006). Recently, the presence of conserved CIRCE elements were also observed on the promoter of small heat shock genes in *L. plantarum* (Spano *et al.*, 2004) and in the *clpL* gene of *L. rhamnosus* (Suokko *et al.*, 2005). These findings are interesting since it broadens the repertories of genes regulated by HrcA what is no longer restricted to the major chaperon families.

The class III of heat shock genes in *Bacillus* are controlled by the class III gene repressor CtsR which binds to a specific direct repeat (CtsR box: a/**ggtcaaa**NaNa/**ggtcaaa**; Derrè *et al.*, 1999). This target sequence was also found upstream from Clp and other heat shock genes (i.e. HtrA, FtsH) of several Gram-positive bacteria, including *List. monocytogenes, Strep. salivarius, Strep. pneumoniae, Strep. pyogenes, Strep. thermophilus, Ent. faecalis, Staph. aureus, Leuconostoc oenos, L. sake, Lact. lactis* and *Clostridium acetobutylicum* (Derrè *et al.*, 1999; Ingmer *et al.*, 1999; Smeds *et al.*, 1998). CtsR homologues were also identified in several of these bacteria, indicating that heat shock regulation by CtsR is highly conserved in Gram-positive bacteria (Derrè *et al.*, 1999).

The regulation of stress responsible genes of Gram-positive bacteria used to be described as HrcA or CtsR dependent. While this distinction is still valid for *B. subtilis* and its closely related species, *B. anthracis, Clost. perfringens* and *List. monocytogenes* however, in the streptococcal group (*Strep. pneumoniae, Strep. pyogenes, Strep. salivarius* and *Lact. lactis*) the HrcA and CtsR regulons partially overlap (Chastanet *et al.*, 2003). Additionally, the genome sequences of *Oen. oeni* and *L. bulgaricus* indicated the presence of only a predominant stress regulator gene, CtsR or HrcA, respectively (Grandvalet *et al.*, 2005). Sapano and Massa (2006) summarized the overlapping regulation of the small heat stress proteins *hsp*18.5 and *hsp*19.5 in *L. plantarum*. Conserved CIRCE elements were observed upstream the *hsp*18.5 and *hsp*19.5. Additionally, the 5′ non-coding region of the *hsp*18.5 gene displayed a putative sequence homologue to the CtsR box (Sapano and Massa, 2006). A similar system of dual regulation by the HrcA and CtsR proteins has been reported for ClpP in *Staph. aureus* (Chastanet and Msadek, 2003).

Future directions

Bacterial adaptation is a natural ongoing phenomenon in the environment, in industrial processes and in healthcare. Initially the responses of lactic acid bacteria to stress conditions were compared to those in well-known model microorganisms such as *B. subtilis* and *E. coli*. However, it soon became clear that while they share some general response mechanisms, some are just inherent to the physiology and the ecology of lactobacilli. The most relevant and striking difference between *Bacillus* and *Lactobacillus* is the absence of a σ^B orthologue while several stress proteins (i.e. DnaK, GroEL, Clp, etc.) and their regulators (HrcA and CtsR) are conserved. It represents a challenge to determine how lactobacilli sense and respond to the various stimulus in the environment. Is it through global transcriptional regulators? Is it mediated by two component systems? Maybe, it is a combination of both. However, only a small fraction of the transcription regulators and their regulatory targets are known. Thus more efforts are necessary to answer these questions.

With more than 15 *Lactobacillus* genomes deposited in public domain (http://www.ncbi.nlm.nih.gov/genomes/lproks.cgi), whole genome comparisons will help the identification of common regulatory elements and/or plasticity regions. A function in niche adaptation or involvement in stress resistance has been shown or proposed for a number of genes located in plasticity zones of other organisms. However, in many cases, their function is unknown. The challenge now is to understand how those genes contribute to microorganism adaptation placed in a physiological context.

References

Alex, L.A., and Simon, M.I. (1994). Protein histidine kinases and signal transduction in prokaryotes and eukaryotes. Trends Genet. *10*, 133–139.

Ananta, E., and Knorr, D. (2004). Evidence on the role of protein biosynthesis in the induction of heat tolerance of *Lactobacillus rhamnosus* GG by pressure pre-treatment. Int. J. Food Microbiol. *96*, 307–313.

Arikado, E., Ishihara, H., Ehara, T., Shibata, C., Saito, H., Kakegawa, T., Igarashi, K., and Kobayashi, H. (1999). Enzyme level of enterococcal F1F0-ATPase is regulated by pH at the step of assembly. Eur. J. Biochem. *259*, 262–268.

Axe, D.D., and Bailey, J.E. (1995). Transport of lactate and acetate through the energized cytoplasmic membrane of *Escherichia coli*. Biotechnol. Bioeng. *47*, 8–19.

Azcarate-Peril, M.A., Altermann, E., Hoover-Fitzula, R.L., Cano, R.J., and Klaenhammer, T.R. (2004) Identification and inactivation of genetic loci involved with *Lactobacillus acidophilus* acid tolerance. Appl. Environ. Microbiol. *70*, 5315–5322.

Azcarate-Peril, M.A., McAuliffe, O., Altermann, E., Lick, S., Russell, W.M., and Klaenhammer, T.R. (2005). Microarray analysis of a two-component regulatory system involved in acid resistance and proteolytic activity in *Lactobacillus acidophilus*. Appl. Environ. Microbiol. *71*, 5794–5804.

Baati, L., Fabre-Gea, C., Auriol, D., and Blanc, P.J. (2000).Study of the cryotolerance of *Lactobacillus acidophilus*: effect of culture and freezing conditions on the viability and cellular protein levels. Int. J. Food Microbiol. 59, 241–247.

Bakker, E.P., and Harold, F.M. (1980). Energy coupling to potassium transport in *Streptococcus faecalis*. Interplay of ATP and the protonmotive force. J. Biol. Chem. 255, 433–440.

Barcelona-Andres, B., Marina, A., and Rubio, V. (2002). Gene structure, organization, expression, and potential regulatory mechanisms of arginine catabolism in *Enterococcus faecalis*. J. Bacteriol. *184*, 6289–6300.

Bateup, J.M., McConnell, M.A., Jenkinson, H.F., and Tannock, G.W. (1995). Comparison of *Lactobacillus* strains with respect to bile salt hydrolase activity, colonization of the gastrointestinal tract, and growth rate of the murine host. Appl. Environ. Microbiol. *61*, 1147–1149.

Begley, M., Gahan, C.G., and Hill, C. (2002). Bile stress response in *Listeria monocytogenes* LO28: Adaptation, cross-protection, and identification of genetic loci involved in bile resistance. Appl. Environ. Microbiol. *68*, 6005–6012.

Begley, M., Hill, C., and Gahan, C.G. (2006). Bile salt hydrolase activity in probiotics. Appl. Environ. Microbiol. 72, 1729–1738.

Bolotin, A., Wincker, P., Mauger, S., Jaillon, O., Malarme, K., Weissenbach, J., Ehrlich, S.D., and Sorokin, A. (2001). The complete genome sequence of the lactic acid bacterium *Lactococcus lactis* ssp. *lactis* IL1403. Genome Res. *11*, 731–753.

Borch, E., Berg, H., and Holst, O. (1991). Heterolactic fermentation by a homofermentative *Lactobacillus* sp. during glucose limitation in anaerobic continuous culture with complete cell recycle. J. Appl. Bacteriol. *71*, 265–269.

Broadbent, J.R., Orberg, C.J., and Wei, L. (1998). Characterization of the *Lactobacillus helveticus groELS* operon. Res. Microbiol. *149*, 247–253.

Broadbent, J.R., Oberg, J.C., Wang, H., and Wie, L. (1997). Attributes of the heat shock response in three species of dairy *Lactobacillus*. Syst. Appl. Microbiol. *20*, 12–19.

Bron, P.A., Marco, M., Hoffer, S.M., Van Mullekom, E., de Vos, W.M., and Kleerebezem, M. (2004). Genetic characterization of the bile salt response in *Lactobacillus plantarum* and analysis of responsive promoters *in vitro* and *in situ* in the gastrointestinal tract. J. Bacteriol. *186*, 7829–7835.

Bron, P.A., Molenaar, D., de Vos, W.M., and Kleerebezem, M. (2006). DNA micro-array-based identification of bile-responsive genes in *Lactobacillus plantarum*. J. Appl. Microbiol. *100*, 728–738.

Cappa, F., Cattivelli, D., and Cocconcelli, P.S. (2005). The *uvrA* gene is involved in oxidative and acid stress responses in *Lactobacillus helveticus* CNBL1156. Res. Microbiol. *156*, 1039–1047.

Castaldo, C., Siciliano, R.A., Muscariello, L., Marasco, R., and Sacco, M. (2006). CcpA affects expression of the *groESL* and *dnaK* operons in *Lactobacillus plantarum*. Microb. Cell Fact. 5, 35.

Chapot-Chartier, M.P., Schouler, C., Lepeuple, A.S., Gripon, J.C., and Chopin, M.C. (1997). Characterization of *cspB*, a cold-shock-inducible gene from *Lactococcus lactis*, and evidence for a family of genes homologous to the *Escherichia coli cspA* major cold shock gene. J. Bacteriol. *179*, 5589–5593.

Chastanet, A., Fert, J., and Msadek, T. (2003). Comparative genomics reveal novel heat shock regulatory mechanisms in *Staphylococcus aureus* and other Gram-positive bacteria. Mol. Microbiol. *47*, 1061–1073.

Chastanet, A., and Msadek, T. (2003). ClpP of *Streptococcus salivarius* is a novel member of the dually regulated class of stress response genes in gram-positive bacteria. J. Bacteriol. *185*, 683–687.

Champomier Verges, M.C., Zuniga, M., Morel-Deville, F., Perez-Martınez, G., Zagorec, M., and Ehrlich, S.D. (1999). Relationships between arginine degradation, pH and survival in *Lactobacillus sakei*. FEMS Microbiol. Lett. *180*, 297–304.

Charteris, W.P., Kelly, P.M., Morelli, L., and Collins, J.K. (2000). Effect of conjugated bile salts on antibiotic susceptibility of bile salt-tolerant *Lactobacillus* and *Bifidobacterium* isolates. J. Food Prot. 63, 1369–1376.

Chikai, T., Nakao, H., and Uchida, K. (1987). Deconjugation of bile acids by human intestinal bacteria implanted in germ free rats. Lipids 22, 669–671.

Christiaens, H., Leer, R.J., Pouwels, P.H., and Verstraete, W. (1992). Cloning and expression of a conjugated bile acid hydrolase gene from *Lactobacillus plantarum* by using a direct plate assay. Appl. Environ. Microbiol. 58, 3792–3798.

Clausen, V.A., Bae, W., Throup, J., Burnham, M.K., Rosenberg, M., and Wallis, N.G. (2003) Biochemical characterization of the first essential two-component signal transduction system from *Staphylococcus aureus* and *Streptococcus pneumoniae*. J. Mol. Microbiol. Biotechnol. 5, 252–260.

Cogan, T.M., O'Dowd, M., and Mellerick, D. (1981) Effects of pH and sugar on acetoin production from citrate by *Leuconostoc lactis*. Appl. Environ. Microbiol. *41*, 1–8.

Conrad, P.B., Miller, D.P., Cielenski, P.R., and de Pablo, J.J. (2000). Stabilization and preservation of *Lactobacillus acidophilus* in saccharide matrices. Cryobiology *41*, 17–24.

Cotter, P.D., Emerson, N., Gahan, C.G., and Hill, C. (1999). Identification and disruption of *lisRK*, a genetic locus encoding a two-component signal transduction system involved in stress tolerance and virulence in *Listeria monocytogenes*. J. Bacteriol. *181*, 6840–6843.

Cotter, P.D., and Hill, C. (2003). Surviving the acid test: responses of gram-positive bacteria to low pH. Microbiol. Mol. Biol. Rev. 67, 429–453.

Couto, J.A., Rozes, N., and Hogg, T. (1996). Ethanol-induced changes in the fatty acid composition of *Lactobacillus hilgardii*, its effect on plasma membrane fluidity and relationship with ethanol tolerance. J. Appl. Bacteriol. *81*, 126–132.

Cox, D.J., and Henick-Kling, T. (1989). Proton-motive force and ATP generation during malolactic fermentation. Am. J. Enol. Vitic. *46*, 319–323.

Csonka, L.N., and Hanson, A.D. (1991). Prokaryotic osmoregulation: genetics and physiology. Annu. Rev. Microbiol. *45*, 569–606.

Cunin, R., Glansdorff, N., Pierard, A., and Stalon, V. (1986). Biosynthesis and metabolism of arginine in bacteria. Microbiol. Rev. *50*, 314–352.

Curatolo, W. (1987). Glycolipids function. Biochim. Biophys. Acta *906*, 137–160.

Darmon, E., Noone, D., Masson, A., Bron, S., Kuipers, O.P., Devine, K.M., and van Dijl, J.M. (2002). A novel class of heat and secretion stress-responsive genes is controlled by the autoregulated CssRS two-component system of *Bacillus subtilis*. J. Bacteriol. *184*, 5661–5671.

Dashkevicz, M.P., and Feighner, S.D. (1989). Development of a differential medium for bile salt hydrolase-active *Lactobacillus* spp. Appl. Environ. Microbiol. 55, 11–16.

Dashper, S.G., and Reynolds, E.C. (1992). pH regulation by *Streptococcus mutans*. J. Dent. Res. *71*, 1159–1165.

De Angelis, M., Bini, L., Pallini, V., Cocconcelli, P.S., and Gobbetti, M. (2001). The acid-stress response in *Lactobacillus sanfranciscensis* CB1. Microbiology *147*, 1863–1873.

De Angelis, M., Di Cagno, R., Huet, C., Crecchio, C., Fox, P.F., and Gobetti, M. (2004). Heat shock response in *Lactobacillus plantarum*. Appl. Environ. Microbiol. *70*, 1336–1346.

De Angelis, M., and Gobbetti, M. (2004). Environmental stress responses in *Lactobacillus*: A review. Proteomics *4*, 106–122.

De Boever, P., and Verstraete, W. (1999). Bile salt deconjugation by *Lactobacillus plantarum* 80 and its implication for bacterial toxicity. J. Appl. Microbiol. *87*, 345–352.

Derre, I., Rapoport, G., and Msadek, T. (1999). CtsR, a novel regulator of stress and heat shock response, controls *clp* and molecular chaperone gene expression in gram-positive bacteria. Mol. Microbiol. *31*, 117–131.

Derzelle, S., Hallet, B., Francis, K.P., Ferain, T., Delcour, J., and Hols, P. (2000). Changes in *cspL*, *cspP*, and *cspC* mRNA abundance as a function of cold shock and growth phase in *Lactobacillus plantarum*. J. Bacteriol. *182*, 5105–5113.

Derzelle, S., Hallet, B., Ferain, T., Delcour, J., and Hols, P. (2003). Improved adaptation to cold-shock, stationary-phase, and freezing stresses in *Lactobacillus plantarum* overproducing cold-shock proteins. Appl. Environ. Microbiol. *69*, 4285–4290.

Derzelle, S., Hallet, B., Ferain, T., Delcour, J., and Hols, P. (2002). Cold shock induction of the *cspL* gene in *Lactobacillus plantarum* involves transcriptional regulation. J. Bacteriol. *184*, 5518–5523.

De Smet, I., Van Hoorde, L., Vande Woestyne, M., Christiaens, H., and Verstraete, W. (1995). Significance of bile salt hydrolytic activities of lactobacilli. J. Appl. Bacteriol. *79*, 292–301.

Dong, Y., Chen, Y.-Y.M., Snyder, J.A., and Burne, R.A. (2002). Isolation and molecular analysis of the gene cluster for the arginine deiminase system from *Streptococcus gordonii* DL1. Appl. Environ. Microbiol. *68*, 5549–5553.

Drews, O., Weiss, W., Reil, G., Parlar, H., Wait, R., and Gorg, A. (2002). High pressure effects step-wise altered protein expression in *Lactobacillus sanfranciscensis*. Proteomics *2*, 765–774.

Driessen, F.M., and de Boer, R. (1989). Fermented milk with selected intestinal bacteria: a healthy trend in new products. Neth. Milk Dairy J. *43*, 367–382.

Duong, T., Barrangou, R., Russell, W.M., and Klaenhammer, T.R. (2006). Characterization of the *tre* locus and analysis of trehalose cryoprotection in *Lactobacillus acidophilus* NCFM. Appl. Environ. Microbiol. 72, 1218–1225.

Earnshaw, R.G., Appleyard, J., and Hurst., R.M. (1995). Understanding physical inactivation processes: combined preservation opportunities using heat, ultrasound and pressure. Int. J. Food Microbiol. *28*, 197–219.

Elkins, E.A., and Savage, D.C. (1998). Identification of genes encoding conjugated bile salt hydrolase and transport in *Lactobacillus johnsonii* 100–100. J. Bacteriol. *180*, 4344–4349.

Fernandez Murga, M.L., Bernik, D., Font de Valdez, G., and Disalvo, A.E. (1999). Permeability and stability properties of membranes formed by lipids extracted from *Lactobacillus acidophilus* grown at different temperatures. Arch. Biochem. Biophys. *364*, 115–121.

Fernandez, M., and Zuniga, M. (2006). Amino acid catabolic pathways of lactic acid bacteria. Crit. Rev. Microbiol. 32, 155–183.

Flahaut, S., Frere, J., Boutibonnes, P., and Auffray, Y. (1996) Comparison of the bile salts and sodium dodecylsulfate stress responses in *Enterococcus faecalis*. Appl. Environ. Microbiol. 62, 2416–2420.

Font de Valdez, G., Savoy de Giori, G., Pesce de Ruiz Holgado, A., and Oliver, G. (1983). Comparative study of the efficiency of some additives in protecting lactic acid bacteria against freeze-drying. Cryobiology *20*, 560–566.

Foster, J.W., and Hall, H.K. (1991). Inducible pH homeostasis and the acid tolerance response of *Salmonella typhimurium*. J. Bacteriol. *173*, 5129–5135.

Frees, D., and Ingmer, H. (1999). ClpP participates in the degradation of misfolded protein in *Lactococcus lactis*. Mol. Microbiol. *31*, 79–87.

Gardiner, G.E., O'Sullivan, E., Kelly, J., Auty, M.A.E., Fitzgerald, G.F., Collins, J.K., Ross, R.P., and Stanton, C. (2000). Comparative survival rates of human-derived probiotic *Lactobacillus paracasei* and *L. salivarius* strains during heat treatment and spray drying. Appl. Environ. Microbiol. *66*, 2605–2612.

Garrigues, C., Loubiere, P., Lindley, N.D., and Cocaign-Bousquet, M. (1997). Control of the shift from homolactic acid to mixed-acid fermentation in *Lactococcus lactis*: predominant role of the NADH/NAD^+ ratio. J. Bacteriol. *179*, 5282–5287.

Gilliland, S.E., and Speck, M.L. (1977). Deconjugation of bile acids by intestinal lactobacilli. Appl. Environ. Microbiol. 33, 15–18.

Gilliland, S.E., and Walker, D.K. (1989). Factors to consider when selecting a culture of *Lactobacillus acidophilus* as a dietary adjunct to produce a hypocholesterolemic effect in humans. J. Dairy Sci. *73*, 905–911.

Glaasker, E., Heuberger, E.H., Konings, W.N., and Poolman, B. (1998a). Mechanism of osmotic activation of the quaternary ammonium compound transporter (QacT) of *Lactobacillus plantarum*. J. Bacteriol. *180*, 5540–5546.

Glaasker, E., Konings, W.N., and Poolman, B. (1996a). Glycine betaine fluxes in *Lactobacillus plantarum* during osmostasis and hyper- and hypo-osmotic shock. J. Biol. Chem. *271*, 10060–10065.

Glaasker, E., Konings, W.N., and Poolman, B. (1996b). Osmotic regulation of intracellular solute pools in *Lactobacillus plantarum*. J. Bacteriol. *178*, 575–582.

Glaasker, E., Tjan, F.S., Ter Steeg, P.F., Konings, W.N., and Poolman, B. (1998b). Physiological response of *Lactobacillus plantarum* to salt and nonelectrolyte stress. J. Bacteriol. *180*, 4718–4723.

Gomez Zavaglia, A., Disalvo, E.A., and De Antoni, G.L. (2000). Fatty acid composition and freeze-thaw resistance in lactobacilli. J. Dairy Res. *67*, 241–247.

Gómez-Zavaglia A., Kociubinski, G., Pérez, P., Disalvo, A., and De Antoni, G. (2002). Effect of bile on the lipid composition and surface properties of bifidobacteria. J. Appl. Microbiol. *93*, 794–799.

Gomez Zavaglia, A., Tymczyszyn, E., De Antoni, G., and Disalvo A.E. (2003). Action of trehalose on the preservation of *Lactobacillus delbrueckii* ssp. *bulgaricus* by heat and osmotic dehydration. J. Appl. Microbiol. 95, 1315–1320.

Gouesbert, G., Jan, G., and Boyaval, P. (2002). Two-dimensional electrophoresis study of *Lactobacillus delbrueckii* subsp. *bulgaricus* thermotolerance. Appl. Environ. Microbiol. *68*, 1055–1063.

Grandvalet, C., Coucheney, F., Beltramo, C., and Guzzo, J. (2005). CtsR is the master regulator of stress response gene expression in *Oenococcus oeni*. J. Bacteriol. *187*, 5614–5623.

Graumann, P., and Marahiel, M.A. (1996). Some like it cold: response of microorganisms to cold shock. Arch. Microbiol. *166*, 293–300.

Graumann, P.L., and Marahiel, M.A. (1998). A superfamily of proteins that contain the cold-shock domain. Trends Biochem. Sci. *23*, 286–290.

Grill, J.P., Cayuela, C., Antoine, J.M., and Schneider, F. (2000). Isolation and characterization of a *Lactobacillus amylovorus* mutant depleted in conjugated bile salt hydrolase activity: relation between activity and bile salt resistance. J. Appl. Microbiol. *89*, 553–563.

Gross, C.A. (1996). Function and regulation of the heat shock proteins. In *Escherichia coli* and *Salmonella*: Cellular and Molecular Biology. Neidhardt F.C., *et al.*, eds, (Washington, DC: American Society for Microbiology), 1382–1399.

Gualerzi, C.O., Giuliodori, A.M., and Pon, C.L. (2003). Transcriptional and post-transcriptional control of cold-shock genes. J. Mol. Biol. *331*, 527–539.

Guedon, E., Serror, P., Ehrlich, S.D., Renault, P., and Delorme, C. (2001). Pleiotropic transcriptional repressor CodY senses the intracellular pool of branched-chain amino acids in *Lactococcus lactis*. Mol. Microbiol. *40*, 1227–1239.

Hanna, M.N., Ferguson, R.J., Li, Y.H., and Cvitkovitch, D.G. (2001). *uvrA* is an acid-inducible gene involved in the adaptive response to low pH in *Streptococcus mutans*. J. Bacteriol. *183*, 5964–5973.

Hartke, A., Bouche, S., Giard, J.-C., Benachour, A., Boutibonnes, P., and Auffray, Y. (1996). The lactic acid stress response of *Lactococcus lactis* subsp. *lactis*. Curr. Microbiol. *33*, 194–199.

Hartl, F.U., Martin, J., and Neupert, W. (1992). Protein folding in the cell: the role of molecular chaperones Hsp70 and Hsp60. Annu. Rev. Biophys. Biomol. Struct. *21*, 293–322.

Hecker, M., Schumann, W., and Volker, U. (1996). Heat-shock and general stress response in *Bacillus subtilis*. Mol. Microbiol. *19*, 417–428.

Helmann, J.D., Winston Wu, F.W., Kobel, P.A., Gamo, F.J., Wilson, M., Morshedi, M.M., Navre, M., and Paddon, C. (2001). Global Transcriptional Response of *Bacillus subtilis* to Heat Shock. J. Bacteriol. *183*, 7318–7328.

Higuchi, T., Hayashi, H., and Abe, K. (1997). Exchange of glutamate and gamma-aminobutyrate in a *Lactobacillus* strain. J. Bacteriol. *179*, 3362–3364.

Hite, B.H. (1899). The effects of pressure on the preservation of milk. West Virginia Agric. Exp. Station Bull. 58, 15–35.

Hofmann, A.F. (1994). Bile acids. In The liver: biology and pathobiology. I.M. Arias, J.L. Boyer, N. Fausto, W.B. Jackoby, D.A. Schachter, and D.A. Shafritz, eds. (New York: Raven Press), pp. 677–718.

Hofmann, A.F., and Mysels, K.J. (1992). Bile acid solubility and precipitation in vitro and in vivo: the role of conjugation, pH and Ca2+ ions. J. Lipid Res. 33, 617–626.

Hoover, D.G., Metrick, C., Papineau, A.M., Farkas, D.F., and Knorr, D. (1989). Biological effects of high hydrostatic pressure on food microorganisms. Food Technol. *43*, 99–107.

Hormann, S., Scheyhing, C., Behr, J., Pavlovic, M., Ehrmann, M., and Vogel, R.F. (2006). Comparative proteome approach to characterize the high-pressure stress response of *Lactobacillus sanfranciscensis* DSM 20451(T). Proteomics 6, 1878–1885.

Hutkins, R.W., and Nannen, N.L. (1993). pH homeostasis in lactic acid bacteria. J. Dairy Sci. *76*, 2354–2365.

Hutkins, R.W., Ellefson, W.L., and Kashket, E.R. (1987). Betaine transport imparts osmotolerance on a strain of *Lactobacillus acidophilus*. Appl. Environ. Microbiol. 53, 2275–2281.

Ingmer, H., Vogensen, F.K., Hammer, K., and Kilstrup, M. (1999). Disruption and analysis of the *clpB*, *clpC*, and *clpE* genes in *Lactococcus lactis*: ClpE, a new Clp family in gram-positive bacteria. J. Bacteriol. *181*, 2075–2083.

Jewell, J.B., and Kashket, E.R. (1991). Osmotically regulated transport of proline by *Lactobacillus acidophilus* IFO 3532. Appl. Environ. Microbiol. *57*, 2829–2833.

Jiang, W., Hou, Y., and Inouye, M. (1997). CspA, the major cold-shock protein of *Escherichia coli*, is an RNA chaperone. J. Biol. Chem. *272*, 196–202.

Jones, P.G., VanBogelen, R.A., and Neidhardt, F.C. (1987). Induction of proteins in response to low temperature in *Escherichia coli*. J. Bacteriol. *169*, 2092–2095.

Kashket, E.R. (1987). Bioenergetics of lactic acid bacteria, cytoplasmic pH and osmotolerance. FEMS Microbiol. Rev. *46*, 233–244.

Kashket, E.R., and Barker, S.L. (1977). Effects of potassium ions on the electrical and pH gradients across the membrane of *Streptococcus lactis* cells. J. Bacteriol. *130*, 1017–1023.

Kets, E.P.W., and de Bont, J.A.M. (1994). Protective effect of betaine on survival of *Lactobacillus plantarum* subjected to drying. FEMS Microbiol. Lett. *116*, 251–256.

Kilstrup, M., Jacobsen, S., Hammer, K., and Vogensen, F.K. (1997). Induction of heat shock proteins DnaK, GroEL, and GroES by salt stress in *Lactococcus lactis*. Appl. Environ. Microbiol. *63*, 1826–1837.

Kim, W.S., and Dunn, N.W. (1997). Identification of a cold shock gene in lactic acid bacteria and the effect of cold shock on cryotolerance. Curr. Microbiol. *35*, 59–63.

Kim, W.S., Perl, L., Park, J.H., Tandianus, J.E., and Dunn, N.W. (2001). Assessment of stress response of the probiotic *Lactobacillus acidophilus*. Curr. Microbiol. *43*, 346–350.

Kleerebezem, M., Boekhorst, J., van Kranenburg, R., Molenaar, D., Kuipers, O.P., Leer, R., Targhini, R., Peters, S.A., Sandbrink, H.M., Fiers, M.W., Stiekema, W., Lankhorst, R.M., Bron, P.A., Hoffer, S.M., Groot, M.N., Kerkhoven, R., de Vries, M., Ursing, B., de Vos, W.M., and Siezen, J. (2003). Complete genome sequence of *Lactobacillus plantarum* strain WCFS1. Proc. Natl. Acad. Sci. USA *100*, 1990–1995.

Knorr, D. (1999). Novel approaches in food-processing technology: new technologies for preserving foods and modifying function. Curr. Opi. Biotech. *10*, 485–491.

Kobayashi, H. (1982). Second system for potassium transport in *Streptococcus faecalis*. J. Bacteriol. *150*, 506–511.

Kobayashi, H., Suzuki, T., Kinoshita, N., and Unemoto, T. (1984). Amplification of the *Streptococcus faecalis* proton-translocating ATPase by a decrease in cytoplasmic pH. J. Bacteriol. *158*, 1157–1160.

Kobayashi, H., Suzuki, T., and Unemoto, T. (1986). Streptococcal cytoplasmic pH is regulated by changes in amount and activity of a proton-translocating ATPase. J. Biol. Chem. *261*, 627–630.

Kobayashi, K., Ogura, M., Yamaguchi, H., Yoshida, K., Ogasawara, N., Tanaka, T., and Fujita, Y. (2001). Comprehensive DNA microarray analysis of *Bacillus subtilis* two-component regulatory systems. J. Bacteriol. *183*, 7365–7370.

Konings, W.N. (2002). The cell membrane and the struggle for life of lactic acid bacteria. Antonie van Leeuwenhoek Int. J. Gen. Mol. Microbiol. *82*, 3–27.

Kullen, M.J., and Klaenhammer, T.R. (1999). Identification of the pH-inducible, proton-translocating F1F0-ATPase (*atpBEFHAGDC*) operon of *Lactobacillus acidophilus* by differential display: gene structure, cloning and characterization. Mol. Microbiol. 33, 1152–1161.

Kurdi, P., Kawanishi, K., Mizutani, K., and Yokota, A. (2006). Mechanism of growth inhibition by free bile acids in lactobacilli and bifidobacteria. J. Bacteriol. *188*, 1979–1986.

Kurdi, P., van Veen H.W., Tanaka, H., Mierau, I., Konings, W.N., Tannock, G.W., Tomita, F., and Yokota, A. (2000). Cholic acid is accumulated spontaneously, driven by membrane delta pH, in many lactobacilli. J. Bacteriol. *182*, 6525–6528.

Laplace, J.M., Sauvageot, N., Harke, A., and Auffray, Y. (1999). Characterization of *Lactobacillus collinoides* response to heat, acid and ethanol treatments. Appl. Microbiol. Biotechnol. *51*, 659–663.

Leslie, S.B., Israeli, E., Lighthart, B., Crowe, J.H., and Crowe, L.M. (1995). Trehalose and sucrose protect both membranes and proteins in intact bacteria during drying. Appl. Environ. Microbiol. *61*, 3592–3597.

Leverrier, P., Dimova, D., Pichereau, V., Auffray, Y., Boyaval, P., and Jan, G. (2003). Susceptibility and adaptive response to bile salts in *Propionibacterium freudenreichii*: physiological and proteomic analysis. Appl. Environ. Microbiol. *69*, 3809–3818.

Lim, E.M., Ehrlich, S.D., and Maguin, E. (2000). Identification of stress inducible proteins in *Lactobacillus delbrueckii* subsp. *bulgaricus*. Electrophoresis *21*, 2557–2561.

Lin, J.J., and Sancar, A. (1992). (A)BC excinuclease: the *Escherichia coli* nucleotide excision repair enzyme. Mol Microbiol. *6*, 2219–2224.

Lindahl, T., and Nyberg, B. (1972). Rate of depurination of native deoxyribonucleic acid. Biochemistry *11*, 3610–3618.

Lonvaud-Funel, A., and Desens, C. (1990). Constitution en acides gras des membranes des bactéries lactiques du vin. Incidences des conditions de culture. Sci. Aliments *10*, 817–829.

Lorca, G.L, Barabote R.D., Winnen, B., Hvorup, R.N., Zlotopolski, V., Nguyen, E., Huang, L.W., Phan, C., and Saier, M.Jr. (2007). Comparative analysis of the transport capabilities encoded within the genome of eleven gram-positive bacteria. Biochim. Biophys. Acta *1765*, 1342–1366.

Lorca, G.L., and Font de Valdez, G. (1998). Temperature adaptation and cryotolerance in *Lactobacillus acidophilus*. Biotech. Lett. *20*, 847–849.

Lorca, G.L., and Font de Valdez, G. (1999). Effect of suboptimal growth temperature and growth phase on resistance of *Lactobacillus acidophilus* to environmental stress. Cryobiology 39, 144–149.

Lorca, G.L., and Font de Valdez, G. (2001a). Acid tolerance mediated by membrane ATPases in *Lactobacillus acidophilus*. Biotech. Lett. *23*, 777–780.

Lorca, G.L., and Font de Valdez., G. (2001b). A-low-pH-inducible stationary-phase acid tolerance response in *Lactobacillus acidophilus* CRL 639. Curr. Microbiol. *42*, 21–25.

Lorca, G.L., Font de Valdez, G., and Ljungh, Å. (2002). Characterization of the protein-synthesis dependent adaptive acid tolerance response in *Lactobacillus acidophilus*. J. Mol. Microbiol. Biotech. *4*, 525–532.

Lorca, G.L., Raya, R.R., Taranto, M.P., and de Valdez, G.F. (1998). Adaptative acid tolerance response in *Lactobacillus acidophilus*. Biotechnol. Lett. *20*, 239–241.

Lorca, G.L., Raya, R.R., Saier, M., and Font de Valdez, G. (2002). Genetic analysis of the *groES-groEL* and *hrcA-grpE-dnaK-dnaJ* operons and their role on the acid tolerance response of *Lactobacillus acidophilus*. Seventh Symposium on Lactic Acid Bacteria-Genetics, Metabolism and Applications, FEMS, The Netherlands, H69.

Lundeen, S.G., and Savage, D.C. (1990). Characterization and purification of bile salt hydrolase from *Lactobacillus* sp. strain 100–100. J. Bacteriol. *172*, 4171–4177.

Lundeen, S.G., and Savage, D.C. (1992). Multiple forms of bile salt hydrolase from *Lactobacillus* sp. strain 100–100. J. Bacteriol. *174*, 7217–7220.

Machado, M.C., Lopez, C.S., Heras, H., and Rivas, E.A. (2004). Osmotic response in *Lactobacillus casei* ATCC 393: biochemical and biophysical characteristics of membrane. Arch. Biochem. Biophys. *422*, 61–70.

Magni, C., De Mendoza, D., Konings, W.N., and Lolkema, J.S. (1999). Mechanism of citrate metabolism in *Lactococcus lactis*: resistance against lactate toxicity at low pH. J. Bacteriol. *181*, 1451–1457.

Makarova, K., Slesarev, A., Wolf, Y., Sorokin, A., Mirkin, B., Koonin, E., Pavlov, A., Pavlova, N., Karamychev, V., Polouchine, N., Shakhova, V., Grigoriev, I., Lou, Y., Rohksar, D., Lucas, S., Huang, K., Goodstein, D.M., Hawkins, T., Plengvidhya, V., Welker, D., Hughes, J., Goh, Y., Benson, A., Baldwin, K., Lee, J.H., Diaz-Muniz, I., Dosti, B., Smeianov, V., Wechter, W., Barabote, R., Lorca, G., Altermann, E., Barrangou, R., Ganesan, B., Xie, Y., Rawsthorne, H., Tamir, D., Parker, C., Breidt, F., Broadbent, J., Hutkins, R., O'Sullivan, D., Steele, J., Unlu, G., Saier, M., Klaenhammer, T., Richardson, P., Kozyavkin, S., Weimer, B., and Mills, D. (2006). Comparative genomics of the lactic acid bacteria. Proc. Natl. Acad. Sci. USA 103, 15611–15616.

Mansilla, M.C., Cybulski, L.E., Albanesi, D., and de Mendoza, D. (2004). Control of membrane lipid fluidity by molecular thermosensors. J. Bacteriol. *186*, 6681–6688.

Marceau, A., Zagorec, M., Chaillou, S., Mera, T., and Champomier-Verges, M.C. (2004). Evidence for involvement of at least six proteins in adaptation of *Lactobacillus sakei* to cold temperatures and addition of NaCl. Appl. Environ. Microbiol. *70*, 7260–7268.

Marquis, R.E., Bender, G.R., Murray, D.R., and Wong, A. (1987). Arginine deiminase system and bacterial adaptation to acid environments. Appl. Environ. Microbiol. *53*, 198–200.

Martin, M.G., Sender, P.D., Peiru, S., de Mendoza, D., and Magni, C. (2004). Acid inducible transcription of the operon encoding the citrate lyase complex of *Lactococcus lactis* biovar *diacetylactis* CRL264. J. Bacteriol. *186*, 5649–5660.

Mayo, B., Derzelle, S., Fernandez, M., Leonard, C., Ferain, T., Hols, P., Suarez, J.E., and Delcour, J. (1997). Cloning and characterization of *cspL* and *cspP*, two cold-inducible genes from *Lactobacillus plantarum*. J, Bacteriol. *179*, 3039–3042.

Mäyrä-Mäkinen, A., and Bigret M. (1998). Industrial use and production of lactic acid bacteria. In Lactic Acid Bacteria, Microbiology and Functional Aspects, S. Salminen and A. von Wright, eds. (New York: Marcel Dekker Inc.), pp. 73–102.

Mazur, P. (1970). Cryobiology: the freezing of biological systems. Science *168*, 939–949.

Mc Auliffe, O., Cano, R.J., and Klaenhammer, T.R. (2005). Genetic analysis of two bile salt hydrolase activities in *Lactobacillus acidophilus* NCFM. Appl. Environ. Microbiol. *71*, 4925–4929.

Morel-Deville, F., Fauvel, F., and Morel, P. (1998). Two-component signal-transducing systems involved in stress responses and vancomycin susceptibility in *Lactobacillus sakei*. Microbiology *144*, 2873–2883.

Moser, S.A., and Savage, D.C. (2001). Bile salt hydrolase activity and resistance to toxicity of conjugated bile salts are unrelated properties in lactobacilli. Appl. Environ. Microbiol. *67*, 3476–3480.

Narberhaus, F., Waldminghaus, T., and Chowdhury, S. (2006). RNA thermometers. FEMS Microbiol. Rev. *30*, 3–16.

Nehme, B., Ganga, M.A., and Lonvaud-Funel, A. (2006). The arginine deiminase locus of *Oenococcus oeni* includes a putative arginyl-tRNA synthetase ArgS2 at its 3'-end. Appl. Microbiol. Biotechnol. *70*, 590–597.

Obis, D., Guillot, A., Gripon, J.C., Renault, P., Bolotin, A., Mistou, M.Y. (1999). Genetic and biochemical characterization of a high-affinity betaine uptake system (BusA) in *Lactococcus lactis* reveals a new functional organization within bacterial ABC transporters. J. Bacteriol. *181*, 6238–6246.

O'Connell-Motherway, M., van Sinderen, D., Morel-Deville, F., Fitzgerald, G.F., Ehrlich, S.D., and Morel, P. (2000). Six putative two-component regulatory systems isolated from *Lactococcus lactis* subsp. *cremoris* MG1363. Microbiology *146*, 935–947.

Olsen, E.B., Russell, J.B., and Henick-Kling, T. (1991). Electrogenic L-malate transport by *Lactobacillus plantarum*: a basis for energy derivation from malolactic fermentation. J. Bacteriol. *173*, 6199–6206.

O'Mahony, L., McCarthy, J., Kelly, P., Hurley, G., Luo, F., Chen, K., O'Sullivan, G.C., Kiely, B., Collins, J.K., Shanahan, F., and Quigley, E.M. (2005). *Lactobacillus* and *bifidobacterium* in irritable bowel syndrome: symptom responses and relationship to cytokine profiles. Gastroenterology *128*, 541–551.

Ouwehand, A.C., Salminen, S., and Isolauri E. (2002). Probiotics: an overview of beneficial effects. Antonie Van Leeuwenhoek *82*, 279–289.

Pallen, M.J., and Wren, B.W. (1997). Micro review: the HtrA family of serine proteases. Mol. Microbiol. *26*, 209–221.

Panoff, J.M., Thammavongs, B., and Gueguen, M. (2000). Cryoprotectants lead to phenotypic adaptation to freeze-thaw stress in *Lactobacillus delbrueckii* ssp. *bulgaricus* CIP 101027T. Cryobiology *40*, 264–269.

Parkinson, J.S. (1993). Signal transduction schemes of bacteria. Cell *73*, 857–871.

Pavlovic, M., Hormann, S., Vogel, R.F., Ehrmann, M.A. (2005). Transcriptional response reveals translation machinery as target for high pressure in *Lactobacillus sanfranciscensis*. Arch. Microbiol. *184*, 11–17.

Penaud, S., Fernandez, A., Boudebbouze, S., Ehrlich, S.D., Maguin, E., and van de Guchte, M. (2006). Induction of heavy-metal-transporting CPX-type ATPases during acid adaptation in *Lactobacillus bulgaricus*. Appl. Environ. Microbiol. *72*, 7445–7454.

Pereira, D.I., McCartney, A.L., and Gibson, G.R. (2003). An in vitro study of the probiotic potential of a bile-salt-hydrolyzing *Lactobacillus fermentum* strain, and determination of its cholesterol-lowering properties. Appl. Environ. Microbiol. *69*, 4743–4752.

Perroud, B., and Le Rudulier, D. (1985). Glycine betaine transport in *Escherichia coli*: osmotic modulation. J. Bacteriol. *161*, 393–401.

Phadtare, S. (2004). Recent developments in bacterial cold-shock response. Curr. Issues Mol. Biol. *6*, 125–136.

Pieterse, B., Leer, R.J., Schuren, F.H., and van der Werf, M.J. (2005). Unravelling the multiple effects of lactic acid stress on *Lactobacillus plantarum* by transcription profiling. Microbiology *151*, 3881–3894.

Piuri, M., Sanchez-Rivas, C., and Ruzal, S.M. (2003). Adaptation to high salt in *Lactobacillus*: role of peptides and proteolytic enzymes. J. Appl. Microbiol. *95*, 372–379.

Piuri, M., Sanchez-Rivas, C., and Ruzal, S.M. (2005). Cell wall modifications during osmotic stress in *Lactobacillus casei*. J. Appl. Microbiol. *98*, 84–95.

Poolman, B., Molenaar, D., Smid, E.J., Ubbink, T., Abee, T., Renault, P.P., and Konings, W.N. (1991). Malolactic fermentation: electrogenic malate uptake and malate/lactate antiport generate metabolic energy. J. Bacteriol. *173*, 6030–6037.

Prasad, J., McJarrow, P., and Gopal, P. (2003). Heat and osmotic stress responses of probiotic *Lactobacillus rhamnosus* HN001 (DR20) in relation to viability after drying. Appl. Environ. Microbiol. *69*, 917–925.

Price, C.W., Fawcett, P., Ceremonie, H., Su, N., Murphy, C.K., and Youngman, P. (2001). Genome-wide analysis of the general stress response in *Bacillus subtilis*. Mol. Microbiol. *41*, 757–774.

Quivey, R.G., Faustoferri, R.C., Clancy, K.A., and Marquis, R.E. (1995). Acid adaptation in *Streptococcus mutans* Ua159 alleviates sensitization to environmental stress due to RecA deficiency. FEMS Microbiol. Lett. *126*, 257–261.

Rhee, S.K., and Pack, M.Y. (1980). Effect of environmental pH on fermentation balance of *Lactobacillus bulgaricus*. J. Bacteriol. *144*, 217–221.

Rollan, G., Lorca, G.L., and Font de Valdez, G. (2003). Arginine catabolism and acid tolerance response in *Lactobacillus reuteri* isolated from sourdough. Food Microbiol. *20*, 313–319.

Ross, R.P., Morgan, S., and Hill, C. (2002). Preservation and fermentation: past, present and future. Int. J. Food Microbiol. *79*, 3–16.

Russell, N.J., Evans, R.I., Ter Steeg, P.F., Hellemons, J., Verheul, A., and Abee, T. (1995). Membranes as a target for stress adaptation. Int. J. Food Microbiol. *28*, 255–261.

Sakamoto, K., Margolles, A., van Veen, H.W., and Konings, W.N. (2001). Hop resistance in the beer spoilage bacterium *Lactobacillus brevis* is mediated by the ATP-binding cassette multidrug transporter HorA. J. Bacteriol. *183*, 5371–5375.

Salema, M., Poolman, B., Lolkema, J.S., Loureiro Dias, M.C., and Konings, W.N. (1994). Uniport of monoanionic L-malate in membrane vesicles from *Leuconostoc oenos*. FEBS Eur. J. Biochem. *124*, 1–7.

Salema, M., Lolkema, J.S., San Romao, M.V., and Loureiro Dias, M.C. (1996). The proton-motive force generated in *Leuconostoc oenos* by L-malate fermentation. J. Bacteriol. *178*, 3127–3132.

Salminen, S., and von Wright, A. (1993). Lactic Acid Bacteria (New York: Marcel Dekker, Inc.).

Sami, M., Yamashita, H., Hirono, T., Kadokura, H., Kitamoto, K., Yoda, K., and Yamasaki, M. (1997). Hop-resistant *Lactobacillus brevis* contains a novel plasmid harbouring a multidrug resistance-like gene. J. Ferment. Bioeng. *84*, 1–6.

Sancar, A., and Rupp, W.D. (1983). A novel repair enzyme: UVRABC excision nuclease of *Escherichia coli* cuts a DNA strand on both sides of the damaged region. Cell *33*, 249–260.

Sanchez, B., de los Reyes-Gavilan, C.G., and Margolles, A. (2006). The F1F0-ATPase of *Bifidobacterium animalis* is involved in bile tolerance. Environ. Microbiol. *8*, 1825–1833.

Sanders, J.W., Leenhouts, K., Burghoorn, J., Brands, J.R., Venema, G., and Kok, J. (1998). A chloride-inducible acid resistance mechanism in *Lactococcus lactis* and its regulation. Mol. Microbiol. *27*, 299–310.

Sauvageot, N., Beaufils, S., Maze, A., Deutscher, J., and Hartke, A. (2006). Cloning and characterization of a gene encoding a cold-shock protein in *Lactobacillus casei*. FEMS Microbiol. Lett. *254*, 55–62.

Scheyhing, C.H., Hormann, S., Ehrmann, M.A., and Vogel, R.F. (2004). Barotolerance is inducible by preincubation under hydrostatic pressure, cold-, osmotic- and acid-stress conditions in *Lactobacillus sanfranciscensis* DSM 20451T. Lett. Appl. Microbiol. *39*, 284–289.

Schmidt, G., Hertel, C., and Hammes, W.P. (1999). Molecular characterisation of the *dnaK* operon of *Lactobacillus sakei* LTH681. Syst. Appl. Microbiol. *22*, 321–328.

Schumann, W. (2003). The *Bacillus subtilis* heat shock stimulon. Cell Stress Chaperones *8*, 207–217.

Schweder, T., and Hecker, M. (2004). Monitoring of stress responses. Adv. Biochem. Eng. Biotechnol. *89*, 47–71.

Serror, P., Dervyn, R., Ehrlich, S.D., and Maguin, E. (2003). *csp*-like genes of *Lactobacillus delbrueckii* ssp. *bulgaricus* and their response to cold shock. FEMS Microbiol. Lett. *226*, 323–330.

Sheehan, V.M., Sleator, R.D., Fitzgerald, G.F., and Hill, C. (2006). Heterologous expression of BetL, a betaine uptake system, enhances the stress tolerance of *Lactobacillus salivarius* UCC118. Appl. Environ. Microbiol. *72*, 2170–2177.

Silva, J., Carvalho, A.S., Ferreira, R., Vitorino, R., Amado, F., Domingues, P., Teixeira, P., and Gibbs, P.A. (2005). Effect of the pH of growth on the survival of *Lactobacillus delbrueckii* subsp. *bulgaricus* to stress conditions during spray-drying. J. Appl. Microbiol. *98*, 775–782.

Skinner, M.M., and Trempy, J.E. (2001). Expression of clpX, an ATPase subunit of the Clp protease, is heat and cold shock inducible in *Lactococcus lactis*. J. Dairy Sci. *84*, 1783–1785.

Smeds, A., Varmanen, P., and Palva, A. (1998). Molecular characterization of a stress-inducible gene from *Lactobacillus helveticus*. J. Bacteriol. *180*, 6148–6153.

Solioz, M., and Odermatt, A. (1995). Copper and silver transport by CopB ATPase in membrane vesicles of *Enterococcus hirae*. J. Biol. Chem. *270*, 9217–9221.

Souzu, H. (1989). Changes in chemical structure and function in *Escherichia coli* cell membranes caused by freeze-thawing. I. Change of lipid state in bilayer vesicles and in the original membrane fragments depending on rate of freezing. Biochim. Biophys. Acta *978*, 105–111.

Souzu, H., Sato, M., and Kojima, T. (1989). Changes in chemical structure and function in Escherichia coli cell membranes caused by freeze-thawing. II. Membrane lipid state and response of cells to dehydration. Biochim. Biophys. Acta *978*, 112–118.

Spano, G., Capozzi, V., Vernile, A., and Massa, S. (2004). Cloning, molecular characterization and expression analysis of two small heat shock genes isolated from wine *Lactobacillus plantarum*. J. Appl. Microbiol. *97*, 774–782.

Spano, G., Beneduce, L., Perrotta, C., and Massa S. (2005). Cloning and characterization of the hsp 18.55 gene, a new member of the small heat shock gene family isolated from wine *Lactobacillus plantarum*. Res. Microbiol. *156*, 219–224.

Spano, G., and Massa, S. (2006). Environmental stress response in wine lactic acid bacteria: beyond *Bacillus subtilis*. Crit. Rev. Microbiol. *32*, 77–86.

Stock, J.B., Surette, M.G., Levit, M., and Stock, A.M. (1995). Two component signal transduction systems: structure-function relationships and mechanisms of catalysis. In Two-component Signal Transduction, J.A. Hoch and T.J. Silhavy, eds. (Washington, DC: American Society for Microbiology), pp. 25–51.

Suokko, A., Savijoki, K., Malinen, E., Palva, A., and Varmanen, P. (2005). Characterization of a mobile *clpL* gene from *Lactobacillus rhamnosus*. Appl. Environ. Microbiol. *71*, 2061–2069.

Suutari, M., and Laakso, S. (1992). Temperature adaptation in *Lactobacillus fermentum*: Interconversions of oleic, vaccenic and dihydrosterculic acids. J. Gen. Microbiol. *138*, 445–450.

Tanaka, H., Doesburg, K., Iwasaki, T., and Mierau, I. (1999). Screening of lactic acid bacteria for bile salt hydrolase activity. J. Dairy Sci. *82*, 2530–2535.

Tannock, G.W., Dashkevicz, M.P., and Feighner, S.D. (1989). Lactobacilli and bile salt hydrolase in the murine intestinal tract. Appl. Environ. Microbiol. *55*, 1848–1851.

Taranto, M.P., and Font de Valdez, G. 1999. Localization and primary characterization of bile salt hydrolase from *Lactobacillus reuteri*. Biotechnol. Lett. *21*, 935–938.

Taranto, M.P., Fernandez Murga, M.L., Lorca, G., and Font de Valdez, G. (2003). Bile salts and cholesterol induce changes in the lipid cell membrane of *Lactobacillus reuteri*. J. Appl. Microbiol. *95*, 86–91.

Taranto, M.P., Medici, M., Perdigon, G., Ruiz Holgado, A.P., and Valdez G.F. (1998). Evidence for hypocholesterolemic effect of *Lactobacillus reuteri* in hypercholesterolemic mice. J. Dairy Sci. *81*, 2336–2340.

Taranto, M.P., Medici, M., Perdigon, G., Ruiz Holgado, A.P., and Valdez, G.F. (2000). Effect of *Lactobacillus reuteri* on the prevention of hypercholesterolemia in mice. J. Dairy Sci. *83*, 401–403.

Taranto, M.P., Perez-Martinez, G., and Font de Valdez, G.M. (2006). Effect of bile acid on the cell membrane functionality of lactic acid bacteria for oral administration. Res. Microbiol. *157*, 720–725.

Taranto, M.P., Sesma, F., Ruiz Holgado, A.P., and Font de Valdez, G. (1997). Bile salt hydrolase plays a key role on cholesterol removal by *Lactobacillus reuteri*. Biotechnol. Lett. *19*, 845–847.

Thanassi, D.G., Cheng, L.W., and Nikaido, H. (1997). Active efflux of bile salts by *Escherichia coli*. J. Bacteriol. *179*, 2512–2518.

Throup, J.P., Koretke, K.K., Bryant, A.P., Ingraham, K.A., Chalker, A.F., Ge, Y., Marra, A., Wallis, N.G., Brown, J.R., Holmes, D.J., Rosenberg, M., and Burnham, M.K. (2000). A genomic analysis of two-component signal transduction in *Streptococcus pneumoniae*. Mol. Microbiol. *35*, 566–576.

Tonon, T., Bourdineaud, J.P., and Lonvaud-Funel, A. (2001). The *arcABC* gene cluster encoding the arginine deiminase pathway of *Oenococcus oeni*, and arginine induction of a CRP-like gene. Res. Microbiol. *152*, 653–661.

Torino, M.I., Taranto, M.P., Sesma, F., and Font de Valdez, G. (2001). Heterofermentative pattern and exopolysaccharide production by *Lactobacillus helveti-*

cus ATCC 15807 in response to environmental pH. J. Appl. Microbiol. *91*, 846–852.

Tymczyszyn, E.E., Gomez-Zavaglia, A., and Disalvo, E.A. (2005). Influence of the growth at high osmolality on the lipid composition, water permeability and osmotic response of *Lactobacillus bulgaricus*. Arch. Biochem. Biophys. *443*, 66–73.

Ueno, Y., Hayakawa, K., Takahashi, S., and Oda, K. (1997). Purification and characterization of glutamate decarboxylase from *Lactobacillus brevis* IFO12005. Biosci. Biotechnol. Biochem. *61*, 1168–1171.

Usman, and Hosono, A. (1999). Bile tolerance, taurocholate deconjugation, and binding of cholesterol by *Lactobacillus gasseri* strains. J. Dairy Sci. *82*, 243–248.

Valdez, G.F., Martos, G., Taranto, M.P., Lorca, G.L., Oliver, G. and Ruiz Holgado, A.P. (1996). Influence of bile on b-galactosidase activity and cell viability of *Lactobacillus reuteri* when subjected to freeze-drying. J. Dairy Sci. *80*, 1955–1958.

van de Guchte, M., Serror, P., Chervaux, C., Smokvina, T., Ehrlich, S.D., and Maguin, E. (2002). Stress responses in lactic acid bacteria. Antonie Van Leeuwenhoek *82*, 187–216.

van Der Heide, T., and Poolman, B. (2000). Glycine betaine transport in *Lactococcus lactis* is osmotically regulated at the level of expression and translocation activity. J. Bacteriol. *182*, 203–206.

Ventura, M., Canchaya, C., Zhang, Z., Bernini, V., Fitzgerald, G.F., and van Sinderen, D. (2006). How high G+C Gram-positive bacteria and in particular bifidobacteria cope with heat stress: protein players and regulators. FEMS Microbiol. Rev. *30*, 734–759.

Vlahcevic, Z.R., Heuman, D.M., and Hylemon, P.B. (1990). Physiology and pathophysiology of enterohepatic circulation of bile acids. In Hepatology. A textbook of liver disease. D. Zakim and T.D. Boyer eds. (Philadelphia, PA: W.B. Saunders Company), pp. 341–377.

Walker, D.C., Girgis, H.S., and Klaenhammer, T.R. (1999). The *groESL* Chaperone operon of *Lactobacillus johnsonii*. Appl. Envir. Microbiol. *65*, 3033–3041.

Wang, Y., Corrieu, G., and Beal, C. (2005). Fermentation pH and temperature influence the cryotolerance of *Lactobacillus acidophilus* RD758. J. Dairy Sci. *88*, 21–29.

Whatmpre, A.M., Chudek, J.A., and Reed, R.H. (1990). The effects of osmotic upshock on the intracellular solute pools of *Bacillus subtilis*. J. Gen. Microbiol. *136*, 2527–2535.

Wouters, J.A., Jeynov, B., Rombouts, F.M., de Vos, W.M., Kuipers, O.P., and Abee, T. (1999). Analysis of the role of 7 kDa cold-shock proteins of *Lactococcus lactis* MG1363 in cryoprotection. Microbiology *145*, 3185–3194.

Wouters J.A., Rombouts, F.M., Kuipers, O.P., de Vos W.M., and Abee, T. (2000). The role of cold-shock proteins in low-temperature adaptation of food-related bacteria. Syst. Appl. Microbiol. *23*, 165–173.

Wouters, J.A., Sanders, J.-W., Kok, J., de Vos, W.M., Kuipers, O.P., and Abee, T. (1998). Clustered organization and transcriptional analysis of a family of five *csp* genes of *Lactococcus lactis* MG1363. Microbiology *144*, 2885–2893.

Wouters, P.C., Glaasker, E., Smelt, J.P. (1998). Effects of high pressure on inactivation kinetics and events related to proton efflux in *Lactobacillus plantarum*. Appl. Environ. Microbiol. *64*, 509–514.

Yamamoto, N., Masujima, Y., and Takano, T. (1996). Reduction of membrane-bound ATPase activity in a *Lactobacillus helveticus* strain with slower growth at low pH. FEMS Microbiol. Lett. *138*, 179–184.

Yamamoto, Y., Higuchi, M., Poole, L.B. and Kamio, Y. (2000). Role of the *dpr* product in oxygen tolerance in *Streptococcus mutans*. J. Bacteriol. *182*, 3740–3747.

Yamanaka, K., Fang, L., and Inouye, M. (1998). The CspA family in *Escherichia coli*: multiple gene duplication for stress adaptation. Mol. Microbiol. *27*, 247–255.

Yamanaka, K., Mitta, M., and Inouye, M. (1999). Mutation analysis of the 5′ untranslated region of the cold shock *cspA* mRNA of *Escherichia coli*. J. Bacteriol. *181*, 6284–6291.

Yokota, A., Veenstra, M., Kurdi, P., van Veen, H.W., and Konings, W.N. (2000). Cholate resistance in *Lactococcus lactis* is mediated by an ATP-dependent multispecific organic anion transporter. J. Bacteriol. *182*, 5196–5201.

Yura, T., Kanemori, M., and Morita, M.T. (2000). The heat shock response: regulation and function. In Bacterial Stress Responses, G. Storz and R. Hengge-Aronis eds (Washington, DC: ASM Press), pp. 3–18.

Zink, R., Walker, C., Schmidt, G., Elli, M., Pridmore, D., and Reniero, R. (2000). Impact of multiple stress factors on the survival of dairy lactobacilli. Sci. Aliments *20*, 119–126.

Zuber, U., and Schumann, W. (1994). CIRCE, a novel heat shock element involved in regulation of heat shock operon dnaK of *Bacillus subtilis*. J. Bacteriol. *176*, 1359–1363.

Zuniga, M., Perez, G., and Gonzalez-Candelas, F. (2002). Evolution of arginine deiminase (ADI) pathway genes. Mol. Phylogenet. Evol. 25, 429–444.

Interactions of *Lactobacillus* with the Immune System

7

Denny Demeria, Julia Ewaschuk and Karen Madsen

Abstract

Lactobacilli have been used for centuries with the intent to produce a health benefit in the human host. Interactions between the host and lactobacilli involve lactobacilli-induced alteration of gene expression in gut epithelial and immune cells, with the subsequent modulation of both mucosal and systemic immune function. Effects of lactobacilli are concentration and species and strain-dependent. Bacteria are critical in establishing a state of immunological homeostasis within the host. Beneficial effects exerted by lactobacilli bacteria in the treatment of human disease may be broadly classified as those effects which arise due to activity in the large intestine and are related to colonization or inhibition of pathogen growth; and those effects which arise in both the small and large intestine, and are related to enhancement of the host immune response and intestinal barrier function.

Introduction

With the introduction of probiotic organisms into the armamentarium for the treatment of human diseases, so to comes the call to examine the relationships between humans and these microorganisms. Of the 10^{14} bacteria that inhabit the human body, a relatively small proportion are capable of causing harm to the human host at any given time. Bacteria within the host maintain a state of homeostasis with the immune system (Fig. 7.1). This chapter will explore the relationships lactobacilli share with the immune system, both at an organ specific level and systemically. Microorganism and host specific anatomical and physiological characteristics which confer adaptability of the microorganism to sustain itself within a specific host will be described. Effects of lactobacilli on macrophages, monocytes and other extraintestinal cell types which are involved with both innate and acquired immunity will be discussed.

Bacterial structural components

Lactobacillus species are Gram-positive facultative anaerobic bacteria that are widely distributed throughout nature. The unique structure of lactobacilli cell walls, including amino acid composition, lipoteichoic acid content and S-layer protein configurations, confers the ability of lactobacilli to adhere to host surfaces and are critical in regulating the host immune response.

Lipoteichoic acid

Although within the genus *Lactobacillus* there exists at least 165 species and subspecies, all lactobacilli share a cell wall structure consistent with most other gram-positive bacteria, and therefore, share many of the potential antigens with those bacteria. Many of these species share a common antigen within the cell membrane, consisting primarily of the 1–3 linked glycerol phosphate unit of the membrane-bound glycerol teichoic acid. Lipoteichoic acid has been shown to elicit both pro- and anti-inflammatory cytokines from immune cells in a strain-dependent fashion (Morath *et al.*, 2001). Further studies indicated that the amount of d-Alanine (d-Ala) within the lactobacilli cell wall was a critical

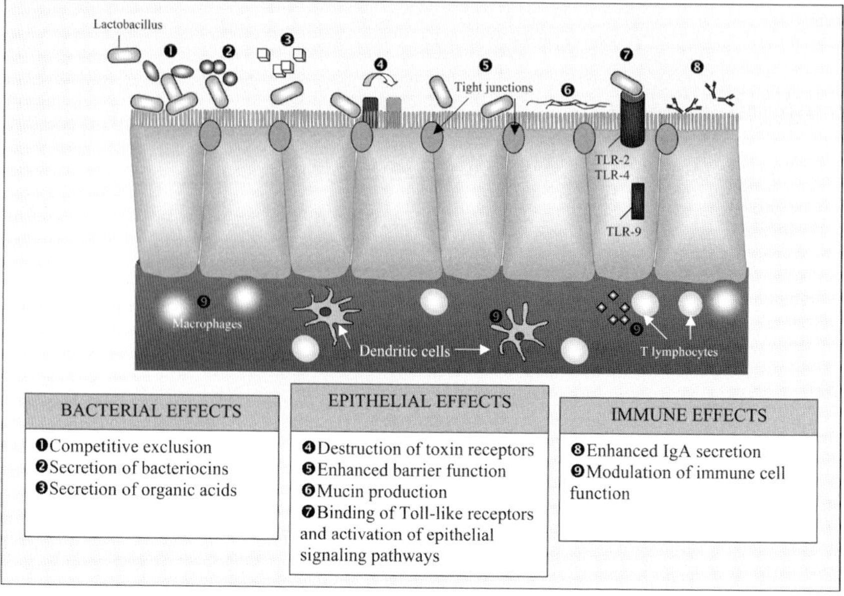

Figure 7.1 Actions of lactobacilli within the host intestinal tract.

determining factor in the ability of a given strain to induce cytokine secretion (Morath *et al.*, 2002). Grangette *et al.* (2005) examined the immunomodulatory effects of a mutant lactobacilli strain, Dlt⁻ *L. plantarum* NCIMB8826 having significantly fewer d-Ala residues than wild type (WT) strains (1.15 vs. 41.7%). It was observed that in isolated peripheral blood mononuclear cells, the Dlt⁻ strain stimulated greater IL-10 secretion and a reduced amount of TNF-α and IFN-γ secretion compared with WT controls. Purified lipoteichoic acids from the *L. plantarum* NCIMB8826 and the Dlt- strains demonstrated similar effects, in that lipoteichoic acid from WT strains elicited a significant secretion of TNF-α, IL-1β, IL-6 and IL-8 while the mutant strain did not elicit any TNF-α response and markedly reduced secretion of the other cytokines. Thus, on the basis of these data, it seems that the cell wall structure, including the amino acid sequence of certain structures such as lipoteichoic acid, play a critical role in determining both the character and magnitude of immune response elicited from host immune cells.

Surface Layers (S-layers)

Numerous protein or glycoprotein subunits, ranging from 25 to 71 kDa, are found in the outermost region of the cell envelope of lactobacillus and other bacteria (Fig. 7.2). These *surface layers* (S-layers) are arranged in a monomolecular crystalline array and have been shown to have numerous functional properties, including acting as virulence factors on pathogens such as *Campylobacter fetus* spp. *fetus* and *Aeromonas salmonicida* (Ebanks *et al.*, 2005); being a depository for surface-exposed enzymes, e.g. *Bacillus stearothermophilus*; determining shape in such organisms as *Thermoproteus tenax* (Messner *et al.*, 1986). The function of the S-layer with respect to immune regulation is of great interest. Avail-Jaaskelainen *et al.* (Avall-Jaaskelainen and Palva, 2005) were the first to review the S-layers of lactobacilli in great detail. They reported that the genetic arrangement of the protein-encoding genes for different strains of lactobacilli is strain-dependent. The amino acid composition of the S-layer consists of predominantly hydrophobic amino acid residues (31.9–38.7%) and an analysis of the secondary protein structure predictions shows that *L. brevis* S-layer proteins demonstrate a high degree of variance compared to each other, while the secondary structures of *L. acidophilus, L. crispatus* and *L. helveticus* share a high degree of similarity. The specific mechanisms controlling S-layer gene expression and regulation in lactobacilli at present are largely unknown. Some of the known genes that encode lactobacilli S-

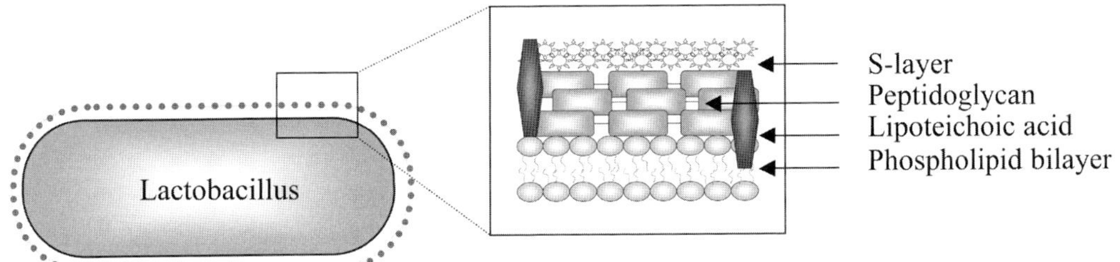

Figure 7.2 Cell wall structures of lactobacilli.

layer proteins include *slp A*, *slp B*, *slp C* and *slp D*. It has been demonstrated that there is variation of the S-layer protein content of *L. brevis* ATCC 14869 under aerated and anaerobic conditions, involving activation of *slp D* transcription by a soluble factor (Jakava-Viljanen *et al.*, 2002). Furthermore, this species was also observed to reversibly vary its colony morphology; the colonies were smooth in anaerobic conditions and had a rough appearance in aerobic conditions. Prior to this, the only confirmed function of the S-layer in lactobacilli was that of adhesion to different host surfaces. Recently, the S-layer protein SlpA of *L. brevis* ATCC 8287 has been shown to function as an adhesin to human intestinal epithelial cell lines and fibronectin (Hynonen *et al.*, 2002). Furthermore, the S-layer protein CbsA of *L. crispatus* mediates adhesion to collagens and laminin, components of the extracellular matrix (Antikainen *et al.*, 2002, Sillanpaa *et al.*, 2000, Toba *et al.*, 1995). Thus, the S-layer appears to have multiple functions; given the variability in structure and their presence throughout the microbial world, many questions remain as to the antigenic potential of this structure.

Effects of *Lactobacillus* on autoimmune disease models

The immune modulatory actions of lactobacilli have been investigated in numerous animal models of autoimmune and inflammatory disease (Matsuzaki, 2007), including arthritis (Kato *et al.*, 1998; Baharav *et al.*, 2004), eczema (Isolauri *et al.*, 2000; Pessi *et al.*, 2000; Viljanen *et al.*, 2005a,b), pouchitis (Gionchetti *et al.*, 2000, Mimura *et al.*, 2004), radiation and NSAID-induced enteropathy (Demirer *et al.*, 2006), ulcerative colitis (Bibiloni *et al.*, 2005), Crohn's disease (Van Gossum, 2007; Marteau *et al.*, 2006; Rolfe *et al.*, 2006), and experimental colitis (Madsen *et al.*, 2001; O'Mahony *et al.*, 2001; Shibolet *et al.*, 2002; Moller *et al.*, 2005). Oral administration of *L. casei* strain Shirota (LcS) has also shown effectiveness in reducing the incidence of diabetes in non-obese diabetic (NOD) mice (Matsuzaki, 2007) and in preventing the onset of type II collagen-induced arthritis in mice (Kato *et al.*, 1998). This was associated with a decrease in the proportion of CD8+ T cells in spleen cells in the LcS-treated animals, and an inhibition of the disappearance of insulin-secreting β-cells in Langerhans islets (Matsuzaki, 2007). The mode of action differs between different lactobacilli strains, and furthermore, the behaviour of a particular strain can differ based upon the host environment and the presence or absence of other bacterial strains.

Lactobacillus interactions with the epithelium

The interactions between lactobacilli and the human host, including colonization and adhesive behaviours, have been well documented (Fernandez, 2003). Human epithelia are composed of both keratinized and mucosal surfaces, each with inherent mechanisms to maintain bacterial populations and to minimize the potential for the risk of bacterial invasion across these barriers. Microbial interactions with the mucosal epithelium (e.g. lung, gut, and genitourinary tract) have demonstrable effects in modulating both mucosal and systemic adaptive and innate immunity. In the intestine, the mucosal epithelium is comprised of epithelial cells, the mucus layer, Paneth cells and goblet cells. The follicle-associated epithelium is a specialized intestinal epithelium lacking mucus, crypts, microvillus or organelles that cover the top of Peyer's patches. Immune

Table 7.1 Biological effects of probiotic bacteria in experimental models

Mechanism of action	Biological effect	Experimental model	Bacterial species	Reference
Epithelial barrier	Enhanced epithelial resistance	IL-10 deficient mouse T84 cells	VSL3	(Madsen *et al.*, 2001)
	Enhanced phosphorylation of actinin and occludin	T84 cells	*Streptococcus thermophilus* *Lactobacillus acidophilus*	(Resta-Lenert and Barrett, 2006)
Antibody production/cell-mediated immunity	Enhanced Ab production Enhanced phagocytic activity Enhanced NK activity	Murine Human F344 rats	*L. rhamnosus* *L. casei* *L. acidophilus 74–2* *L. johnsonii La1* *L. kefiranofaciens*	(Fang *et al.*, 2000) (Gill and Rutherfurd, 2001) (Klein, 2007) (Kaburagi, 2007) (Vinderola, 2006)
Cell signalling	Modulation of NF-kB pathway	HT29 cells T84 cells	Non-pathogenic Salmonella strains	(Neish *et al.*, 2000) (O'Hara *et al.*, 2006)
Apoptosis	Prevention of apoptosis	YAMC epithelial cells	*L. rhamnosus LGG*	(Yan and Polk, 2002)
	Induction of apoptosis	Jurkat cells	*brevis* *Streptococcus thermophilus*	(Di Marzio *et al.*, 2001)
Mucin secretion	Up-regulation of MUC3 mRNA and secretion	HT29 cells human	*Lactobacillus rhamnosus*	(Mack *et al.*, 2003)
Antioxidative	Inhibited linoleic acid peroxidation Scavenge DPPH free radicals		*Lactobacillus fermentum* *L. acidophilus* *Bifidobacterium longum*	(Kullisaar *et al.*, 2002)

Adhesive properties	Prevention of pathogenic strains from adhering to epithelial cells Disease-specific adhesion Strain-specific adhesion	Caco-2 HT29 Intestine 407 Human mucosa	*L. casei* Shirota *L. rhamnosus* *L. rhamnosus GG* *L. reuteri* *Lactobacillus acidophilus MJLA1*	(Fernandez *et al.*, 2003) (Huang and Adams, 2003) (Schaer-Zammaretti and Ubbink, 2003)
Metabolic processes	Formation of short-chain fatty acids Energy homeostasis	Rat mouse	*L. plantarum 299v* *L. acidophilus NCFB 1748* *L. paracasei ssp. Paracasei F19*	(Nerstedta, 2007)
Antimicrobial	Production of Organic acids Hydrogen peroxide Bacteriocins Reduction of luminal pH	Caco-2 cells C3/He/Oujco mice	*Enterococcus faecium* *L. acidophilus* *L. paracasei* *Streptococcus bovis*	(Coconnier *et al.*, 2000) (Caridi, 2002) (Eijsink *et al.*, 2002)
Cytokine production	Increased proinflammatory and/or Decreased proinflammatory	Murine splenocytes Human PBMC IL-10 deficient mouse Murine J774.1 cells HT29 cells	*L. reuteri* *L. breve* *L. rhamnosus* GG *L. casei* *L. salivarius*	(Perdigon *et al.*, 2002) (Lammers *et al.*, 2002) (von der Weid *et al.*, 2001) (Lammers *et al.*, 2003) (Morita *et al.*, 2002b, Morita *et al.*, 2002a, Michail and Abernathy, 2002) (O'Hara *et al.*, 2006) (Vinderola, 2006)
Oral tolerance	Suppression of humoral and cellular responses	Murine	*L.paracasei* *L. johnsonii* *Bifidobacterium lactis*	(Prioult *et al.*, 2003)

cells (dendritic, T and B cells, macrophages) are dispersed throughout the epithelium and lamina propria. It is clear from experimental models that *Lactobacillus* strains differ greatly in their mechanisms of action, and that a single mechanism of action is not responsible for their clinical effects (Table 7.1). It has been demonstrated that orally administered *Lactobacillus* strains are incorporated into M cells in Peyer's patches, and are subsequently found in intracellular spaces of lymphocytes and macrophages adjacent to M cells (Kang and Conway, 2006).

Effects on epithelial barrier

Lactobacilli stimulate several epithelial cell protective responses, including enhancement of barrier function (Madsen *et al.*, 2001; Resta-Lenert and Barrett, 2006), mucin synthesis and secretion (Mack *et al.*, 2003; Otte and Podolsky, 2004), and cell survival (Yan and Polk, 2002). Intestinal epithelial cells are joined together by a series of intercellular junctions along their lateral membranes. At the uppermost apical surface, the belt-like apical junctional complex (AJC) connects epithelial cells, composed of a tight junction (TJ) and an adherens junction (AJ). Numerous TJ proteins have been described, including claudins, occludins, cingulin, junctional adhesion molecules (JAMs), the coxsackie adenovirus receptor, catenin, and E-cadherin (Utech *et al.*, 2006).

Lactobacillus spp. can influence intestinal barrier function through modulations of tight junctions. Madsen *et al.* (2001) have shown that the probiotic combination VSL#3, which contains several species of *Lactobacillus* as well as *Bifidobacteria*, was effective in enhancing barrier function both *in vitro* in cultured cells and *in vivo*. They also found that secreted bioactive factors from strains found in the VSL3 mixture were effective in preventing *Salmonella* invasion into epithelial cells (Madsen *et al.*, 2001). *Streptococcus thermophilus* and *Lactobacillus acidophilus* ATCC4356 have been shown to enhance phosphorylation of actinin and occludin in the tight junction and inhibit the invasion of *E. coli* into human intestinal epithelial cell lines (Resta-Lenert and Barrett, 2006). These effects were associated with modulation of ERK-1 and 2, p38, and PI3K, which are associated with the MAPK-dependent pathways. Montalto *et al.* (2004) found that heat-killed spent culture supernatant from *L. acidophilus* strain LB was able to prevent an acetylsalicylic acid-induced disruption of tight junction structure and reduced ZO-1 expression.

L. rhamnosus GG has been shown to modulate cell proliferation and apoptosis (Ichikawa *et al.*, 1999) via the release of specific protein factors (Yan, 2007). These soluble proteins activated the Akt/protein kinase B pathway (anti-apoptotic) in a phosphatidylinositol-3′-kinase (PI3K)-dependent manner and inactivated the pro-apoptotic p38 mitogen-activated protein kinase signalling pathway. These soluble proteins released from LGG have also been shown to induce cytoprotective heat shock protein synthesis in intestinal epithelial cells (Tao *et al.*, 2006).

Immunomodulatory effects on epithelial cells

Epithelial–microbial interactions generally involve toll-like receptors (TLR). The TLR family is a group of pattern-recognition receptors that are evolutionary conserved germline-encoded transmembrane proteins which recognize conserved motifs referred to as microbe-associated molecular patterns (MAMPs) (Harris *et al.*, 2006). TLRs detect multiple MAMPs, including lipopolysaccharide (LPS) by TLR4; lipoproteins and lipoteichoic acids by TLR2; flagellin by TLR5; CpG DNA by TLR9; dsRNA by TLR3; and ssRNA by TLR7. Gut epithelial cells express TLR1-TLR9, as well as the nucleotide-binding oligomerization domain (NOD) molecules, Nod1 and Nod2 (Otte and Podolsky, 2004). In addition to ligand specificity, individual TLRs differ in their expression patterns and the signal transduction pathways they activate appear to be cell-type dependent. The downstream effects of TLR activation include induction of host defence genes, which serve to stimulate expression of chemokines, cytokines, antigen-presenting molecules and other molecules involved in the process of immune signal transduction (Harris *et al.*, 2006). The specific mechanisms and TLR-signalling pathways involved in the interaction between lactobacillus and the surface epithelium are currently being elucidated. Lactobacilli modulate epithelial barrier function possibly through interactions with TLR2, which recognizes bacterial lipoproteins, zymosan and lipoteichoic acid (Kalliomaki and Walker, 2005). However, *Lactobacillus* can also interact with TLR9 through their DNA sequences. Bacterial DNA is recognized in a differential manner by epithelial cells (Jijon *et al.*, 2004); with pathogenic strains evoking a phosphorylation of the ERK pathway and activation of AP-1 (Akhtar *et al.*, 2003), and probiotic strains modulating the NF-κB pathway (Jijon *et al.*, 2004).

Kim *et al.* (2006) examined the effects of *L. casei* ATCC27139 and *L. casei* ATCC27139-J1R on mRNA expression levels of TLRs 1–9 in murine splenocytes exposed to *Listeria monocytogenes*. The former had a longer up-regulation of TLR-2 (up to 12 h) versus the latter (2 h with rapid decline by 4 h); there was no up-regulation of any other TLR. These data, therefore, support the role of lactobacillus in stimulating different TLRs in variable ways to produce differential patterns of cytokine secretion and resultant specific T-cell response. Given the breadth and novelty of the TLR family, further studies need to be performed to elucidate the specific relationships that lactobacillus shares with these receptors.

Lactobacilli also modulate cytokine release from epithelial cells in a dose- and strain-dependent manner. Zhang et al (Zhang *et al.*, 2005) investigated the ability of live and dead LGG to attenuate TNFα-induced IL-8 secretion from CaCo2 cells. They found that at doses of 10^{6-8} cfu/ml, IL-8 secretion was reduced; however, at higher concentrations, LGG actually increased IL-8 secretion above that seen with TNFα alone (Zhang *et al.*, 2005), further supporting the idea that determining the correct dosage is critical for successful therapeutic usage. Other studies have shown *L. reuteri* to reduce a TNFα-induced IL-8 secretion and also to up-regulate the levels of nerve growth factor (Ma *et al.*, 2004). Jijon *et al.* (2004) demonstrated that bacterial DNA from VSL3 (four lactobacilli, three bifidobacteria, one streptococcus) was able to decrease IL-8 secretion, delay NFκB activation, and stabilize IκB levels.

Nemeth *et al.* (2006) further sought to explain the potential beneficial effects at the epithelial-microbial interface utilizing a *Salmonella* model. They examined the effects of *L. casei* subsp. *casei* 2756, *L. curvatus* 2775, and *L. plantarum* 2142 as well as their spent culture supernatants (SCS) on *Salmonella enteritidis* 857 growth, interleukin (IL)-8 and Hsp70 synthesis in undifferentiated crypt-like and differentiated villus-like Caco-2 cells. When administered systemically, it was observed that lactobacilli suppressed the synthesis of inflammatory cytokines like IL-8 (Nemeth *et al.*, 2006). They hypothesized that the benefits of probiotics may be related to the induced expression of heat shock proteins (Hsp). Functional studies by Tao *et al.* (2006) confirmed that conditioned media from *Lactobacillus* GG induced heat shock protein expression in intestinal epithelial cells, and that LGG-CM modulated cellular signalling pathways including the activation of MAP kinase.

Effects in human studies

Although observations of the effects of lactobacilli on mucosal and systemic immune function are predominantly available in animal models at present, some studies have sought to assess the effects of probiotic supplementation on human immune function. A limited number of *Lactobacillus* strains have been tested for their ability to prevent and/or treat allergy in infants. A combination of *L. rhamnosus* 19070–2 and *L. reuteri* DSM 122460 significantly reduced the clinical scoring of atopic dermatitis in 1- to 13-year-old children with a positive skin prick test (Rosenfeldt *et al.*, 2003). Taylor et al (Taylor *et al.*, 2006) determined the effects of probiotics on the development of the innate immune system during infancy in humans. Children of mothers with documented allergies (clinical history of asthma, allergic rhinitis or eczema plus a positive skin prick test) were enrolled. They were administered probiotic (3×10^9 CFU *L. acidophilus* LAVRI-A1) or placebo for a period of 6 months. A trend for higher TNF-α levels in the children receiving *L. acidophilus* LAVRI-A1 ($P = 0.47$) was observed, along with alterations in IL-6, IL-10, TNF-α and IFN-γ levels. Another study used *L. rhamnosus* GG in at-risk infants. Pregnant allergic mothers were given *L. rhamnosus* GG or placebo from 2 to 4 weeks before their delivery date in a randomized double-blind trial. After birth, the children received *L. rhamnosus*

GG for 6 months. After 4 years, 46% of the children in the placebo group had developed atopic eczema, while only 26% in the probiotic group had symptoms (Kalliomaki *et al.*, 2003). Another study used a combination of *L. rhamnosus* GG, *L. rhamnosus* LC705, *Bifidobacterium breve*, and *Proprionbacterium freudenreichii* ssp. *shermanii* in a similar randomized, double-blind placebo-controlled trial in pregnant mothers and their high-risk infants (Kukkonen *et al.*, 2007). In this study, probiotics showed no overall preventative effect on allergic diseases in infants by 2 years of age, but did reduce eczema and IgE-associated eczema. Thus, at this time, although probiotic therapy appears to be a promising approach in the treatment and prevention of the onset of allergies in infants, further studies are necessary to determine the most efficacious strains to use and the best time to initiate treatment. A more detailed knowledge of the intestinal microflora of allergic and healthy infants also may provide insight into the mechanisms underlying individual responses to probiotic-based therapies.

Studies in adults have demonstrated *Lactobacillus*-induced effects on unspecific cellular immune responses as indicated by increased phagocytic activity of granulocytes and monocytes (Klein, 2008). This was accompanied by decreased serum triacylglycerols (Klein, 2008). Effects of orally administered heat-killed *Lactobacillus plantarum* on acquired immunity have also been observed in healthy adults (Hirose *et al.*, 2006).

Interactions with dendritic cells and T cells

Dendritic cells (DCs) represent the primary antigen-presenting cell population of the intestine. It has been shown that through the expression of pattern recognition receptors and TLRs, DCs may distinguish between different microbial strains. *Lactobacillus* sp alter DC maturation and cytokine release in a concentration and strain-dependent fashion (O'Mahony *et al.*, 2006). Different strains of *Lactobacillus* can induce dendritic cells to drive either a Th1, Th2, or Th3 response by mucosal T cells. This response depends on the state of stimulation of the DC and which cytokines and other co-factors are present. For example, whereas *L. casei* appears to elicit a predominant immunostimulatory response with an increased production of IL-12 and increased IFN-γ secretion by T-cells and natural killer (NK) cells, *L. reuteri* appears to have opposing effects, in that IL-12 production is suppressed and the antigen–MHC class II complex is repressed, thereby decreasing inflammation (Heufler *et al.*, 1996). Thus, the presence and concentration of each specific type of bacteria present within the lumen of the gastrointestinal tract will elicit specific cytokine patterns and immune responses, in part on the basis of their interactions with dendritic cells.

To assess the differential response of DCs to both pathogenic and commensal microorganisms in the state of disease, O'Mahony *et al.* (2006) examined the *in vitro* responses of human mononuclear cell and DC populations obtained from mesenteric lymph nodes (MLNCs) and peripheral blood mononuclear cells from patients with inflammatory bowel disease (IBD). This study matched the cytokine response profiles elicited by *L. salivarius* UCC118 and *B. infantis* 35624 against those seen with *Salmonella typhimurium* UK1. MLN cells exposed to *L. salivarius* and *B. infantis* responded with enhanced IL-10 secretion, while *S. typhimurium* significantly induced IL-12 and TNF-α secretion. TGF-β was upregulated only by *B. infantis* and not *L. salivarius*. Furthermore, the authors observed that there was also a differential cytokine response depending on the cellular environment or compartment. There was a significant increase in IL-12 and TNF-α secretion by peripheral blood mononuclear cells exposed to *L. salivarius*; this effect was not observed in MLNCs. Conversely, IL-10 secretion was significantly greater by MLCs as opposed to peripheral blood mononuclear cells induced by *L. salivarius*. Therefore, these results indicate that each strain of lactobacillus, and likely all probiotic organisms, is processed by dendritic cells differently, both within the intestine and systemically, to produce variable T-cell responses and perhaps serves to maintain immune homeostasis within a given host. Overall, probiotic bacteria interactions with dendritic cells leads to an immunoregulatory state, rather than aggressive immune responses.

Regulatory T cells (T_{reg}) appear to function in the suppression or regulation of effector T cell function through production of cytokines such as IL-10 and/or TGFβ, or through receptor-ligand interactions such as CTLA-4 (Paust and Cantor, 2005). To date, several $CD4^+$ T regulatory subsets have been identified, including Tr1, which produces high levels of IL-1; Th3, which produces TGFβ; and $CD4^+CD25^+$ T cells that mediate suppression in a contact-dependent manner through cytotoxic T cell-associated antigen 4 (Paust and Cantor, 2005). Induction of T_{reg} cells by certain *Lactobacillus* strains may explain how probiotics given by the subcutaneous route may be effective not only against inflammation in the gut, but also in the treatment of arthritis (Sheil *et al.*, 2004).

Interactions with macrophages and monocytes

Aside from DCs, other immune cells within the gut have also been shown to interact with probiotic bacteria. Macrophages represent a major source of cytokine production and scavenging of microbes and other antigens. They exist in the lamina propria in close proximity to Peyer's patches and line inter-epithelial layers. It has been demonstrated that direct communication exists between infiltrating monocytes and commensal organisms at the onset of or during the process of active inflammation of the intestine. These interactions are critical in the pathogenesis of certain inflammatory diseases (Mahida, 2000; Saniabadi *et al.*, 2003; Watanabe *et al.*, 2003). Kim *et al.* (2006) examined the cytokine immune response by macrophages in response to *L. rhamnosus* GG and *L. rhamnosus* GR-1, with particular attention being paid to the downstream signal transduction pathways activated by these two organisms. Both organisms were less potent inducers of TNF-α secretion than pathogenic *E. coli* GR-12 and *Enterococcus faecalis* 1311. Furthermore, stimulated macrophages released granulocyte colony stimulating factor (G-CSF), a potent TNFα suppressing agent. STAT-3 was strongly activated by GR-1-CM, which preferentially induced phosphorylation of STAT-3 at a tyrosine residue (Y705). *E. coli*-CM, however, induced phosphorylation at both Y705 and TNF-α serine (S727) amino acid residues; a pathway mediated via mitogen-activated protein kinases (MAPK) ERK1 and ERK2 (Jain *et al.*, 1998, Tian and An, 2004, Wierenga *et al.*, 2003). Utilizing macrophages deficient in both IL-10 ($\text{GR-1-CM}^{\text{IL-10-/-}}$) and G-CSF receptor ($\text{G-CSFR}^{-/-}$), the authors demonstrated that G-CSF is the predominant factor responsible for the phosphorylation events at Y705, which in turn is mediated via Janus kinase 2 (JAK2) (Kim *et al.*, 2006).

Conclusion

Currently, there is much interest in understanding both the organ specific mechanisms of immune modulation conferred by probiotics, as well as the systemic repercussions. There is increasing clinical evidence that the utility of probiotics has extended to not only include gastrointestinal luminal disease such as inflammatory bowel disease and gastrointestinal infections, but that the extra-intestinal effects induced by oral or systemic administration of probiotics have been utilized with varying degrees of success in other disorders including chronic liver disease (Loguercio *et al.*, 2005), multiple organ dysfunction syndrome (Alberda, 2007), and autoimmune disease (Marteau *et al.*, 2001, Matsuzaki, 2007). Our current understanding of interactions between lactobacilli and the mucosal and systemic immune system suggests that these interactions are necessary for the development of intestinal immune homeostasis. Immune responses to lactobacilli depend on the species under study, the concentration of bacteria utilized and the immunological compartment studied. As our understanding of the commensal organisms that reside within the human host and the complex reactions that exist between microbes and their host increases, the clinical utility of these bacteria will be improved. On the basis of the work completed to date, future studies may do well to carefully characterize the specific populations and concentrations of specific bacteria within the host, both in states of disease and health.

References

Akhtar, M., Watson, J.L., Nazli, A., and McKay, D.M. (2003). Bacterial DNA evokes epithelial IL-8 production by a MAPK-dependent, NF-kappaB-independent pathway. FASEB J. *17*, 1319–1321.

Alberda, C., Gramlich L., Meddings J., Field C., Mccargar L., Kutsogiannis D., Fedorak R., and Madsen K. (2007). Effects of probiotic therapy in critically ill patients: a randomized, double-blind, placebo-controlled trial. Am. J. Clin. Nutr. 85, 816–823.

Antikainen, J., Anton, L., Sillanpaa, J., and Korhonen, T.K. (2002). Domains in the S-layer protein CbsA of *Lactobacillus crispatus* involved in adherence to collagens, laminin and lipoteichoic acids and in self-assembly. Mol. Microbiol. *46*, 381–394.

Avall-Jaaskelainen, S., and Palva, A. (2005). *Lactobacillus* surface layers and their applications. FEMS Microbiol. Rev. 29, 511–529.

Baharav, E., Mor, F., Halpern, M. and Weinberger, A. (2004). *Lactobacillus* GG bacteria ameliorate arthritis in Lewis rats. J. Nutr. *134*, 1964–1969.

Bibiloni, R., Fedorak, R.N., Tannock, G.W., Madsen, K. L., Gionchetti, P., Campieri, M., De Simone, C., and Sartor, R.B. (2005). VSL#3 probiotic-mixture induces remission in patients with active ulcerative colitis. Am. J. Gastroenterol. *100*, 1539–1546.

Caridi, A. (2002). Selection of *Escherichia coli*-inhibiting strains of *Lactobacillus paracasei* subsp. *paracasei*. J. Indust. Microbiol. Biotechnol. *29*, 303–308.

Coconnier, M.H., Lievin, V., Lorrot, M., and Servin, A.L. (2000). Antagonistic activity of *Lactobacillus acidophilus* LB against intracellular *Salmonella enterica* serovar *Typhimurium* infecting human enterocyte-like Caco-2/TC-7 cells. Appl. Environ. Microbiol. *66*, 1152–1157.

Demirer, S., Aydintug, S., Aslim, B., Kepenekci, I., Sengul, N., Evirgen, O., Gerceker, D., Andrieu, M.N., Ulusoy, C., and Karahuseyinoglu, S. (2006). Effects of probiotics on radiation-induced intestinal injury in rats. Nutrition *22*, 179–186.

Di Marzio, L., Russo, F. P., D'Alo, S., Biordi, L., Ulisse, S., Amicosante, G., De Simone, C., and Cifone, M.G. (2001). Apoptotic effects of selected strains of lactic acid bacteria on a human T leukemia cell line are associated with bacterial arginine deiminase and/or sphingomyelinase activities. Nutr. Cancer 40, 185–196.

Ebanks, R.O., Goguen, M., Mckinnon, S., Pinto, D.M., and Ross, N.W. (2005). Identification of the major outer membrane proteins of *Aeromonas salmonicida*. Dis. Aquat. Organ. 68, 29–38.

Eijsink, V.G., Axelsson, L., Diep, D.B., Havarstein, L.S., Holo, H., and Nes, I.F. (2002). Production of class II bacteriocins by lactic acid bacteria; an example of biological warfare and communication. Antonie van Leeuwenhoek *81*, 639–654.

Fang, H., Elina, T., Heikki, A., and Seppo, S. (2000). Modulation of humoral immune response through probiotic intake. *FEMS Immunol. Med. Microbiol. 29*, 47–52.

Fernandez, M.F., Boris, S., and Barbes, C. (2003). Probiotic properties of human lactobacilli strains to be used in the gastrointestinal tract. J. Appl. Microbiol. *94*, 449–455.

Gill, H.S., and Rutherfurd, K.J. (2001) Immune enhancement conferred by oral delivery of *Lactobacillus rhamnosus* HN001 in different milk-based substrates. J. Dairy Res. *68*, 611–616.

Gionchetti, P., Rizzello, F., Venturi, A., Brigidi, P., Matteuzzi, D., Bazzocchi, G., Poggioli, G., Miglioli, M., and Campieri, M. (2000) Oral bacteriotherapy as maintenance treatment in patients with chronic pouchitis: a double-blind, placebo-controlled trial [see comment]. Gastroenterology *119*, 305–309.

Grangette, C., Nutten, S., Palumbo, E., Morath, S., Hermann, C., Dewulf, J., Pot, B., Hartung, T., Hols, P., and Mercenier, A. (2005). Enhanced antiinflammatory capacity of a *Lactobacillus plantarum* mutant synthesizing modified teichoic acids. Proc. Natl. Acad. Sci. USA *102*, 10321–1036.

Harris, G., Kuolee, R., and Chen, W. (2006). Role of Toll-like receptors in health and diseases of gastrointestinal tract. World J. Gastroenterol. *12*, 2149–2160.

Heufler, C., Koch, F., Stanzl, U., Topar, G., Wysocka, M., Trinchieri, G., Enk, A., Steinman, R.M., Romani, N., and Schuler, G. (1996) Interleukin-12 is produced by dendritic cells and mediates T helper 1 development as well as interferon-gamma production by T helper 1 cells. Eur. J. Immunol. *26*, 659–668.

Hirose, Y., Murosaki, S., Yamamoto, Y., Yoshikai, Y., and Tsuru, T. (2006). Daily intake of heat-killed *Lactobacillus plantarum* L-137 augments acquired immunity in healthy adults. J. Nutr. *136*, 3069–3073.

Huang, Y., and Adams, M.C. (2003) An in vitro model for investigating intestinal adhesion of potential dairy propionibacteria probiotic strains using cell line C2BBe1. Lett. Appl. Microbiol. *36*, 213–216.

Hynonen, U., Westerlund-Wikstrom, B., Palva, A. and Korhonen, T.K. (2002). Identification by flagellum display of an epithelial cell- and fibronectin-binding function in the SlpA surface protein of *Lactobacillus brevis*. J. Bacteriol. *184*, 3360–3367.

Ichikawa, H., Kuroiwa, T., Inagaki, A., Shineha, R., Nishihira, T., Satomi, S., and Sakata, T. (1999). Probiotic bacteria stimulate gut epithelial cell proliferation in rat. Dig. Dis. Sci. *44*, 2119–2223.

Isolauri, E., Arvola, T., Sutas, Y., Moilanen, E., and Salminen, S. (2000). Probiotics in the management of atopic eczema. Clin. Exp. Allergy *30*, 1604–10.

Jain, N., Zhang, T., Fong, S.L., Lim, C.P., and Cao, X. (1998). Repression of Stat3 activity by activation of mitogen-activated protein kinase (MAPK). Oncogene *17*, 3157–67.

Jakava-Viljanen, M., Avall-Jaaskelainen, S., Messner, P., Sleytr, U.B., and Palva, A. (2002). Isolation of three new surface layer protein genes (slp) from *Lactobacillus brevis* ATCC 14869 and characterization of the change in their expression under aerated and anaerobic conditions. J. Bacteriol. *184*, 6786–6795.

Jijon, H., Backer, J., Diaz, H., Yeung, H., Thiel, D., Mckaigney, C., De Simone, C., and Madsen, K. (2004). DNA from probiotic bacteria modulates murine and human epithelial and immune function. Gastroenterology *126*, 1358–1373.

Kaburagi, T., Yamano, T., Fukushima Y., Yoshino H., Mito N., and Sato, K. (2007). Effect of *Lactobacillus johnsonii* La1 on immune function and serum albumin in aged and malnourished aged mice. Nutrition 2, 342–350.

Kalliomaki, M., Salminen, S., Poussa, T., Arvilommi, H., and Isolauri, E. (2003). Probiotics and prevention of atopic disease: 4-year follow-up of a randomised placebo-controlled trial. Lancet *361*, 1869–1871.

Kalliomaki, M.A., and Walker, W.A. (2005). Physiologic and pathologic interactions of bacteria with gastrointestinal epithelium. Gastroenterol. Clin. North Am. *34*, 383–399, vii.

Kang, S.S., and Conway, P.L. (2006). Characteristics of the adhesion of PCC *Lactobacillus fermentum* VRI 003 to Peyer's patches. FEMS Microbiol. Lett. *261*, 19–24.

Kato, I., Endo-Tanaka, K., and Yokokura, T. (1998). Suppressive effects of the oral administration of *Lactobacillus casei* on type II collagen-induced arthritis in DBA/1 mice. Life Sci. 63, 635–644.

Kim, Y. G., Ohta, T., Takahashi, T., Kushiro, A., Nomoto, K., Yokokura, T., Okada, N., and Danbara, H. (2006) Probiotic *Lactobacillus casei* activates innate immunity via NF-kappaB and p38 MAP kinase signaling pathways. Microbes Infect. 8, 994–1005.

Klein, A., Friedrich U., Vogelsang H., and Jahreis, G. (2008). *Lactobacillus acidophilus* 74–2 and *Bifidobacterium animalis* sub sp *lactis* DGCC 420 modulate unspecific cellular immune response in healthy adults. Eur. J. Clin. Nutr. *62*, 584–593.

Kukkonen, K., Savilahti, E., Haahtela, T., Juntunen-Backman, K., Korpela, R., Poussa, T., Tuure, T., and Kuitunen, M. (2007). Probiotics and prebiotic galacto-oligosaccharides in the prevention of allergic diseases: a randomized, double-blind, placebo-controlled trial. J. Allergy Clin. Immunol. *119*, 192–198.

Kullisaar, T., Zilmer, M., Mikelsaar, M., Vihalemm, T., Annuk, H., Kairane, C., and Kilk, A. (2002). Two antioxidative lactobacilli strains as promising probiotics. Int. J. Food Microbiol. 72, 215–224.

Lammers, K. M., Brigidi, P., Vitali, B., Gionchetti, P., Rizzello, F., Caramelli, E., Matteuzzi, D., and Campieri, M. (2003). Immunomodulatory effects of probiotic bacteria DNA: IL-1 and IL-10 response in human peripheral blood mononuclear cells. FEMS Immunol. Med. Microbiol. *38*, 165–72.

Lammers, K.M., Helwig, U., Swennen, E., Rizzello, F., Venturi, A., Caramelli, E., Kamm, M.A., Brigidi, P., Gionchetti, P., and Campieri, M. (2002). Effect of probiotic strains on interleukin 8 production by HT29/19A cells. Am. J. Gastroenterol. *97*, 1182–1186.

Loguercio, C., Federico, A., Tuccillo, C., Terracciano, F., D'Auria, M. V., De Simone, C., and Del Vecchio Blanco, C. (2005). Beneficial effects of a probiotic VSL#3 on parameters of liver dysfunction in chronic liver diseases. J. Clin. Gastroenterol. 39, 540–3.

Ma, D., Forsythe, P., and Bienenstock, J. (2004). Live *Lactobacillus reuteri* is essential for the inhibitory effect on tumor necrosis factor alpha-induced interleukin-8 expression. Infect. Immun. 72, 5308–14.

Mack, D. R., Ahrne, S., Hyde, L., Wei, S. and Hollingsworth, M.A. (2003). Extracellular MUC3 mucin secretion follows adherence of *Lactobacillus* strains to intestinal epithelial cells in vitro. Gut 52, 827–833.

Madsen, K., Cornish, A., Soper, P., Mckaigney, C., Jijon, H., Yachimec, C., Doyle, J., Jewell, L., and De Simone, C. (2001). Probiotic bacteria enhance murine and human intestinal epithelial barrier function. Gastroenterology *121*, 580–591.

Mahida, Y.R. (2000) The key role of macrophages in the immunopathogenesis of inflammatory bowel disease. Inflamm. Bowel Dis. *6*, 21–33.

Marteau, P., Lemann, M., Seksik, P., Laharie, D., Colombel, J., Bouhnik, Y., Cadiot, G., Soule, J.C., Bourreille, A., Metman, E., Lerebours, E., Carbonnel, F., Dupas, J.L., Veyrac, M., Coffin, B., Moreau, J., Abitbol, V., Blum-Sperisen, S., and Mary, J.Y. (2006). Ineffectiveness of *Lactobacillus johnsonii* LA1 for prophylaxis of postoperative recurrence in Crohn's disease: a randomised, double blind, placebo controlled GETAID trial [see comment]. Gut 55, 842–847.

Marteau, P. R., De Vrese, M., Cellier, C.J. and Schrezenmeir, J. (2001). Protection from gastrointestinal diseases with the use of probiotics. Am. J. Clin. Nutr. 73, 430S–436S.

Matsuzaki, T., Takagi A., Ikemura H., Matsuguchi T., and Yokokura, T. (2007). Intestinal microflora: probiotics and autoimmunity. J. Nutr. *137*, 798S–802S.

Messner, P., Pum, D., Sara, M., Stetter, K. O. and Sleytr, U. B. (1986) Ultrastructure of the cell envelope of the archaebacteria *Thermoproteus tenax* and *Thermoproteus neutrophilus*. J. Bacteriol. *166*, 1046–1054.

Michail, S. and Abernathy, F. (2002). *Lactobacillus plantarum* reduces the in vitro secretory response of intestinal epithelial cells to enteropathogenic *Escherichia coli* infection. J. Pediatr. Gastroenterol. Nutr. 35, 350–355.

Mimura, T., Rizzello, F., Helwig, U., Poggioli, G., Schreiber, S., Talbot, I. C., Nicholls, R. J., Gionchetti, P., Campieri, M., and Kamm, M.A. (2004). Once daily high dose probiotic therapy (VSL#3) for maintaining remission in recurrent or refractory pouchitis. Gut 53, 108–114.

Moller, P.L., Paerregaard, A., Gad, M., Kristensen, N.N., and Claesson, M. H. (2005). Colitic scid mice fed *Lactobacillus* spp. show an ameliorated gut histopathology and an altered cytokine profile by local T cells. Inflammatory Bowel Disease *11*, 814–819.

Montalto, M., Maggiano, N., Ricci, R., Curigliano, V., Santoro, L., Di Nicuolo, F., Vecchio, F.M., Gasbarrini, A., and Gasbarrini, G. (2004). *Lactobacillus acidophilus* protects tight junctions from aspirin damage in HT-29 cells. Digestion *69*, 225–228.

Morath, S., Geyer, A., and Hartung, T. (2001) Structure-function relationship of cytokine induction by lipoteichoic acid from *Staphylococcus aureus*. J. Exp. Med. *193*, 393–3907.

Morath, S., Stadelmaier, A., Geyer, A., Schmidt, R.R., and Hartung, T. (2002). Synthetic lipoteichoic acid from Staphylococcus aureus is a potent stimulus of cytokine release. J. Exp. Med. *195*, 1635–1640.

Morita, H., He, F., Fuse, T., Ouwehand, A.C., Hashimoto, H., Hosoda, M., Mizumachi, K., and Kurisaki, J. (2002a). Adhesion of lactic acid bacteria to caco-2 cells and their effect on cytokine secretion. Microbiol. Immunol. *46*, 293–297.

Morita, H., He, F., Fuse, T., Ouwehand, A.C., Hashimoto, H., Hosoda, M., Mizumachi, K., and Kurisaki, J. (2002b). Cytokine production by the murine macrophage cell line J774.1 after exposure to lactobacilli. Biosci. Biotechnol. Biochem. *66*, 1963–1966.

Neish, A. S., Gewirtz, A. T., Zeng, H., Young, A.N., Hobert, M.E., Karmali, V., Rao, A.S. and Madara, J. L. (2000). Prokaryotic regulation of epithelial responses by inhibition of IkappaB-alpha ubiquitination [comment]. Science *289*, 1560–1563.

Nemeth, E., Fajdiga, S., Malago, J., Koninkx, J., Tooten, P., and Van Dijk, J. (2006). Inhibition of *Salmonella*-induced IL-8 synthesis and expression of Hsp70 in enterocyte-like Caco-2 cells after exposure to non-starter lactobacilli. Int. J. Food Microbiol. *112*, 266–274.

Nerstedta, N.E., Ohlsona, K., Håkanssona, J., Svenssona, L., Löwenadlera, B., Svenssona, U., and Mahlapuu, M. (2007). Administration of *Lactobacillus* evokes coordinated changes in the intestinal expression profile of genes regulating energy homeostasis and immune phenotype in mice. Br. J. Nutr. *97*(6), 1117–1127.

O'Hara, A.M., O'Regan, P., Fanning, A., O'Mahony, C., Macsharry, J., Lyons, A., Bienenstock, J., O'Mahony, L. and Shanahan, F. (2006). Functional modulation of human intestinal epithelial cell responses by *Bifidobacterium infantis* and *Lactobacillus salivarius*. Immunology *118*, 202–215.

O'Mahony, L., Feeney, M., O'Halloran, S., Murphy, L., Kiely, B., Fitzgibbon, J., Lee, G., O'Sullivan, G., Shanahan, F. and Collins, J.K. (2001). Probiotic impact on microbial flora, inflammation and tumour development in IL-10 knockout mice. Aliment. Pharmacol. Ther. *15*, 1219–1225.

O'Mahony, L., O'Callaghan, L., Mccarthy, J., Shilling, D., Scully, P., Sibartie, S., Kavanagh, E., Kirwan, W.O., Redmond, H.P., Collins, J.K., and Shanahan, F. (2006). Differential cytokine response from dendritic cells to commensal and pathogenic bacteria in different lymphoid compartments in humans. Am. J. Physiol. Gastrointest. Liver Physiol. *290*, G839–45.

Otte, J.M., and Podolsky, D.K. (2004). Functional modulation of enterocytes by gram-positive and gram-negative microorganisms. Am. J. Physiol. Gastrointest. Liver Physiol. *286*, G613–G626.

Paust, S., and Cantor, H. (2005). Regulatory T cells and autoimmune disease. Immunol. Rev. *204*, 195–207.

Perdigon, G., Maldonado Galdeano, C., Valdez, J. C. and Medici, M. (2002). Interaction of lactic acid bacteria with the gut immune system. Eur. J. Clin. Nutr. *56* (Suppl. 4), S21–S26.

Pessi, T., Sutas, Y., Hurme, M., and Isolauri, E. (2000). Interleukin-10 generation in atopic children following oral *Lactobacillus rhamnosus* GG. Clin. Exp. Allergy 1804–1808.

Prioult, G., Fliss, I., and Pecquet, S. (2003). Effect of probiotic bacteria on induction and maintenance of oral tolerance to beta-lactoglobulin in gnotobiotic mice. Clin. Diagn. Labo. Immun. *10*, 787–92.

Resta-Lenert, S., and Barrett, K.E. (2006). Probiotics and commensals reverse TNF-alpha- and IFN-gamma-induced dysfunction in human intestinal epithelial cells. Gastroenterology *130*, 731–46.

Rolfe, V.E., Fortun, P.J., Hawkey, C.J. and Bath-Hextall, F. (2006). Probiotics for maintenance of remission in Crohn's disease. Cochrane Database of Systematic Reviews, CD004826.

Rosenfeldt, V., Benfeldt, E., Nielsen, S.D., Michaelsen, K.F., Jeppesen, D.L., Valerius, N.H. and Paerregaard, A. (2003). Effect of probiotic *Lactobacillus* strains in children with atopic dermatitis. J. Allergy Clin. Immunol. *111*, 389–395.

Saniabadi, A.R., Hanai, H., Takeuchi, K., Umemura, K., Nakashima, M., Adachi, T., Shima, C., Bjarnason, I., and Lofberg, R. (2003). Adacolumn, an adsorptive carrier based granulocyte and monocyte apheresis device for the treatment of inflammatory and refractory diseases associated with leukocytes. Ther. Apher. Dial. *7*, 48–59.

Schaer-Zammaretti, P. and Ubbink, J. (2003). Imaging of lactic acid bacteria with AFM – elasticity and adhesion maps and their relationship to biological and structural data. Ultramicroscopy *97*, 199–208.

Sheil, B., Mccarthy, J., O'Mahony, L., Bennett, M.W., Ryan, P., Fitzgibbon, J. J., Kiely, B., Collins, J.K. and Shanahan, F. (2004). Is the mucosal route of administration essential for probiotic function? Subcutaneous administration is associated with attenuation of murine colitis and arthritis.[see comment][erratum appears in Gut 2004 Aug; 53(8), 1216]. Gut *53*, 694–700.

Shibolet, O., Karmeli, F., Eliakim, R., Swennen, E., Brigidi, P., Gionchetti, P., Campieri, M., Morgenstern, S., and Rachmilewitz, D. (2002) Variable response to probiotics in two models of experimental colitis in rats. Inflammatory Bowel Diseases *8*, 399–406.

Sillanpaa, J., Martinez, B., Antikainen, J., Toba, T., Kalkkinen, N., Tankka, S., Lounatmaa, K., Keranen, J., Hook, M., Westerlund-Wikstrom, B., Pouwels, P.H., and Korhonen, T.K. (2000). Characterization of the collagen-binding S-layer protein CbsA of *Lactobacillus crispatus*. J. Bacteriol. *182*, 6440–6450.

Tao, Y., Drabik, K. A., Waypa, T. S., Musch, M. W., Alverdy, J. C., Schneewind, O., Chang, E.B. and and Petrof, E.O. (2006). Soluble factors from *Lactobacillus* GG activate MAPKs and induce cytoprotective heat shock proteins in intestinal epithelial cells [erratum appears in Am. J. Physiol. Cell Physiol. 2006 291(1): C194]. Am. J. Physiol. Cell Physiol. *290*, C1018–C1030.

Taylor, A., Hale, J., Wiltschut, J., Lehmann, H., Dunstan, J.A. and Prescott, S.L. (2006). Evaluation of the effects of probiotic supplementation from the neonatal period on innate immune development in infancy. Clin. Exp. Allergy *36*, 1218–1226.

Tian, Z. J. and An, W. (2004). ERK1/2 contributes negative regulation to STAT3 activity in HSS-transfected HepG2 cells. Cell Res. *14*, 141–147.

Toba, T., Virkola, R., Westerlund, B., Bjorkman, Y., Sillanpaa, J., Vartio, T., Kalkkinen, N., and Korhonen, T.K. (1995). A collagen-binding S-layer protein in *Lactobacillus crispatus*. Appl. Environ. Microbiol. *61*, 2467–2471.

Utech, M., Bruwer, M., and Nusrat, A. (2006). Tight junctions and cell–cell interactions. Methods Mol. Biol. *341*, 185–195.

Van Gossum, A., Dewit O., Louis E., De Hertogh G., Baert F., Fontaine F., Devos M., Enslen M., Paintin M., and Franchimont, D. (2007). Multicenter randomized-controlled clinical trial of probiotics on early endoscopic recurrence of Crohn's disease after ileo-caecal resection. Inflammatory Bowel Disease *13*(2), 135–142.

Viljanen, M., Pohjavuori, E., Haahtela, T., Korpela, R., Kuitunen, M., Sarnesto, A., Vaarala, O., and Savilahti, E. (2005a). Induction of inflammation as a possible mechanism of probiotic effect in atopic eczema-dermatitis syndrome. J. Allergy Clin. Immunol. *115*, 1254–1259.

Viljanen, M., Savilahti, E., Haahtela, T., Juntunen-Backman, K., Korpela, R., Poussa, T., Tuure, T. and Kuitunen, M. (2005b). Probiotics in the treatment of atopic eczema/dermatitis syndrome in infants: a double-blind placebo-controlled trial [see comment]. Allergy *60*, 494–500.

Vinderola, G., Perdigon, G., Duarte, J., Farnworth, E., and Matar, C. (2006). Effects of the oral administration of the exopolysaccharide produced by *Lactobacillus kefiranofaciens* on the gut mucosal immunity. Cytokine *36*, 254–260.

Von Der Weid, T., Bulliard, C., and Schiffrin, E. J. (2001). Induction by a lactic acid bacterium of a population of CD4(+) T cells with low proliferative capacity that produce transforming growth factor beta and interleukin-10. Clin. Diagn. Lab. Immunol. *8*, 695–701.

Watanabe, N., Ikuta, K., Okazaki, K., Nakase, H., Tabata, Y., Matsuura, M., Tamaki, H., Kawanami, C., Honjo, T. and Chiba, T. (2003). Elimination of local macrophages in intestine prevents chronic colitis in interleukin-10-deficient mice. Dig. Dis. Sci. *48*, 408–414.

Wierenga, A. T., Vogelzang, I., Eggen, B. J. and Vellenga, E. (2003). Erythropoietin-induced serine 727 phosphorylation of STAT3 in erythroid cells is mediated by a MEK-, ERK-, and MSK1-dependent pathway. Exp. Hematol. *31*, 398–405.

Yan, F., Cao, H., Cover, T., Whitehead, R., Washington, M., and Polk, D.B. (2007) Soluble proteins produced by probiotic bacteria regualte intestinal epithelial cell survival and growth. Gastroenterology *132*, 562–575.

Yan, F., and Polk, D.B. (2002). Probiotic bacterium prevents cytokine-induced apoptosis in intestinal epithelial cells. J. Biol. Chem. *277*, 50959–50965.

Zhang, L., Li, N., Caicedo, R., and Neu, J. (2005). Alive and dead *Lactobacillus rhamnosus* GG decrease tumor necrosis factor-alpha-induced interleukin-8 production in Caco-2 cells. J. Nutr. *135*, 1752–1756.

Lactic Acid Bacteria: Probiotics With Anti-Cancer Activities

Chandra Iyer and James Versalovic

Abstract

Beneficial bacteria include *Lactobacillus* and *Bifidobacterium* spp. and other lactic acid bacteria (LAB) commonly known as probiotics. LAB possesses numerous potential therapeutic properties including anti-inflammatory and anti-cancer activities and other features of interest. In recent years, studies with *in vitro* cell culture and animal models that clearly demonstrated protective effects of LAB for anti-tumour and anti-cancer effects. Dietary administration of LAB alleviated the risks of certain types of cancers and suppressed colonic tumour incidence, volume and multiplicity induced by various carcinogens in different animal models. Oral administration of LAB effectively reduced DNA adduct formation, ameliorated DNA damage and prevented putative preneoplastic lesions such as aberrant crypt foci induced by chemical carcinogens in the gastrointestinal (GI) tract of various animal models. LAB also increased the latency period and survival rates in test animals when challenged with carcinogenic agents. Reports also indicated that LAB cultures administered to animals inhibited liver, colon, bladder and mammary tumours, highlighting potential systemic effects of probiotics with anti-neoplastic activities.

Evidence from multiple human trials supports the notion that selected LAB strains may have anti-cancer properties. Although some studies suggested that certain LAB did not yield evidence of anti-cancer activities (Kampman *et al.*, 1994a,b), other reports favour the role of LAB in cancer prevention. Consumption of large quantities of dairy products containing *Lactobacillus* or *Bifidobacterium* has been reported to yield a lower incidence of colon cancer (Ishikawa *et al.*, 2005; Roller *et al.*, 2007; Takeda and Okumura, 2007). Oral administration of *Lactobacillus casei* demonstrated a marked suppressive effect on the urinary mutagenicity arising from ingestion of fried ground beef in humans (Hayatsu and Hayatsu, 1993; Hirayama and Rafter, 2000; Lidbeck *et al.*, 1992). *Lactobacillus* spp. also reduced the specific activities of faecal enzymes (microbial enzymes that are associated with carcinogen production in the gut) in human volunteer studies (Ayebo *et al.*, 1980; Brigidi *et al.*, 2001; Goldin and Gorbach, 1984b; Goldin *et al.*, 1980; Ling *et al.*, 1992; Ouwehand *et al.*, 2002; Spanhaak *et al.*, 1998). However, epidemiological studies supporting the association between fermented dairy products containing LAB and cancer prevention are very limited. An epidemiological study performed in Finland demonstrated an inverse relationship between the frequency of yogurt consumption and fermented milk products and the incidence of colon cancer, despite a high fat intake (Maclennan and Jensen, 1977). A relationship between yogurt consumption and reduced risk of colon cancer in a general population has also been reported (Boutron *et al.*, 1996; Malhotra, 1977; Peters *et al.*, 1992; Young and Wolf, 1988). In another case–control study, LAB have been shown to reduce the incidence of breast cancer in women (Le *et al.*, 1986; van't Veer *et al.*, 1989).

Potential mechanisms of action

Although numerous scientific reports support the anti-cancer and antitumorigenic activities of LAB, the mechanisms of action are still being explored. Proposed mechanisms of action include alteration of intestinal microbiota and its metabolic activities, alteration of physico-chemical conditions in the GI tract, binding and degradation of potential carcinogens, production of antitumorigenic or antimutagenic compounds, enhancement of the host's immune response, anti-inflammatory activities and induction of apoptosis. Different LAB strains may be exerting effects at different stages of carcinogenesis. Potential anti-cancer mechanisms of LAB are shown in Fig. 8.1 and discussed in more detail below.

Modulation of intestinal microbiota and associated metabolic activities

The GI tract represents a complex ecosystem comprised of hundreds of microbial species which can be broadly categorized into two major groups, (i) potentially pathogenic microbes or (ii) health-promoting, beneficial microorganisms. A third group may include transient or indigenous organisms, which lack pathogenic or beneficial effects. Bacterial pathogens may predispose an individual to enteric infections, inflammatory bowel disease (IBD), or cancer of the GI tract (Ellmerich *et al.*, 2000; Fox and Wang, 2007; Geypens *et al.*, 1997; Hope *et al.*, 2005; Horie *et al.*, 1999; Huycke and Gaskins, 2004; McGarr *et al.*, 2005; Swidsinski *et al.*, 1998). Animal studies have shown that specific members of the commensal microbiota may produce potently mutagenic metabolites (Carman *et al.*, 1988). Certain commensal bacteria may promote chemically induced carcinogenesis by increasing the rate of tumour progression (Onoue *et al.*, 1997). Evidence from germ-free mice supports the importance of the intestinal microbiota in the induction of intestinal inflammation and subsequent cancer (Engle *et al.*, 2002; Kado *et al.*, 2001; Sellon *et al.*, 1998; Takaku *et al.*, 1998). Studies by Balish and Warner (2002) have shown that certain commensal bacteria are capable of inducing IBD, dysplasia and adenocarcinoma in IL-10 knockout mice, whereas no pathology was observed in their germ-free counterparts. Overall, these studies provide compelling evidence that certain commensal microbiota may play an important role in the development of colorectal cancer.

On the other hand, beneficial bacteria such as lactobacilli and bifidobacteria, may promote intestinal health by modulating the gut microbiota and intestinal community-associated metabolic activities (Bengmark, 2002; Guarner, 2006; Isolauri *et al.*, 2004; Ljungh and Wadstrom, 2006; Shi and Walker, 2004; Versalovic and Relman, 2006). Lactobacilli and bifidobacteria may modulate metabolic activities of the gut microbiota by three possible mechanisms: (Commane *et al.*, 2005)

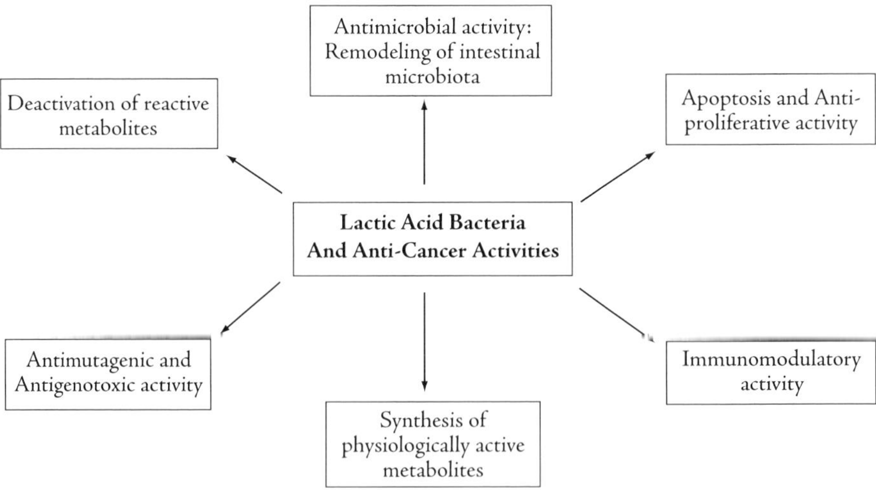

Figure 8.1 Potential anti-cancer mechanisms of LAB. The diagram illustrates potential mechanisms of action of probiotic LAB in the context of cancer.

- competing with and displacing members of the gut microbiota;
- production of antimicrobial agents (such as bacteriocins and bacteriocin-like components);
- production of short-chain fatty acids (SCFAs) and other metabolites.

Dietary supplementation with LAB have beneficial effects on the intestinal microecology by suppression of potentially pathogenic and proinflammatory intestinal microorganisms such as *Clostridium*, *Bacteroides* and coliforms (Ayebo *et al.*, 1980; Doron and Gorbach, 2006; Forestier *et al.*, 2001; Johnson-Henry *et al.*, 2004; Johnson-Henry *et al.*, 2005; Olivares *et al.*, 2006; Shahani and Ayebo, 1980; Sullivan and Nord, 2005). Bifidobacterial isolates of human origin adhered to human enterocytes (Caco-2 cells) and inhibited epithelial cell invasion by various intestinal pathogens (Bernet *et al.*, 1993). Adherence of *Lactobacillus acidophilus* prevented binding of different enteric pathogens to human intestinal epithelial cells (IECs) (Bernet *et al.*, 1994; Chauviere *et al.*, 1992). Reports also indicate that lactobacilli have differential capabilities with respect to attachment to human IECs (Kleeman and Klaenhammer, 1982). Diminished epithelial barrier integrity may be associated with neoplastic progression and LAB may improve barrier function *in vivo* (Garcia-Urkia *et al.*, 2002; Ko *et al.*, 2007; Madsen, 2001; Parassol *et al.*, 2005; Qin *et al.*, 2005; Resta-Lenert and Barrett, 2003; Zareie *et al.*, 2006). Interestingly, feeding animals with non-digestible polysaccharides (i.e. prebiotics) selectively stimulated the proliferation of beneficial bacteria such as LAB in the GI tract (Gibson and Roberfroid, 1995). Potential prebiotics include inulin, fructo-oligosaccharides, galacto-oligosaccharides, and xylo-oligosaccharides (Swennen *et al.*, 2006). Oral administration of prebiotics yielded anti-carcinogenic effects in multiple studies (Femia *et al.*, 2002; Hsu *et al.*, 2004; Hughes and Rowland, 2001; Pool-Zobel, 2005; Reddy, 1998; Taper and Roberfroid, 2005). Probiotics with prebiotics may be mixed to form synergistic combinations such as synbiotics. When combined with prebiotics, probiotics may deliver more potent anti-carcinogenic effects (Challa *et al.*, 1997; Femia *et al.*, 2002; Gallaher and Khil, 1999; Geier *et al.*, 2006; Le Leu *et al.*, 2005; Rafter *et al.*, 2007; Roller *et al.*, 2007; Roller *et al.*, 2004; Rowland *et al.*, 1998).

LAB produce antimicrobial agents such as bacteriocins that target pathogenic microbes and may remodel intestinal communities (Corr *et al.*, 2007; Ennahar *et al.*, 1999; Forestier *et al.*, 2001; Rodriguez *et al.*, 2002; Talarico and Dobrogosz, 1989; Vincent *et al.*, 1959). Shifts in the intestinal bacterial communities may result in reduced inflammation, cell proliferation or cancer risk. The major classes of bacteriocins produced by LAB include: (i) lantibiotics, (ii) small heat-stable peptides, (iii) large heat-labile proteins, and (iv) proteins whose activity requires the presence of carbohydrate or lipid moieties (Klaenhammer, 1993). The best-studied LAB-derived bacteriocin is nisin A, a 34-residue antibacterial peptide and lantibiotic that is produced by several strains of *Lactococcus lactis* and strongly inhibits the growth of a wide range of gram-positive bacteria (Hurst, 1967; Jung *et al.*, 1992). Nisin A demonstrated potential in treating peptic ulcers by inhibiting the growth and colonization by a gastric pathogen, *Helicobacter pylori* (Delves-Broughton *et al.*, 1996). The inhibitory effect of nisin A on *H. pylori* highlights the potential of bacteriocins to suppress cancer-promoting bacterial pathogens.

Large bowel cancer may be influenced by alterations or reductions of intestinal pH, thereby suppressing the growth of putrefactive bacteria (Modler *et al.*, 1990). In addition, commensal microbiota are involved in fermentation reactions yielding SCFAs such as lactate and acetate, reducing intestinal pH and providing energy sources (butyrate) for colonocytes (Cummings and Macfarlane, 2002). The primary SCFA metabolites of LAB are aceto-acetate, acetate and lactate. Lactate and aceto-acetate may serve as substrates for other bacterial components of the microbiota and may be degraded into other SCFAs such as butyrate and propionate. Luminal SCFAs, in particular butyrate, are potential anti-carcinogenic agents within the gut. A positive correlation between presence of SCFAs (e.g. butyrate and propionate) and crypt development in the colonic mucosa has been reported in *in vitro* models (Lupton and Kurtz, 1993). Addition of prebiotics such as fructo-oligosaccharides, galacto-oligosaccharides, and xylo-oligosaccharides induced saccharolytic bacteria to produce butyrate (Manning and Gibson, 2004). Dietary alterations with prebiotics may facilitate a shift from proteolytic to saccharolytic fermentation in clostridia and *Bacteroides* species. This beneficial shift results in increased production of SCFAs, thus lowering the colonic pH. In rats given inulin-containing diets with or without *Bifidobacterium longum*, a reduction in colonic pH was observed (Rowland *et al.*, 1998), although some studies did not detect a significant change in intestinal pH (Abdelali *et al.*, 1995; Bartram *et al.*, 1994; Figler *et al.*, 2006). In another case–control study, oral administration of LAB significantly reduced the faecal pH of patients associated with colonic adenomas (Biasco *et al.*, 1991).

Deactivation of reactive metabolites by LAB

Three different mechanisms have been proposed by which LAB deactivate reactive metabolites:

- modification of potentially carcinogenic enzymes;
- altering bile composition;
- anti-oxidative activities.

Modification of potentially carcinogenic enzymes

Experimental animal and human studies show that LAB supplementation can modify bacterial enzymatic activities in the GI tract. Many chemical compounds are detoxified by glucuronide formation in the liver before entering the intestine via the bile. The bacterial enzyme β-glucuronidase has the ability to hydrolyse many glucuronides and may thus liberate carcinogenic aglycones in the intestinal lumen. Several other bacterial enzymes such as β-glycosidases, azo- and nitroreductases, arylsulphatases, and alcohol dehydrogenases have been implicated in carcinogenesis by facilitating release of carcinogens in the intestinal tract (McBain and Macfarlane, 1998; Rowland, 2004). Interestingly, LAB supplementation in rodents significantly decreased faecal enzymatic activities that are associated with carcinogen production (Abdelali *et al.*, 1995; de Moreno de LeBlanc and Perdigon, 2005; Goldin and Gorbach, 1984b; Kulkarni and Reddy, 1994; Rowland *et al.*, 1998). In addition, LAB increased the colonic enzyme NADPH-cytochrome P-450 reductase activities; that are involved in the metabolism of carcinogens in animals (Pool-Zobel *et al.*, 1996). Renner and Munzner (1991) reported that a multicomponent probiotic mixture (containing different LAB strains) suppressed mutagenicity by suppressing potentially carcinogenic enzymes. Fermented milk products containing single or multiple LAB reduced the specific activities of faecal enzymes in human subjects (Ayebo *et al.*, 1980; Bouhnik, 2001; Goldin *et al.*, 1980; Ling *et al.*, 1994; Marteau *et al.*, 1990; Spanhaak *et al.*, 1998). Goldin and Gorbach (1984b) studied the effects of feeding two different *L. acidophilus* strains on activities of three bacterial enzymes including β-glucuronidase, nitroreductase and azoreductase in 21 healthy volunteers. Both strains had similar effects and caused significant declines in the specific activities of three enzymes in all subjects after a 10-day feeding trial. However, enzymatic functions recovered within days of stopping *Lactobacillus* feeding, suggesting that continuous consumption of LAB is necessary to maintain the benefits. Similarly, Ling *et al.* (1992) observed modified faecal enzyme activities in elderly volunteers consuming a *Lactobacillus rhamnosus* (LGG) preparation. Significant reductions in glycolic acid hydrolase, β-glucuronidase and urease activities were recorded. *Lactobacillus acidophilus* contributed to the detoxification of various carcinogens such as polycyclic aromatic hydrocarbons and heterocyclic aromatic amines (Goldin and Gorbach, 1984a). The ingestion of *B. longum* significantly inhibited ornithine decarboxylase (ODC) activity, a rate-limiting enzyme crucial in polyamine biosynthetic pathway (Singh *et al.*, 1997). LAB consumption significantly reduced faecal ammonia (a putative tumour promoter) concentrations in human subjects (De Preter *et al.*, 2006). Overall, these animal and human studies indicate that oral feeding of certain LAB strains to mammals can result in reductions of quantities of faecal enzyme activities and other carcinogens that may be involved in neoplastic progression. Studies have suggested that alterations in intestinal enzymatic functions may reduce cancer risk in humans, and LAB strains may indirectly suppress cancer development by reducing host enzymatic functions.

Altering the bile composition

Bile acids exert a cytotoxic effect upon intestinal epithelial cells, resulting in enhanced cell proliferation and a potentially higher probability of colonic cancer development (Cheng and Raufman, 2005; Nagengast *et al.*, 1995). Cytotoxic effects have been attributed to elevated concentrations of bile acids in the colon. Bile acids in the colon are initially deconjugated and undergo a dehydroxylation reaction of the 7α-hydroxyl group, forming potentially detrimental secondary bile acids such as deoxycholic and lithocholic acids. Modulation of the intestinal microbiota through LAB consumption affected the activity of 7α-dehydroxylase (Kitahara *et al.*, 2000). De Boever *et al.* (2000) reported that treatment with *L. reuteri* activated secondary bile acids and protected against toxic effects, potentially by binding bile salts. Administration of *L. acidophilus*-fermented milk supplements to colon cancer patients for 6 weeks resulted in trends towards lower concentrations of soluble bile acids in faeces (Lidbeck *et al.*, 1992). Anticancer activities of LAB may have resulted from decreased levels of bile acids in the intestine.

Anti-oxidative activities

Reactive oxygen intermediates (ROIs) are derivatives of molecular oxygen and commonly include superoxide radicals, hydrogen peroxide, hypochlorous acid, singlet oxygen, and the hydroxyl radical. ROIs are produced during cell metabolism and may react with lipids or proteins to generate intermediates that cause DNA damage. Base modifications and DNA strand breakage, as a result of ROI production results in chromosomal instability in the form of mutations, deletions, sister chromatid exchanges and chromosomal translocations (Marnett, 2000; Shacter and Weitzman, 2002). Increasing evidence supports the role of ROIs in the development of colorectal cancer (Owen *et al.*, 2000). Evidence from scientific studies shows that metabolic activities of the colonic microbiota may be responsible for production of ROIs, resulting in oxidative DNA damage in colonocytes. The generation of ROIs was demonstrated in the faecal matrix. The lack of ROIs present in autoclaved faeces led to the suggestion that ROIs are largely bacterial metabolites (Babbs, 1990). Soluble polysaccharides derived from *L. acidophilus* exert significant anti-oxidative activities on a variety of cancer-derived cell lines (Choi *et al.*, 2006; Kullisaar *et al.*, 2002; Lin and Chang, 2000).

Antitumorigenic, antimutagenic and antigenotoxic activities

Numerous studies have described the anticarcinogenic potential of LAB strains (Biffi *et al.*, 1997; Kato *et al.*, 1994; Perdigon *et al.*, 2002; Rachid *et al.*, 2006; Reddy *et al.*, 1973). Dietary administration of *Lactobacillus casei* strain Shirota inhibited methylcholanthracene-induced tumour development in mice (Matsuzaki and Chin, 2000; Takagi *et al.*, 2001). *Lactobacillus* spp. interfered with the initiation or early promotional stages of chemical carcinogens such as 1,2-dimethylhydrazine (DMH) and azoxymethane (AOM) (Goldin *et al.*, 1996; Rowland *et al.*, 1998). Heat-killed cells of *L. acidophilus* demonstrated profound inhibitory effects on cancer cell growth (Kim, 2002). Rachid *et al.* (2006) reported a significant reduction in colonic tumour growth at 2 and 7 days following feeding of *Lactobacillus helveticus*. Lim *et al.* (2002) were able to inhibit the subcutaneous development of an implanted bladder tumour cell line in mice fed LGG, compared with control mice. In double-blind studies of human cancer patients fed *L. casei* preparations, Aso *et al.* (1995) reported the suppression of bladder tumour recurrence. *B. longum* supplementation reduced the extent of colon and liver carcinogenesis induced by 2-amino-3-methylimidazo [4,5-f] quinoline as well as AOM-induced colon cancer in rats (Reddy and Rivenson, 1993; Singh *et al.*, 1997). Fermented milk containing *L. casei* inhibited the growth of tumour cells injected into mice (Kato *et al.*, 1981). Reports from other studies showed that repeated intralesional or subcutaneous injections of *B. infantis* inhibited the growth of Meth-A tumour cells transplanted subcutaneously into syngeneic BALB/c mice, resulting in tumour regression in the mice (Kohwi *et al.*, 1978). Sekine *et al.* (1985) reported that a single subcutaneous injection of peptidoglycan isolated from *B. infantis* significantly suppressed tumour progression by up to 70% in mice.

Aberrant crypt foci (ACF) are putative precursor lesions that may develop into colonic adenomas and carcinomas. ACF were consistently observed in experimentally induced colon carcinogenesis in laboratory animals and in the intestinal mucosa of patients with colon cancer (McLellan *et al.*, 1991; Pretlow *et al.*, 1992). ACF often contain mutations in the *APC* gene and *ras* oncogene, and these mutations may contribute to multistep progression to colonic adenocarcinoma (Vivona *et al.*, 1993). Inhibitors of ACF reduced the incidence of colonic tumours in laboratory animals (Wargovich *et al.*, 1996). *Bifidobacterium longum* reduced the formation of ACF (Reddy, 1998; Singh *et al.*, 1997). Ingestion of *B. longum* significantly inhibited the expression of the Ras-p21 oncoprotein (Singh *et al.*, 1997). Tien *et al.* (2006) reported that *L. casei* pretreatment up-regulated tumour suppressor genes, *sel-1* and *gp96*, by microarray analysis.

Ability of LAB to bind to various food mutagens including 3-amino-1-methyl 5 h pyrido [4,3-b] indole acetate (Trp-P-2), 2-amino-3-methylimidazo [4,5-f] quinoline (IQ), 2-amino-3,8-demethylimidazo [4,5-f] quinoxaline and 2-amino-1-methyl-6-phenylimidazo [4,5-b] pyridine have been reported by different groups (Hosono *et al.*, 1986; Orrhage *et al.*, 1994; Renner and Munzner, 1991). Studies also showed that the mutagenicity of Trp-P-2 for *S. typhimurium* was inhibited by the addition of *L. casei* to the reaction mixture, indicating that bound Trp-P-2 caused fewer mutations (Morotomi and Mutai, 1986). Zhang and Ohta (1993) showed that freeze-dried cells of LAB significantly reduced the absorption of Trp-P-1 from the small intestine in rats, resulting in decreased levels of Trp-P-1 in peripheral blood. LAB (*L. acidophilus* and *B. longum*) intake reduced uptake of the food mutagen Trp-P-2 and its metabolites in various tissues in mice (Orrhage *et al.*, 2002). Results from *in vitro* studies suggested that LAB supplements influence the uptake or excretion of mutagens by adsorption within the intestine. Carcinogens were bound by intestinal bacteria or LAB, but the level of binding was reported to be pH-dependent. *In vivo* evidence that LAB bind carcinogens is less conclusive. Oral administration of *L. rhamnosus* and *L. casei* to humans reduced the quantities of urinary mutagens such as heterocyclic amines and aflatoxin B (1)-N (7)-guanine (AFB-N (7)-guanine), thus suggesting the possibility that these compounds are not being absorbed in the intestine (El-Nezami *et al.*, 2006; Hayatsu and Hayatsu, 1993). In another study, the antimutagenicity of milk fermented with *L. helveticus* was related to the proteolytic activity of this strain (Matar *et al.*, 1997). All of these experiments suggest that probiotic consumption may affect the initiation phase of carcinogenesis, by decreasing epithelial cell exposure to activated faecal carcinogens.

Pool-Zobel *et al.* (1993) demonstrated the abilities of *L. casei* Shirota to inhibit DNA damage in the colons of rats exposed to the mutagen *N*-methyl-*N*-nitro, *N*-nitrosoguanidine. A subsequent study by the same group confirmed the antigenotoxic effects for different species of lactobacilli in rats challenged with the colonic carcinogen DMH (Pool-Zobel *et al.*, 1996). Bolognani *et al.* (1997) found that oral administration of LAB with heterocyclic amine carcinogens IQ and Trp-P-2 and with the polycyclic aromatic hydrocarbon benzo-α-pyrene did not reduce the hepatotoxicity and failed to lower blood or hepatic concentrations of these chemicals. The lack of effect was attributed to the reversible nature of carcinogen binding under conditions of neutral or alkaline pH.

Immunomodulatory activities

LAB directly modulates cytokine networks and innate immunity by multiple mechanisms and effects were dose-and strain-dependent (Yasui *et al.*, 1999). LAB stimulated immunoglobulin A (IgA) production, macrophage activation and phagocytosis, antigen uptake in Peyer's patches, proinflammatory cytokine (TNF-α, IFN-γ, and IL-12) secretion, and regulatory cytokine (IL-4, IL-10) production (Christensen *et al.*, 2002, Gill *et al.*, 2001; Gorbach, 2000; Hessle *et al.*, 1999; Kato *et al.*, 1994; Matar *et al.*, 2001; Pessi *et al.*, 2000; Tejada-Simon *et al.*, 1999). Kim *et al.* (2006) reported that both nuclear factor kappa-B (NF-κB) and Mitogen-Activated Protein Kinase (p38 MAPK) signalling pathways play important roles in the augmentation of innate immunity by probiotic *L. casei*. Probiotic strain LGG activated NF-κB and signal transducer and activator of transcription (STAT) signalling pathways in human macrophages (Miettinen *et al.*, 2000).

Cytotoxicity mediated by natural killer (NK) cells, plays an important role in inhibiting experimental tumour metastasis, and diminished NK cell activities result in a high incidence of tumour occurrence (Takeuchi *et al.*, 2001). Studies reported increased NK cell activity and proinflammatory responses with the administration of probiotic strains (Lim *et al.*, 2002; Matsuzaki and Chin, 2000; Takagi *et al.*, 2001). Dietary intervention studies in human volunteers revealed increased NK cell activity in response to probiotic consumption (Gill *et al.*, 2001; Nagao *et al.*, 2000; Takeda and Okumura, 2007). NK cell activity in treatment groups of mice were observed, and treated animals demonstrated delayed tumour development in comparison to the control group. As expected, NK cell-deficient mice did not show delayed tumour development in response to treatment by dietary intervention with LAB. Various other tumour-infiltrating lymphocytes, including cytotoxic T cells and lymphokine-activated killer cells, may be potential effectors of antitumor immunity and may prevent neoplastic progression (Hadden, 1999). IFN-γ is a proinflammatory cytokine and has been reported as a key effector molecule in the immune response against solid tumours. Studies reported an increased IFN-γ activity with oral administration of probiotic strains (Aattour *et al.*, 2002; Raitano and Korc, 1993; Yasutake *et al.*, 1999; Yokokura *et al.*, 1984).

LAB may induce adaptive immune responses by enhancing proliferation or function of regulatory T-cells (T_{reg}) (Di Giacinto *et al.*, 2005). *Lactobacillus* strains stimulated the proliferation of T-cells that produce immuno-regulatory cytokines such as IL-10 and TGF-β (von der Weid *et al.*, 2001). Commensal bacteria may stimulate production of TGF- β by intestinal epithelial cells, and TGF- β induces expression of IL-10 by dendritic cells and macrophages. TGF- β and IL-10 favour the differentiation of T_{reg} cells at mucosal inductive sites (Guarner *et al.*, 2006). T_{reg} cells may be associated with reduced cancer risk.

Anti-inflammatory activities

The link between chronic inflammation and cancer has been well established (Aggarwal *et al.*, 2006; Balkwill and Mantovani, 2001; Coussens and Werb, 2002). Excessive Th1 cell responses such as TNF-α and IL-1 in tissues support tumour development and growth (Miles *et al.*, 1994). Evidence from animal models has shown that dysregulation of T cell responses results in chronic intestinal inflammation, and increased risk of colorectal cancers. Individuals with IBD have an increased risk of colorectal neoplasia (Campbell *et al.*, 2001; Vagefi and Longo, 2005). Oral administration of probiotic strain, LGG, suppressed TNF-α production in murine macrophages by a contact-independent mechanism (Pena and Versalovic, 2003). Anti-cancer effects of *L. helveticus* were mediated by decreased production of the proinflammatory cytokines, in particular IL-6 (de Moreno de LeBlanc *et al.*, 2006; Rachid *et al.*, 2006). Cell culture studies have demonstrated that *Lactobacillus* and *Bifidobacterium* spp. suppressed production of proinflammatory cytokine, IL-8, by human IECs derived from colorectal cancers (Bai *et al.*, 2006; Ma *et al.*, 2004; Zhang *et al.*, 2005). Borruel *et al.* (2002) observed inhibition of TNF-α production in human ileal specimens obtained from patients with Crohn's disease and treated *ex vivo* with *L. casei* and *L. bulgaricus*. However, differences in TNF-α quantities were not observed in non-inflamed mucosa. In a study of IL-10 knockout mice, O'Mahony *et al.* (2001) found that *L. salivarus* diminished colonic inflammation and tumour development in mice. These results indicate that down-regulation of Th1-associated inflammation by probiotics is associated with reduced risk of various carcinomas.

NF-κB is a family of Rel domain-containing cytoplasmic proteins that regulate the expression of many genes whose products are involved in inflammation, cell proliferation and tumorigenesis. NF-κB-regulated targets include anti-apoptotic genes (e.g. cIAP1/2, survivin, TRAF, Bcl-2, and Bcl-xL), Cox-2, MMP-9 genes encoding adhesion molecules, chemokines, and inflammatory cytokines; and cell cycle regulatory genes (e.g. cyclin D1 and c-Myc) (Aggarwal, 2004; Karin and Greten, 2005; Pacifico and Leonardi, 2006). Consistent with its regulatory role, constitutive activation of NF-κB is frequently observed in different types of cancers (Karin and Greten, 2005) and has been correlated with resistance to radiation- and chemotherapeutic agent-induced apoptosis (Aggarwal, 2004). Consumption of different LAB strains may yield anti-cancer effects partly as a result of anti-inflammatory capacities conferred by LAB (Borruel *et al.*, 2002; Christensen *et al.*, 2002; Rachid *et al.*, 2006). LAB strains may differentially modulate cytokine production. The probiotic combination VSL#3 conferred anti-inflammatory activities by inhibiting NF-κB signalling and inducing heat shock proteins in colonic epithelial cells through proteasome inhibition (Petrof *et al.*, 2004). Tien *et al.* (2006) reported that anti-inflammatory effects of *L. casei* were mediated by inhibition of NF-κB signalling, particularly via stabilization of Iκ-Bα. Riedel *et al.* (2006) reported that six of eight bifidobacteria inhibited lipopolysaccharide- (LPS-) induced NF-κB activation in a dose-and strain-dependent manner. However, NF-κB activation in response to challenge with TNF-α was not affected by any of the tested bifidobacteria, indicating that the inhibitory effects of bifidobacteria were specific for LPS-induced inflammation in IECs. Other commensal bacteria have been reported to diminish proinflammatory cytokine production by inhibiting NF-κB activation. Avirulent strains of *Salmonella* inhibited ubiquitination of the NF-κB inhibitory subunit, Iκ-Bα, (Neish *et al.*, 2000) and *Bacteroides thetaiotaomicron* modulated NF-κB signalling by facilitating the nuclear export of RelA (NF-κB subunit) via peroxisome proliferation activated receptor gamma (PPAR-γ) (Kelly *et al.*, 2004).

Induction of apoptosis and anti-proliferative activities

The inducible transcription factor NF-κB enhances cell viability by activating genes that counteract cell death pathways (apoptosis) (Sen, 2006). Apoptosis is a complex and active cellular process by which individual cells are triggered to undergo self-destruction (programmed cell death) (Chinnaiyan and Dixit, 1996). Promotion of apoptosis may inhibit cancer progression. The balance between cell proliferation and cell death is important in order to maintain equilibrium with respect to cell proliferation and cancer development. Yogurt feeding increased the number of apoptotic cells in a colon cancer model (Rachid *et al.*, 2002). Incubation with LAB samples resulted in time-dependent apoptosis of Jurkat cells, in contrast to relative sparing of primary human peripheral blood-derived lymphocytes (Di Marzio *et al.*, 2001). The apoptosis-inducing ability of *Streptococcus thermophilus* preparations has been reported to be attributed to high levels of neutral sphingomyelinase activity and the resultant generation of ceramide, a known apoptotic death messenger in Jurkat cells. Apoptosis induced by *L. brevis* could be associated with high levels of arginine deiminase activity, and consequent down-regulation of polyamine synthesis in Jurkat cells (Di Marzio *et al.*, 2001). In contrary, Yan and Polk (2002) reported that a different strain of probiotic bacteria (LGG) prevented cytokine-induced apoptosis in human and mouse IECs and may facilitate epithelial cell recovery following acute injury.

Biffi *et al.* (1997) reported that soluble components of milk fermented by five different LAB species possessed anti-proliferative effects in different cancer cell lines. SCFAs have been implicated in regulation of apoptosis and cellular differentiation (Augenlicht *et al.*, 1999; Basson *et al.*, 2000; Hague *et al.*, 1993; Mariadason *et al.*, 2000). At a molecular level, butyrate affects gene expression via phosphorylation and acetylation of histone proteins (Archer and Hodin, 1999). Certain probiotics may modify the ratio of SCFAs in the colon resulting in increased butyrate levels. Butyrate and propionate stimulate crypt cell proliferation in the caecum to a smaller extent (Gamet *et al.*, 1992). Butyrate has the potential to arrest the growth of neoplastic colonocytes, induce apoptosis and inhibits preneoplastic proliferation induced by certain tumour promoters (Williams *et al.*, 2003). Conjugated linoleic acids (CLA) are a group of isomers of linoleic acid (LA) possessing anti-inflammatory and anticarcinogenic properties, and CLAs can be produced from LA by certain bacterial strains. Ewaschuk *et al.* (2006) reported that conditioned media, containing probiotic-produced CLA, reduced viability and induced apoptosis of HT-29 and Caco-2 cells. Soluble polysaccharide fractions of heat-killed cells of *L. acidophilus* were found to specifically inhibit HT-29 cell proliferation (Fichera and Giese, 1994; Kim, 2002). This effect was attributed, at least in part, to the induction of apoptosis. Prebiotics such as chicory-derived fructans induced apoptosis and inhibited cancer cell proliferation in animal models (Hughes and Rowland, 2001).

Overexpression of anti-apoptotic members of the Bcl-2 family, such as Bcl-2 and Bcl-xL, has been implicated in cancer chemoresistance, whereas high levels of pro-apoptotic proteins, such as Bax, promote apoptosis and sensitize tumour cells to various anti-cancer therapies (Gardner, 2004). de Moreno de Leblanc *et al.* (2007) demonstrated that milk fermented with *L. helveticus* reduced Bcl-2 expression and enhanced apoptosis in a cancer model. Similar reduction in Bcl-2 gene expression by LAB was also reported by another group (Carol *et al.*, 2006). Studies from our lab also demonstrated that *L. reuteri* induces apoptosis in human myeloid cells via inhibition of NF-κB signalling (Iyer *et al.*, 2008).

Cox-2 and i-NOS are up-regulated in human and experimental colonic neoplasms (DuBois *et al.*, 1996; Kojima *et al.*, 1999) a phenomenon which has been linked to apoptosis resistance, DNA damage, mutation, increased proliferation, oxidative stress, increased tumour vascularity and metastasis potential. Inhibition of Cox-2 and i-NOS activities diminished chemically induced carcinogenesis (Watanabe *et al.*, 2000). Oral administration of *Lactobacillus salivarius* reduced i-NOS production in rat colitis models. Femia *et al.* (2002) reported that both i-NOS and Cox-2 expression were slightly reduced in tumours of rats treated with prebiotics, probiotics and synbiotics.

Conclusions

Individual LAB strains or species appear to be capable of delivering anticarcinogenic effects. LAB modulate the gut microbiota, regulate immune responses, bind directly to and degrade carcinogens, produce metabolites that directly result in carcinogen detoxification, suppress inflammation and induce apoptosis. Many chemotherapeutic agents are inherently limited because of systemic cytotoxicity and adverse effects. However, LAB may be selected as natural strains with specific functions and used as oral therapeutic agents, with a relatively high degree of safety and efficacy. Although probiotics appear to be beneficial in the context of cancer and cell proliferation, it will require a greater understanding of mechanisms of probiosis and identification of probiotic strains with several desirable features. Although the inhibition of carcinogenesis correlates with changes in immune function, knowledge of immune responses to LAB consumption currently are inadequate. The strain-dependent variability of biological functions within a probiotic species complicates our attempts to understand the role of immunity in probiotic-mediated anticarcinogenesis. Further work is needed to assess long-term effects of probiotics on host immunity and any correlation with reduced susceptibility to carcinogenesis. The continued development of techniques in cellular and molecular biology may ultimately lead to a greater understanding of the mechanisms of probiosis, cancer prevention or treatment with beneficial LAB.

Acknowledgements

We would like to thank Tiffany Morgan and Karla Sternberg for helping to prepare this manuscript. This work was supported by grants from the Crohn's and Colitis Foundation of America First Award, Moran Foundation and National Institutes of Health grant award RO1 DK065075.

References

Aattour, N., Bouras, M., Tome, D., Marcos, A., and Lemonnier, D. (2002). Oral ingestion of lactic-acid bacteria by rats increases lymphocyte proliferation and interferon-gamma production. Br. J. Nutr. *87*, 367–373.

Abdelali, H., Cassand, P., Soussotte, V., Daubeze, M., Bouley, C., and Narbonne, J.F. (1995). Effect of dairy products on initiation of precursor lesions of colon cancer in rats. Nutr. Cancer *24*, 121–132.

Aggarwal, B.B. (2004). Nuclear factor-kappaB: the enemy within. Cancer Cell *6*, 203–208.

Aggarwal, B.B., Shishodia, S., Sandur, S.K., Pandey, M.K., and Sethi, G. (2006). Inflammation and cancer: how hot is the link? Biochem. Pharmacol. *72*, 1605–1621.

Archer, S.Y., and Hodin, R.A. (1999). Histone acetylation and cancer. Curr. Opin. Genet. Dev. *9*, 171–174.

Aso, Y., Akaza, H., Kotake, T., Tsukamoto, T., Imai, K., and Naito, S. (1995). Preventive effect of a *Lactobacillus casei* preparation on the recurrence of superficial bladder cancer in a double-blind trial. The BLP Study Group. Eur. Urol. *27*, 104–109.

Augenlicht, L.H., Anthony, G.M., Church, T.L., Edelmann, W., Kucherlapati, R., Yang, K., Lipkin, M., and Heerdt, B.G. (1999). Short-chain fatty acid metabolism, apoptosis, and Apc-initiated tumorigenesis in the mouse gastrointestinal mucosa. Cancer Res. *59*, 6005–6009.

Ayebo, A.D., Angelo, I.A., and Shahani, K.M. (1980). Effect of ingesting *Lactobacillus acidophilus* milk upon fecal flora and enzyme activity in humans. Milch Wissenschaft 35, 730–733.

Babbs, C.F. (1990). Free radicals and the etiology of colon cancer. Free Radic. Biol. Med. *8*, 191–200.

Bai, A.P., Ouyang, Q., Xiao, X.R., and Li, S.F. (2006). Probiotics modulate inflammatory cytokine secretion from inflamed mucosa in active ulcerative colitis. Int. J. Clin. Pract. *60*, 284–288.

Balish, E., and Warner, T. (2002). *Enterococcus faecalis* induces inflammatory bowel disease in interleukin-10 knockout mice. Am. J. Pathol. *160*, 2253–2257.

Balkwill, F., and Mantovani, A. (2001). Inflammation and cancer: back to Virchow? Lancet *357*, 539–545.

Bartram, H.P., Scheppach, W., Gerlach, S., Ruckdeschel, G., Kelber, E., and Kasper, H. (1994). Does yogurt enriched with *Bifidobacterium longum* affect colonic microbiology and fecal metabolites in healthy subjects? Am. J. Clin. Nutr. 59, 428–432.

Basson, M.D., Liu, Y.W., Hanly, A.M., Emenaker, N.J., Shenoy, S.G., and Gould Rothberg, B.E. (2000). Identification and comparative analysis of human colonocyte short-chain fatty acid response genes. J. Gastrointest. Surg. *4*, 501–512.

Bengmark, S. (2002). Gut microbial ecology in critical illness: is there a role for prebiotics, probiotics, and synbiotics? Curr. Opin. Crit. Care *8*, 145–151.

Bernet, M.F., Brassart, D., Neeser, J.R., and Servin, A.L. (1993). Adhesion of human bifidobacterial strains to cultured human intestinal epithelial cells and inhibition of enteropathogen–cell interactions. Appl. Environ. Microbiol. 59, 4121–4128.

Bernet, M.F., Brassart, D., Neeser, J.R., and Servin, A.L. (1994). *Lactobacillus acidophilus* LA 1 binds to cultured human intestinal cell lines and inhibits cell attachment and cell invasion by enterovirulent bacteria. Gut *35*, 483–489.

Biasco, G., Paganelli, G.M., Brandi, G., Brillanti, S., Lami, F., Callegari, C., and Gizzi, G. (1991). Effect of *Lactobacillus acidophilus* and *Bifidobacterium bifidum* on rectal cell kinetics and fecal pH. Ital. J. Gastroenterol. 23, 142.

Biffi, A., Coradini, D., Larsen, R., Riva, L., and Di Fronzo, G. (1997). Antiproliferative effect of fermented milk on the growth of a human breast cancer cell line. Nutr. Cancer *28*, 93–99.

Bolognani, F., Rumney, C.J., and Rowland, I.R. (1997). Influence of carcinogen binding by lactic acid-producing bacteria on tissue distribution and in vivo mutagenicity of dietary carcinogens. Food Chem. Toxicol. 35, 535–545.

Borruel, N., Carol, M., Casellas, F., Antolin, M., de Lara, F., Espin, E., Naval, J., Guarner, F., and Malagelada, J.R. (2002). Increased mucosal tumour necrosis factor alpha production in Crohn's disease can be down-regulated ex vivo by probiotic bacteria. Gut *51*, 659–664.

Bouhnik, Y. (2001). [Can we modify colonic microflora and to what end?]. Gastroenterol. Clin. Biol. *25*, C101–104.

Boutron, M.C., Faivre, J., Marteau, P., Couillault, C., Senesse, P., and Quipourt, V. (1996). Calcium, phosphorus, vitamin D., dairy products and colorectal carcinogenesis: a French case–control study. Br. J. Cancer *74*, 145–151.

Brigidi, P., Vitali, B., Swennen, E., Bazzocchi, G., and Matteuzzi, D. (2001). Effects of probiotic administration upon the composition and enzymatic activity of human fecal microbiota in patients with irritable bowel syndrome or functional diarrhea. Res. Microbiol. *152*, 735–741.

Campbell, B.J., Yu, L.G., and Rhodes, J.M. (2001). Altered glycosylation in inflammatory bowel disease: a possible role in cancer development. Glycoconj. J. *18*, 851–858.

Carman, R.J., Van Tassell, R.L., Kingston, D.G., Bashir, M., and Wilkins, T.D. (1988). Conversion of IQ, a dietary pyrolysis carcinogen to a direct-acting mutagen by normal intestinal bacteria of humans. Mutat. Res. *206*, 335–342.

Carol, M., Borruel, N., Antolin, M., Llopis, M., Casellas, F., Guarner, F., and Malagelada, J.R. (2006). Modulation of apoptosis in intestinal lymphocytes by a probiotic bacteria in Crohn's disease. J. Leukoc. Biol. *79*, 917–922.

Challa, A., Rao, D.R., Chawan, C.B., and Shackelford, L. (1997). *Bifidobacterium longum* and lactulose suppress azoxymethane-induced colonic aberrant crypt foci in rats. Carcinogenesis *18*, 517–521.

Chauviere, G., Coconnier, M.H., Kerneis, S., Fourniat, J., and Servin, A.L. (1992). Adhesion of human *Lactobacillus acidophilus* strain LB to human enterocyte-like Caco-2 cells. J. Gen. Microbiol. *138 Pt 8*, 1689–1696.

Cheng, K., and Raufman, J.P. (2005). Bile acid-induced proliferation of a human colon cancer cell line is mediated by transactivation of epidermal growth factor receptors. Biochem. Pharmacol. *70*, 1035–1047.

Chinnaiyan, A.M., and Dixit, V.M. (1996). The cell-death machine. Curr. Biol. 6, 555–562.

Choi, S.S., Kim, Y., Han, K.S., You, S., Oh, S., and Kim, S.H. (2006). Effects of *Lactobacillus* strains on cancer cell proliferation and oxidative stress in vitro. Lett Appl. Microbiol. *42*, 452–458.

Christensen, H.R., Frokiaer, H., and Pestka, J.J. (2002). Lactobacilli differentially modulate expression of cytokines and maturation surface markers in murine dendritic cells. J. Immunol. *168*, 171–178.

Commane, D., Hughes, R., Shortt, C., and Rowland, I. (2005). The potential mechanisms involved in the anti-carcinogenic action of probiotics. Mutat. Res. *591*, 276–289.

Corr, S.C., Li, Y., Riedel, C.U., O'Toole, P.W., Hill, C., and Gahan, C.G. (2007). Bacteriocin production as a mechanism for the antiinfective activity of *Lactobacillus salivarius* UCC118. Proc. Natl. Acad. Sci. USA *104*, 7617–7621.

Coussens, L.M., and Werb, Z. (2002). Inflammation and cancer. Nature *420*, 860–867.

Cummings, J.H., and Macfarlane, G.T. (2002). Gastrointestinal effects of prebiotics. Br. J. Nutr. *87 Suppl 2*, S145–151.

De Boever, P., Wouters, R., Verschaeve, L., Berckmans, P., Schoeters, G., and Verstraete, W. (2000). Protective effect of the bile salt hydrolase-active *Lactobacillus reuteri* against bile salt cytotoxicity. Appl. Microbiol. Biotechnol. *53*, 709–714.

de Moreno de LeBlanc, A., Matar, C., Farnworth, E., and Perdigon, G. (2006). Study of cytokines involved in the prevention of a murine experimental breast cancer by kefir. Cytokine *34*, 1–8.

de Moreno de Leblanc, A., Matar, C., Farnworth, E., and Perdigon, G. (2007). Study of immune cells involved in the antitumor effect of kefir in a murine breast cancer model. J. Dairy Sci. *90*, 1920–1928.

de Moreno de LeBlanc, A., and Perdigon, G. (2005). Reduction of beta-glucuronidase and nitroreductase activity by yoghurt in a murine colon cancer model. Biocell *29*, 15–24.

De Preter, V., Coopmans, T., Rutgeerts, P., and Verbeke, K. (2006). Influence of long-term administration of lactulose and *Saccharomyces boulardii* on the colonic generation of phenolic compounds in healthy human subjects. J. Am. Coll. Nutr. *25*, 541–549.

Delves-Broughton, J., Blackburn, P., Evans, R.J., and Hugenholtz, J. (1996). Applications of the bacteriocin, nisin. Antonie Van Leeuwenhoek *69*, 193–202.

Di Giacinto, C., Marinaro, M., Sanchez, M., Strober, W., and Boirivant, M. (2005). Probiotics ameliorate recurrent Th1-mediated murine colitis by inducing IL-10 and IL-10-dependent TGF-beta-bearing regulatory cells. J. Immunol. *174*, 3237–3246.

Di Marzio, L., Russo, F.P., D'Alo, S., Biordi, L., Ulisse, S., Amicosante, G., De Simone, C., and Cifone, M.G. (2001). Apoptotic effects of selected strains of lactic acid bacteria on a human T leukemia cell line are associated with bacterial arginine deiminase and/or sphingomyelinase activities. Nutr. Cancer *40*, 185–196.

Doron, S., and Gorbach, S.L. (2006). Probiotics: their role in the treatment and prevention of disease. Expert Rev. Anti. Infect. Ther. *4*, 261–275.

DuBois, R.N., Radhika, A., Reddy, B.S., and Entingh, A.J. (1996). Increased cyclooxygenase-2 levels in carcinogen-induced rat colonic tumors. Gastroenterology *110*, 1259–1262.

El-Nezami, H.S., Polychronaki, N.N., Ma, J., Zhu, H., Ling, W., Salminen, E.K., Juvonen, R.O., Salminen, S.J., Poussa, T., and Mykkanen, H.M. (2006). Probiotic supplementation reduces a biomarker for increased risk of liver cancer in young men from Southern China. Am. J. Clin. Nutr. *83*, 1199–1203.

Ellmerich, S., Scholler, M., Duranton, B., Gosse, F., Galluser, M., Klein, J.P., and Raul, F. (2000). Promotion of intestinal carcinogenesis by *Streptococcus bovis*. Carcinogenesis *21*, 753–756.

Engle, S.J., Ormsby, I., Pawlowski, S., Boivin, G.P., Croft, J., Balish, E., and Doetschman, T. (2002). Elimination of colon cancer in germ-free transforming growth factor beta 1-deficient mice. Cancer Res, *62*, 6362–6366.

Ennahar, S., Sonomoto, K., and Ishizaki, A. (1999). Class IIa bacteriocins from lactic acid bacteria: antibacterial activity and food preservationJ. Biosci. Bioeng. *87*, 705–716.

Ewaschuk, J.B., Walker, J.W., Diaz, H., and Madsen, K.L. (2006). Bioproduction of conjugated linoleic acid by probiotic bacteria occurs in vitro and in vivo in mice. J. Nutr. *136*, 1483–1487.

Femia, A.P., Luceri, C., Dolara, P., Giannini, A., Biggeri, A., Salvadori, M., Clune, Y., Collins, K.J., Paglierani, M., and Caderni, G. (2002). Antitumorigenic activity of the prebiotic inulin enriched with oligofructose in combination with the probiotics *Lactobacillus rhamnosus* and *Bifidobacterium lactis* on azoxymethane-induced colon carcinogenesis in rats. Carcinogenesis *23*, 1953–1960.

Fichera, G.A., and Giese, G. (1994). Non-immunologically-mediated cytotoxicity of *Lactobacillus casei* and its derivative peptidoglycan against tumor cell lines. Cancer Lett. *85*, 93–103.

Figler, M., Mozsik, G., Schaffer, B., Gasztonyi, B., Acs, P., Szili, B., Rab, R., and Szakaly, S. (2006). Effect of special Hungarian probiotic kefir on faecal microflora. World J. Gastroenterol. *12*, 1129–1132.

Forestier, C., De Champs, C., Vatoux, C., and Joly, B. (2001). Probiotic activities of *Lactobacillus casei rhamnosus*: in vitro adherence to intestinal cells and antimicrobial properties. Res. Microbiol. *152*, 167–173.

Fox, J.G., and Wang, T.C. (2007). Inflammation, atrophy, and gastric cancer. J. Clin. Invest. *117*, 60–69.

Gallaher, D.D., and Khil, J. (1999). The effect of synbiotics on colon carcinogenesis in rats. J. Nutr. *129*, 1483S-1487S.

Gamet, L., Daviaud, D., Denis-Pouxviel, C., Remesy, C., and Murat, J.C. (1992). Effects of short-chain fatty acids on growth and differentiation of the human colon-cancer cell line HT29. Int. J. Cancer 52, 286–289.

Garcia-Urkia, N., Asensio, A.B., Zubillaga Azpiroz, I., Zubillaga Huici, P., Vidales, C., Garcia-Arenzana, J.M., Aldazabal, P., and Eizaguirre, I. (2002). [Beneficial effects of *Bifidobacterium lactis* in the prevention of bacterial translocation in experimental short bowel syndrome]. Cir. Pediatr. *15*, 162–165.

Gardner, C.R. (2004). Anticancer drug development based on modulation of the Bcl-2 family core apoptosis mechanism. Expert Rev. Anticancer Ther. *4*, 1157–1177.

Geier, M.S., Butler, R.N., and Howarth, G.S. (2006). Probiotics, prebiotics and synbiotics: a role in chemoprevention for colorectal cancer? Cancer Biol. Ther. *5*, 1265–1269.

Geypens, B., Claus, D., Evenepoel, P., Hiele, M., Maes, B., Peeters, M., Rutgeerts, P., and Ghoos, Y. (1997). Influence of dietary protein supplements on the formation of bacterial metabolites in the colon. Gut *41*, 70–76.

Gibson, G.R., and Roberfroid, M.B. (1995). Dietary modulation of the human colonic microbiota: introducing the concept of prebiotics. J. Nutr. *125*, 1401–1412.

Gill, H.S., Cross, M.L., Rutherfurd, K.J., and Gopal, P.K. (2001). Dietary probiotic supplementation to enhance cellular immunity in the elderly. Br. J. Biomed. Sci. *58*, 94–96.

Goldin, B.R., and Gorbach, S.L. (1984a). Alterations of the intestinal microflora by diet, oral antibiotics, and *Lactobacillus*: decreased production of free amines from aromatic nitro compounds, azo dyes, and glucuronides. J. Natl. Cancer Inst. *73*, 689–695.

Goldin, B.R., and Gorbach, S.L. (1984b). The effect of milk and *Lactobacillus* feeding on human intestinal bacterial enzyme activity. Am. J. Clin. Nutr. *39*, 756–761.

Goldin, B.R., Gualtieri, L.J., and Moore, R.P. (1996). The effect of *Lactobacillus GG* on the initiation and promotion of DMH-induced intestinal tumors in the rat. Nutr. Cancer *25*, 197–204.

Goldin, B.R., Swenson, L., Dwyer, J., Sexton, M., and Gorbach, S.L. (1980). Effect of diet and *Lactobacillus acidophilus* supplements on human fecal bacterial enzymes. J. Natl. Cancer Inst. *64*, 255–261.

Gorbach, S.L. (2000). Probiotics and gastrointestinal health. Am. J. Gastroenterol. *95*, S2–S4.

Guarner, F. (2006). Enteric flora in health and disease. Digestion *73* (Suppl. 1), 5–12.

Guarner, F., Bourdet-Sicard, R., Brandtzaeg, P., Gill, H.S., McGuirk, P., van Eden, W., Versalovic, J., Weinstock, J.V., and Rook, G.A. (2006). Mechanisms of disease: the hygiene hypothesis revisited. Nat. Clin. Pract. Gastroenterol. Hepatol. *3*, 275–284.

Hadden, J.W. (1999). The immunology and immunotherapy of breast cancer: an update. Int. J. Immunopharmacol. *21*, 79–101.

Hague, A., Manning, A.M., Hanlon, K.A., Huschtscha, L.I., Hart, D., and Paraskeva, C. (1993). Sodium butyrate induces apoptosis in human colonic tumour cell lines in a p53-independent pathway: implications for the possible role of dietary fibre in the prevention of large-bowel cancer. Int. J. Cancer 55, 498–505.

Hayatsu, H., and Hayatsu, T. (1993). Suppressing effect of *Lactobacillus casei* administration on the urinary mutagenicity arising from ingestion of fried ground beef in the human. Cancer Lett. *73*, 173–179.

Hessle, C., Hanson, L.A., and Wold, A.E. (1999). Lactobacilli from human gastrointestinal mucosa are strong stimulators of IL-12 production. Clin. Exp. Immunol. *116*, 276–282.

Hirayama, K., and Rafter, J. (2000). The role of probiotic bacteria in cancer prevention. Microbes Infect. *2*, 681–686.

Hope, M.E., Hold, G.L., Kain, R., and El-Omar, E.M. (2005). Sporadic colorectal cancer–role of the commensal microbiota. FEMS Microbiol. Lett. *244*, 1–7.

Horie, H., Kanazawa, K., Okada, M., Narushima, S., Itoh, K., and Terada, A. (1999). Effects of intestinal bacteria on the development of colonic neoplasm: an experimental study. Eur. J. Cancer Prev. *8*, 237–245.

Hosono, A., Kashina, T., and Kada, T. (1986). Antimutagenic properties of lactic acid-cultured milk on chemical and fecal mutagens. J. Dairy Sci. *69*, 2237–2242.

Hsu, C.K., Liao, J.W., Chung, Y.C., Hsieh, C.P., and Chan, Y.C. (2004). Xylooligosaccharides and fructooligosaccharides affect the intestinal microbiota and precancerous colonic lesion development in rats. J. Nutr. *134*, 1523–1528.

Hughes, R., and Rowland, I.R. (2001). Stimulation of apoptosis by two prebiotic chicory fructans in the rat colon. Carcinogenesis 22, 43–47.

Hurst, A. (1967). Function of nisin and nisin-like basic proteins in the growth cycle of *Streptococcus lactis*. Nature *214*, 1232–1234.

Huycke, M.M., and Gaskins, H.R. (2004). Commensal bacteria, redox stress, and colorectal cancer: mechanisms and models. Exp. Biol. Med. (Maywood) *229*, 586–597.

Ishikawa, H., Akedo, I., Otani, T., Suzuki, T., Nakamura, T., Takeyama, I., Ishiguro, S., Miyaoka, E., Sobue, T., and Kakizoe, T. (2005). Randomized trial of dietary fiber and *Lactobacillus casei* administration for prevention of colorectal tumors. Int. J. Cancer *116*, 762–767.

Isolauri, E., Salminen, S., and Ouwehand, A.C. (2004). Microbial-gut interactions in health and disease. Probiotics. Best Prac. Res. Clin. Gastroenterol. *18*, 299–313.

Iyer, C., Kosters, A., Sethi, G., Kunnamakkara, A.B., Aggarwal, B.B., and Versalovic, J. (2008). Probiotic *Lactobacillus reuteri* promotes TNf-induced apoptosis in human myeloid leukemia-derived cells by modulation of Nf-kappaB and MAPK signalling. Cell. Microbiol. *10*, 1442–1452.

Johnson-Henry, K.C., Mitchell, D.J., Avitzur, Y., Galindo-Mata, E., Jones, N.L., and Sherman, P.M. (2004). Probiotics reduce bacterial colonization and gastric inflammation in H. pylori-infected mice. Dig. Dis. Sci. *49*, 1095–1102.

Johnson-Henry, K.C., Nadjafi, M., Avitzur, Y., Mitchell, D.J., Ngan, B.Y., Galindo-Mata, E., Jones, N.L., and Sherman, P.M. (2005). Amelioration of the effects of *Citrobacter rodentium* infection in mice by pretreatment with probiotics. J. Infect. Dis. *191*, 2106–2117.

Jung, D.S., Bodyfelt, F.W., and Daeschel, M.A. (1992). Influence of fat and emulsifiers on the efficacy of nisin in inhibiting *Listeria monocytogenes* in fluid milk. J. Dairy Sci. *75*, 387–393.

Kado, S., Uchida, K., Funabashi, H., Iwata, S., Nagata, Y., Ando, M., Onoue, M., Matsuoka, Y., Ohwaki, M., and Morotomi, M. (2001). Intestinal microflora are necessary for development of spontaneous adenocarcinoma of the large intestine in T-cell receptor beta chain and p53 double-knockout mice. Cancer Res. *61*, 2395–2398.

Kampman, E., Goldbohm, R.A., van den Brandt, P.A., and van 't Veer, P. (1994a). Fermented dairy products, calcium, and colorectal cancer in The Netherlands Cohort Study. Cancer Res. *54*, 3186–3190.

Kampman, E., van 't Veer, P., Hiddink, G.J., van Aken-Schneijder, P., Kok, F.J., and Hermus, R.J. (1994b). Fermented dairy products, dietary calcium and colon cancer: a case-control study in The Netherlands. Int. J. Cancer *59*, 170–176.

Karin, M., and Greten, F.R. (2005). NF-kappaB: linking inflammation and immunity to cancer development and progression. Nat. Rev. Immunol. *5*, 749–759.

Kato, I., Endo, K., and Yokokura, T. (1994). Effects of oral administration of *Lactobacillus casei* on antitumor responses induced by tumor resection in mice. Int. J. Immunopharmacol. *16*, 29–36.

Kato, I., Kobayashi, S., Yokokura, T., and Mutai, M. (1981). Antitumor activity of *Lactobacillus casei* in mice. Gann *72*, 517–523.

Kelly, D., Campbell, J.I., King, T.P., Grant, G., Jansson, E.A., Coutts, A.G., Pettersson, S., and Conway, S. (2004). Commensal anaerobic gut bacteria attenuate inflammation by regulating nuclear-cytoplasmic shuttling of PPAR-gamma and RelA. Nat. Immunol. 5, 104–112.

Kim, J.Y., Woo, H.J., Kim, Y.S., Lee, J.H. (2002). Screening for antiproliferative effects of cellular components from latic acid bacteria against human cancer cell lines. Biotechnol. Lett. *24*, 1431–1436.

Kim, Y.G., Ohta, T., Takahashi, T., Kushiro, A., Nomoto, K., Yokokura, T., Okada, N., and Danbara, H. (2006). Probiotic *Lactobacillus casei* activates innate immunity via NF-kappaB and p38 MAP kinase signaling pathways. Microbes Infect. *8*, 994–1005.

Kitahara, M., Takamine, F., Imamura, T., and Benno, Y. (2000). Assignment of *Eubacterium* sp. VPI 12708 and related strains with high bile acid 7alpha-dehydroxylating activity to *Clostridium scindens* and proposal of *Clostridium hylemonae* sp. nov., isolated from human faeces. Int. J. Syst. Evol. Microbiol. *50 Pt 3*, 971–978.

Klaenhammer, T.R. (1993). Genetics of bacteriocins produced by lactic acid bacteria. FEMS Microbiol. Rev. *12*, 39–85.

Kleeman, E.G., and Klaenhammer, T.R. (1982). Adherence of *Lactobacillus* species to human fetal intestinal cells. J. Dairy Sci. *65*, 2063–2069.

Ko, J.S., Yang, H.R., Chang, J.Y., and Seo, J.K. (2007). *Lactobacillus plantarum* inhibits epithelial barrier dysfunction and interleukin-8 secretion induced by tumor necrosis factor-alpha. World J. Gastroenterol. *13*, 1962–1965.

Kohwi, Y., Imai, K., Tamura, Z., and Hashimoto, Y. (1978). Antitumor effect of *Bifidobacterium infantis* in mice. Gann *69*, 613–618.

Kojima, M., Morisaki, T., Tsukahara, Y., Uchiyama, A., Matsunari, Y., Mibu, R., and Tanaka, M. (1999). Nitric oxide synthase expression and nitric oxide production in human colon carcinoma tissue. J. Surg. Oncol. *70*, 222–229.

Kulkarni, N., and Reddy, B.S. (1994). Inhibitory effect of *Bifidobacterium longum* cultures on the azoxymethane-induced aberrant crypt foci formation and fecal bacterial beta-glucuronidase. Proc. Soc. Exp. Biol. Med. *207*, 278–283.

Kullisaar, T., Zilmer, M., Mikelsaar, M., Vihalemm, T., Annuk, H., Kairane, C., and Kilk, A. (2002). Two antioxidative lactobacilli strains as promising probiotics. Int. J. Food Microbiol. *72*, 215–224.

Le Leu, R.K., Brown, I.L., Hu, Y., Bird, A.R., Jackson, M., Esterman, A., and Young, G.P. (2005). A synbiotic combination of resistant starch and *Bifidobacterium lactis* facilitates apoptotic deletion of carcinogen-damaged cells in rat colon. J. Nutr. *135*, 996–1001.

Le, M.G., Moulton, L.H., Hill, C., and Kramar, A. (1986). Consumption of dairy produce and alcohol in a case-control study of breast cancer. J. Natl. Cancer Inst. *77*, 633–636.

Lidbeck, A., Nord, C.E., Gustafsson, J.A., and Rafter, J. (1992). Lactobacilli, anticarcinogenic activities and human intestinal microflora. Eur. J. Cancer Prev. *1*, 341–353.

Lim, B.K., Mahendran, R., Lee, Y.K., and Bay, B.H. (2002). Chemopreventive effect of *Lactobacillus rhamnosus* on growth of a subcutaneously implanted bladder cancer cell line in the mouse. Jpn. J. Cancer Res. *93*, 36–41.

Lin, M.Y., and Chang, F.J. (2000). Antioxidative effect of intestinal bacteria *Bifidobacterium longum* ATCC 15708 and *Lactobacillus acidophilus* ATCC 4356. Dig. Dis. Sci. *45*, 1617–1622.

Ling, W.H., Hanninen, O., Mykkanen, H., Heikura, M., Salminen, S., and Von Wright, A. (1992). Colonization and fecal enzyme activities after oral *Lactobacillus* GG administration in elderly nursing home residents. Ann. Nutr. Metab. *36*, 162–166.

Ling, W.H., Korpela, R., Mykkanen, H., Salminen, S., and Hanninen, O. (1994). *Lactobacillus* strain GG supplementation decreases colonic hydrolytic and reductive enzyme activities in healthy female adults. J. Nutr. *124*, 18–23.

Ljungh, A., and Wadstrom, T. (2006). Lactic acid bacteria as probiotics. Curr. Issues Intest. Microbiol. *7*, 73–89.

Lupton, J.R., and Kurtz, P.P. (1993). Relationship of colonic luminal short-chain fatty acids and pH to in vivo cell proliferation in rats. J. Nutr. *123*, 1522–1530.

Ma, D., Forsythe, P., and Bienenstock, J. (2004). Live *Lactobacillus reuteri* is essential for the inhibitory effect on tumor necrosis factor alpha-induced interleukin-8 expression. Infect. Immun. *72*, 5308–5314.

Maclennan, R., and Jensen, O.M. (1977). Dietary fibre, transit-time, faecal bacteria, steroids, and colon cancer in two Scandinavian populations. Report from the International Agency for Research on Cancer Intestinal Microecology Group. Lancet *2*, 207–211.

Madsen, K.L. (2001). The use of probiotics in gastrointestinal disease. Can. J. Gastroenterol. *15*, 817–822.

Malhotra, S.L. (1977). Dietary factors in a study of cancer colon from Cancer Registry, with special reference to the role of saliva, milk and fermented milk products and vegetable fibre. Med. Hypotheses 3, 122–126.

Manning, T.S., and Gibson, G.R. (2004). Microbial-gut interactions in health and disease. Prebiotics. Best Pract. Res. Clin. Gastroenterol. *18*, 287–298.

Mariadason, J.M., Rickard, K.L., Barkla, D.H., Augenlicht, L.H., and Gibson, P.R. (2000). Divergent phenotypic patterns and commitment to apoptosis of Caco-2 cells during spontaneous and butyrate-induced differentiation. J. Cell Physiol. *183*, 347–354.

Marnett, L.J. (2000). Oxyradicals and DNA damage. Carcinogenesis *21*, 361–370.

Marteau, P., Pochart, P., Flourie, B., Pellier, P., Santos, L., Desjeux, J.F., and Rambaud, J.C. (1990). Effect of chronic ingestion of a fermented dairy product containing *Lactobacillus acidophilus* and *Bifidobacterium bifidum* on metabolic activities of the colonic flora in humans. Am. J. Clin. Nutr. 52, 685–688.

Matar, C., Nadathur, S.S., Bakalinsky, A.T., and Goulet, J. (1997). Antimutagenic effects of milk fermented by *Lactobacillus helveticus* L89 and a protease-deficient derivative. J. Dairy Sci. *80*, 1965–1970.

Matar, C., Valdez, J.C., Medina, M., Rachid, M., and Perdigon, G. (2001). Immunomodulating effects of milks fermented by *Lactobacillus helveticus* and its non-proteolytic variant. J. Dairy Res. *68*, 601–609.

Matsuzaki, T., and Chin, J. (2000). Modulating immune responses with probiotic bacteria. Immunol. Cell Biol. *78*, 67–73.

McBain, A.J., and Macfarlane, G.T. (1998). Ecological and physiological studies on large intestinal bacteria in relation to production of hydrolytic and reductive enzymes involved in formation of genotoxic metabolites. J. Med. Microbiol. *47*, 407–416.

McGarr, S.E., Ridlon, J.M., and Hylemon, P.B. (2005). Diet, anaerobic bacterial metabolism, and colon cancer: a review of the literature. J. Clin. Gastroenterol. *39*, 98–109.

McLellan, E.A., Medline, A., and Bird, R.P. (1991). Sequential analyses of the growth and morphological characteristics of aberrant crypt foci: putative preneoplastic lesions. Cancer Res. *51*, 5270–5274.

Miettinen, M., Lehtonen, A., Julkunen, I., and Matikainen, S. (2000). Lactobacilli and streptococci activate NF-kappa B and STAT signaling pathways in human macrophages. J. Immunol. *164*, 3733–3740.

Miles, D.W., Happerfield, L.C., Naylor, M.S., Bobrow, L.G., Rubens, R.D., and Balkwill, F.R. (1994). Expression of tumour necrosis factor (TNF alpha) and its receptors in benign and malignant breast tissue. Int. J. Cancer 56, 777–782.

Modler, G.W., McKellar, R.C., and M., Y. (1990). Bifidobacteria and bifidogenic factors. Can. Inst. Food Sci. Technol. Journ. 23 29–41.

Morotomi, M., and Mutai, M. (1986). In vitro binding of potent mutagenic pyrolysates to intestinal bacteria. J. Natl. Cancer Inst. *77*, 195–201.

Nagao, F., Nakayama, M., Muto, T., and Okumura, K. (2000). Effects of a fermented milk drink containing *Lactobacillus casei* strain Shirota on the immune system in healthy human subjects. Biosci. Biotechnol. Biochem. *64*, 2706–2708.

Nagengast, F.M., Grubben, M.J., and van Munster, I.P. (1995). Role of bile acids in colorectal carcinogenesis. Eur. J. Cancer *31A*, 1067–1070.

Neish, A.S., Gewirtz, A.T., Zeng, H., Young, A.N., Hobert, M.E., Karmali, V., Rao, A.S., and Madara, J.L. (2000). Prokaryotic regulation of epithelial responses by inhibition of IkappaB-alpha ubiquitination. Science *289*, 1560–1563.

O'Mahony, L., Feeney, M., O'Halloran, S., Murphy, L., Kiely, B., Fitzgibbon, J., Lee, G., O'Sullivan, G., Shanahan, F., and Collins, J.K. (2001). Probiotic impact on microbial flora, inflammation and tumour development in IL-10 knockout mice. Aliment. Pharmacol. Ther. *15*, 1219–1225.

Olivares, M., Diaz-Ropero, M.A., Gomez, N., Lara-Villoslada, F., Sierra, S., Maldonado, J.A., Martin, R., Lopez-Huertas, E., Rodriguez, J.M., and Xaus, J. (2006). Oral administration of two probiotic strains, *Lactobacillus gasseri* CECT5714 and *Lactobacillus coryniformis* CECT5711, enhances the intestinal function of healthy adults. Int. J. Food Microbiol. *107*, 104–111.

Onoue, M., Kado, S., Sakaitani, Y., Uchida, K., and Morotomi, M. (1997). Specific species of intestinal bacteria influence the induction of aberrant crypt foci by 1,2-dimethylhydrazine in rats. Cancer Lett. *113*, 179–186.

Orrhage, K., Sillerstrom, E., Gustafsson, J.A., Nord, C.E., and Rafter, J. (1994). Binding of mutagenic heterocyclic amines by intestinal and lactic acid bacteria. Mutat. Res. *311*, 239–248.

Orrhage, K.M., Annas, A., Nord, C.E., Brittebo, E.B., and Rafter, J.J. (2002). Effects of lactic acid bacteria on the uptake and distribution of the food mutagen Trp-P-2 in mice. Scand. J. Gastroenterol. *37*, 215–221.

Ouwehand, A., Isolauri, E., and Salminen, S. (2002). The role of the intestinal microflora for the development of the immune system in early childhood. Eur. J. Nutr. *41* Suppl 1, I32–37.

Owen, R.W., Spiegelhalder, B., and Bartsch, H. (2000). Generation of reactive oxygen species by the faecal matrix. Gut *46*, 225–232.

Pacifico, F., and Leonardi, A. (2006). NF-kappaB in solid tumors. Biochem. Pharmacol. *72*, 1142–1152.

Parassol, N., Freitas, M., Thoreux, K., Dalmasso, G., Bourdet-Sicard, R., and Rampal, P. (2005). *Lactobacillus casei* DN-114 001 inhibits the increase in paracellular permeability of enteropathogenic *Escherichia coli*-infected T84 cells. Res. Microbiol. *156*, 256–262.

Pena, J.A., and Versalovic, J. (2003). *Lactobacillus rhamnosus* GG decreases TNF-alpha production in lipopolysaccharide-activated murine macrophages by a contact-independent mechanism. Cell. Microbiol. 5, 277–285.

Perdigon, G., de Moreno de LeBlanc, A., Valdez, J., and Rachid, M. (2002). Role of yoghurt in the prevention of colon cancer. Eur. J. Clin. Nutr. *56* (Suppl. 3), S65–S68.

Pessi, T., Sutas, Y., Hurme, M., and Isolauri, E. (2000). Interleukin-10 generation in atopic children following oral *Lactobacillus rhamnosus* GG. Clin. Exp. Allergy *30*, 1804–1808.

Peters, R.K., Pike, M.C., Garabrant, D., and Mack, T.M. (1992). Diet and colon cancer in Los Angeles County, California. Cancer Causes Control 3, 457–473.

Petrof, E.O., Kojima, K., Ropeleski, M.J., Musch, M.W., Tao, Y., De Simone, C., and Chang, E.B. (2004). Probiotics inhibit nuclear factor-kappaB and induce heat shock proteins in colonic epithelial cells through proteasome inhibition. Gastroenterology *127*, 1474–1487.

Pool-Zobel, B.L. (2005). Inulin-type fructans and reduction in colon cancer risk: review of experimental and human data. Br. J. Nutr. *93* (Suppl. 1), S73–S90.

Pool-Zobel, B.L., Bertram, B., Knoll, M., Lambertz, R., Neudecker, C., Schillinger, U., Schmezer, P., and Holzapfel, W.H. (1993). Antigenotoxic properties of lactic acid bacteria in vivo in the gastrointestinal tract of rats. Nutr. Cancer *20*, 271–281.

Pool-Zobel, B.L., Neudecker, C., Domizlaff, I., Ji, S., Schillinger, U., Rumney, C., Moretti, M., Vilarini, I., Scassellati-Sforzolini, R., and Rowland, I. (1996). *Lactobacillus-* and *Bifidobacterium*-mediated antigenotoxicity in the colon of rats. Nutr. Cancer *26*, 365–380.

Pretlow, T.P., O'Riordan, M.A., Somich, G.A., Amini, S.B., and Pretlow, T.G. (1992). Aberrant crypts correlate with tumor incidence in F344 rats treated with azoxymethane and phytate. Carcinogenesis *13*, 1509–1512.

Qin, H.L., Shen, T.Y., Gao, Z.G., Fan, X.B., Hang, X.M., Jiang, Y.Q., and Zhang, H.Z. (2005). Effect of *Lactobacillus* on the gut microflora and barrier function of the rats with abdominal infection. World J. Gastroenterol. *11*, 2591–2596.

Rachid, M., Matar, C., Duarte, J., and Perdigon, G. (2006). Effect of milk fermented with a *Lactobacillus helveticus* R389(+) proteolytic strain on the immune system and on the growth of 4T1 breast cancer cells in mice. FEMS Immunol. Med. Microbiol. *47*, 242–253.

Rachid, M.M., Gobbato, N.M., Valdez, J.C., Vitalone, H.H., and Perdigon, G. (2002). Effect of yogurt on the inhibition of an intestinal carcinoma by increasing cellular apoptosis. Int. J. Immunopathol. Pharmacol. *15*, 209–216.

Rafter, J., Bennett, M., Caderni, G., Clune, Y., Hughes, R., Karlsson, P.C., Klinder, A., O'Riordan, M., O'Sullivan, G.C., Pool-Zobel, B., *et al.* (2007). Dietary synbiotics reduce cancer risk factors in polypectomized and colon cancer patients. Am. J. Clin. Nutr. *85*, 488–496.

Raitano, A.B., and Korc, M. (1993). Growth inhibition of a human colorectal carcinoma cell line by interleukin 1 is associated with enhanced expression of gamma-interferon receptors. Cancer Res. *53*, 636–640.

Reddy, B.S. (1998). Prevention of colon cancer by pre- and probiotics: evidence from laboratory studies. Br. J. Nutr. *80*, S219–223.

Reddy, B.S., and Rivenson, A. (1993). Inhibitory effect of *Bifidobacterium longum* on colon, mammary, and liver carcinogenesis induced by 2-amino-3-methylimidazo[4,5-f]quinoline, a food mutagen. Cancer Res. *53*, 3914–3918.

Reddy, G.V., Shahani, K.M., and Banerjee, M.R. (1973). Inhibitory effect of yogurt on Ehrlich Ascites tumor-cell proliferation. J. Natl. Cancer Inst. *50*, 815–817.

Renner, H.W., and Munzner, R. (1991). The possible role of probiotics as dietary antimutagen. Mutat. Res. *262*, 239–245.

Resta-Lenert, S., and Barrett, K.E. (2003). Live probiotics protect intestinal epithelial cells from the effects of infection with enteroinvasive *Escherichia coli* (EIEC). Gut 52, 988–997.

Riedel, C.U., Foata, F., Philippe, D., Adolfsson, O., Eikmanns, B.J., and Blum, S. (2006). Anti-inflammatory effects of bifidobacteria by inhibition of LPS-induced NF-kappaB activation. World J. Gastroenterol. *12*, 3729–3735.

Rodriguez, J.M., Martinez, M.I., and Kok, J. (2002). Pediocin PA-1, a wide-spectrum bacteriocin from lactic acid bacteria. Crit. Rev. Food Sci. Nutr. *42*, 91–121.

Roller, M., Clune, Y., Collins, K., Rechkemmer, G., and Watzl, B. (2007). Consumption of prebiotic inulin enriched with oligofructose in combination with the probiotics *Lactobacillus rhamnosus* and *Bifidobacterium lactis* has minor effects on selected immune parameters in polypectomised and colon cancer patients. Br. J. Nutr. *97*, 676–684.

Roller, M., Pietro Femia, A., Caderni, G., Rechkemmer, G., and Watzl, B. (2004). Intestinal immunity of rats with colon cancer is modulated by oligofructose-enriched inulin combined with *Lactobacillus rhamnosus* and *Bifidobacterium lactis*. Br J. Nutr. *92*, 931–938.

Rowland, I. (2004). Probiotics and colorectal cancer risk. Br. J. Nutr. *91*, 805–807.

Rowland, I.R., Rumney, C.J., Coutts, J.T., and Lievense, L.C. (1998). Effect of *Bifidobacterium longum* and inulin on gut bacterial metabolism and carcinogen-induced aberrant crypt foci in rats. Carcinogenesis *19*, 281–285.

Sekine, K., Toida, T., Saito, M., Kuboyama, M., Kawashima, T., and Hashimoto, Y. (1985). A new morphologically characterized cell wall preparation (whole peptidoglycan) from *Bifidobacterium infantis* with a higher efficacy on the regression of an established tumor in mice. Cancer Res. *45*, 1300–1307.

Sellon, R.K., Tonkonogy, S., Schultz, M., Dieleman, L.A., Grenther, W., Balish, E., Rennick, D.M., and Sartor, R.B. (1998). Resident enteric bacteria are necessary for development of spontaneous colitis and immune system activation in interleukin-10-deficient mice. Infect. Immun. *66*, 5224–5231.

Sen, R. (2006). Control of B lymphocyte apoptosis by the transcription factor NF-kappaB. Immunity *25*, 871–883.

Shacter, E., and Weitzman, S.A. (2002). Chronic inflammation and cancer. Oncology (Williston Park) *16*, 217–226, 229; discussion 230–212.

Shahani, K.M., and Ayebo, A.D. (1980). Role of dietary lactobacilli in gastrointestinal microecology. Am. J. Clin. Nutr. *33*, 2448–2457.

Shi, H.N., and Walker, A. (2004). Bacterial colonization and the development of intestinal defences. Can. J. Gastroenterol. *18*, 493–500.

Singh, J., Rivenson, A., Tomita, M., Shimamura, S., Ishibashi, N., and Reddy, B.S. (1997). *Bifidobacterium longum*, a lactic acid-producing intestinal bacterium inhibits colon cancer and modulates the intermediate biomarkers of colon carcinogenesis. Carcinogenesis *18*, 833–841.

Spanhaak, S., Havenaar, R., and Schaafsma, G. (1998). The effect of consumption of milk fermented by *Lactobacillus casei* strain Shirota on the intestinal microflora and immune parameters in humans. Eur. J. Clin. Nutr. 52, 899–907.

Sullivan, A., and Nord, C.E. (2005). Probiotics and gastrointestinal diseases. J. Intern. Med. *257*, 78–92.

Swennen, K., Courtin, C.M., and Delcour, J.A. (2006). Non-digestible oligosaccharides with prebiotic properties. Crit. Rev. Food Sci. Nutr. *46*, 459–471.

Swidsinski, A., Khilkin, M., Kerjaschki, D., Schreiber, S., Ortner, M., Weber, J., and Lochs, H. (1998). Association between intraepithelial *Escherichia coli* and colorectal cancer. Gastroenterology *115*, 281–286.

Takagi, A., Matsuzaki, T., Sato, M., Nomoto, K., Morotomi, M., and Yokokura, T. (2001). Enhancement of natural killer cytotoxicity delayed murine carcinogenesis by a probiotic microorganism. Carcinogenesis 22, 599–605.

Takaku, K., Oshima, M., Miyoshi, H., Matsui, M., Seldin, M.F., and Taketo, M.M. (1998). Intestinal tumorigenesis in compound mutant mice of both Dpc4 (Smad4) and Apc genes. Cell *92*, 645–656.

Takeda, K., and Okumura, K. (2007). Effects of a fermented milk drink containing *Lactobacillus casei* strain Shirota on the human NK-cell activity. J. Nutr. *137*, 791S–793S.

Takeuchi, M., Nagai, S., Nakajima, A., Shinya, M., Tsukano, C., Asada, H., Yoshikawa, K., Yoshimura, M., and Izumi, T. (2001). Inhibition of lung natural killer cell activity by smoking: the role of alveolar macrophages. Respiration *68*, 262–267.

Talarico, T.L., and Dobrogosz, W.J. (1989). Chemical characterization of an antimicrobial substance produced by *Lactobacillus reuteri*. Antimicrob. Agents Chemother. *33*, 674–679.

Taper, H.S., and Roberfroid, M.B. (2005). Possible adjuvant cancer therapy by two prebiotics – inulin or oligofructose. In Vivo *19*, 201–204.

Tejada-Simon, M.V., Ustunol, Z., and Pestka, J.J. (1999). Ex vivo effects of lactobacilli, streptococci, and bifidobacteria ingestion on cytokine and nitric oxide production in a murine model. J. Food Prot. *62*, 162–169.

Tien, M.T., Girardin, S.E., Regnault, B., Le Bourhis, L., Dillies, M.A., Coppee, J.Y., Bourdet-Sicard, R., Sansonetti, P.J., and Pedron, T. (2006). Anti-inflammatory effect of *Lactobacillus casei* on Shigella-infected human intestinal epithelial cells. J. Immunol. *176*, 1228–1237.

Vagefi, P.A., and Longo, W.E. (2005). Colorectal cancer in patients with inflammatory bowel disease. Clin. Colorectal Cancer *4*, 313–319.

van't Veer, P., Dekker, J.M., Lamers, J.W., Kok, F.J., Schouten, E.G., Brants, H.A., Sturmans, F., and Hermus, R.J. (1989). Consumption of fermented milk products and breast cancer: a case-control study in The Netherlands. Cancer Res. *49*, 4020–4023.

Versalovic, J., and Relman, D. (2006). How bacterial communities expand functional repertoires. PLoS Biol. *4*, e430.

Vincent, J.G., Veomett, R.C., and Riley, R.F. (1959). Antibacterial activity associated with *Lactobacillus acidophilus*. J. Bacteriol. *78*, 477–484.

Vivona, A.A., Shpitz, B., Medline, A., Bruce, W.R., Hay, K., Ward, M.A., Stern, H.S., and Gallinger, S. (1993). K-ras mutations in aberrant crypt foci, adenomas and adenocarcinomas during azoxymethane-induced colon carcinogenesis. Carcinogenesis *14*, 1777–1781.

von der Weid, T., Bulliard, C., and Schiffrin, E.J. (2001). Induction by a lactic acid bacterium of a population of CD4(+) T cells with low proliferative capacity that produce transforming growth factor beta and interleukin-10. Clin. Diagn. Lab. Immunol. *8*, 695–701.

Wargovich, M.J., Chen, C.D., Jimenez, A., Steele, V.E., Velasco, M., Stephens, L.C., Price, R., Gray, K., and Kelloff, G.J. (1996). Aberrant crypts as a biomarker for colon cancer: evaluation of potential chemopreventive agents in the rat. Cancer Epidemiol. Biomarkers Prev. *5*, 355–360.

Watanabe, K., Kawamori, T., Nakatsugi, S., and Wakabayashi, K. (2000). COX-2 and iNOS, good targets for chemoprevention of colon cancer. Biofactors *12*, 129–133.

Williams, E.A., Coxhead, J.M., and Mathers, J.C. (2003). Anti-cancer effects of butyrate: use of micro-array technology to investigate mechanisms. Proc. Nutr. Soc. *62*, 107–115.

Yan, F., and Polk, D.B. (2002). Probiotic bacterium prevents cytokine-induced apoptosis in intestinal epithelial cells. J. Biol. Chem. *277*, 50959–50965.

Yasui, H., Shida, K., Matsuzaki, T., and Yokokura, T. (1999). Immunomodulatory function of lactic acid bacteria. Antonie Van Leeuwenhoek *76*, 383–389.

Yasutake, N., Matsuzaki, T., Kimura, K., Hashimoto, S., Yokokura, T., and Yoshikai, Y. (1999). The role of tumor necrosis factor (TNF)-alpha in the antitumor effect of intrapleural injection of *Lactobacillus casei* strain Shirota in mice. Med. Microbiol. Immunol. *188*, 9–14.

Yokokura, T., Kato, I., Matsuzaki, T., Mutai, M., and Satoh, H. (1984). [Antitumor activity of *Lactobacillus casei* YIT 9018 (LC 9018) – effect of administration route]. Gan To Kagaku Ryoho *11*, 2427–2433.

Young, T.B., and Wolf, D.A. (1988). Case–control study of proximal and distal colon cancer and diet in Wisconsin. Int. J. Cancer *42*, 167–175.

Zareie, M., Johnson-Henry, K., Jury, J., Yang, P.C., Ngan, B.Y., McKay, D.M., Soderholm, J.D., Perdue, M.H., and Sherman, P.M. (2006). Probiotics prevent bacterial translocation and improve intestinal barrier function in rats following chronic psychological stress. Gut *55*, 1553–1560.

Zhang, L., Li, N., Caicedo, R., and Neu, J. (2005). Alive and dead *Lactobacillus rhamnosus* GG decrease tumor necrosis factor-alpha-induced interleukin-8 production in Caco-2 cells. J. Nutr. *135*, 1752–1756.

Zhang, X.B., and Ohta, Y. (1993). Microorganisms in the gastrointestinal tract of the rat prevent absorption of the mutagen-carcinogen 3-amino-1,4-dimethyl-5H-pyrido(4,3-b)indole. Can. J. Microbiol. *39*, 841–845.

Lactobacillus in the Gastrointestinal Tract

9

John Keohane, Kieran Ryan and Fergus Shanahan

Abstract

Probiotics are living organisms which when consumed have beneficial health benefits outside their inherent nutritional effects. Much has been written in both the medical literature and lay press about their potential benefits, some of it well validated by randomized controlled trials but other claims remain unsubstantiated. There is a growing body of evidence for the role of probiotics in gastrointestinal infections, irritable bowel syndrome and inflammatory bowel disease which will be discussed further in this chapter.

The lactobacillus group of bacteria is the best studied species of probiotic and this chapter concentrates on its role in the management of gastrointestinal disease.

Introduction

Lactic acid bacteria can be broadly defined as bacteria that produce lactic acid as a main end product of carbohydrate metabolism. Probably the most widely known and best studied lactic acid bacteria are the *Lactobacteriaceae*, of which the genus *Lactobacillus* is the biggest group containing over 80 species (Satokari *et al.*, 2003). The nutritional requirements of lactobacilli are reflected in their habitats which are rich in carbohydrate containing substrates, and they are most often found in plant material, in fermented or spoiled foodstuffs, or in association with animals. It has been suggested that the somewhat unusual diversity of the lactobacilli may be a result of their ability to grow in such diverse environments (Canchaya *et al.*, 2006). Although lactobacilli have long been used by humans for many purposes (for example in the fermentation of vegetable, dairy and meat products), in recent times the ability of these bacteria to confer certain beneficial health effects has been of greatest interest (Bernardeau *et al.*, 2006; O'Hara and Shanahan, 2006; O'Sullivan *et al.*, 2005). Lactobacilli are part of the normal gastrointestinal (GI) microbiota of humans and many other animals, and are also found to a lesser extent in the human genital and respiratory tracts. Historically the commensal microbiota of the GI tract was thought to be somewhat benign but experiments in more recent times using germ-free animal models have shown that the normal microbiota has profound effects on both host physiology and immunology (Backhed *et al.*, 2004; Mazmanian *et al.*, 2005; Xu and Gordon, 2003).

At birth the GI tract of vertebrates is sterile, but rapidly becomes colonized by bacteria (Tannock, 1995). In the first months of life the gut microbiota changes and increases in diversity as better colonizers replace early colonizers (Edwards and Parrett, 2002). As the gut flora evolves new metabolic pathways are created allowing subsequent colonization by new species. Over time all niches in the gut will eventually become colonized by a well-adapted sub-set of bacteria (Berg, 1996). Precisely which bacteria will become long-term colonizers depends on the biochemical capability of the micro-organisms, the micro-environment determined by the host cells and the available foodstuffs (Backhed *et al.*,

2005; Kassen and Rainey, 2004). In an adult, the organisms which inhabit an individual's GI tract are extremely stable and it is very difficult to introduce a new species (Zoetendal *et al.*, 1998; Vanhoutte *et al.*, 2006). New rRNA-based metagenomic approaches and advanced in fluorescent *in situ* hybridization (FISH) have helped to expand our knowledge of the diversity of bacterial species in the human gut (Zoetendal *et al.*, 2004; Manichanh *et al.*, 2006; Eckburg *et al.*, 2005). For example, it has been estimated that 1000 different species can inhabit the intestine of an individual at any one time (Xu and Gordon, 2003). It has been estimated that bacteria account for 35–50% of the volume of the contents of the human colon and have a collective metabolism equivalent to that of the liver (O'Hara and Shanahan, 2006). The dominant groups which each make up about 30% of the total bacteria in intestinal mucous and faeces are the Cytophaga-Flavobacterium-Bacteroides and the Firmicutes (the *Clostridium* and *Eubacteria*) (Backhed *et al.*, 2005; Eckburg *et al.*, 2005). As part of this gut microbiota, it is estimated that lactobacilli number 10^3–10^6/ml in the oral cavity, 10^3/ml in the stomach, and 10^4/ml in the duodenum and jejunum (Walter J, 2005). The number of lactobacilli increases in the intestine with up to 10^8/ml in the ileum and 10^9/ml in the colon (Walter J, 2005). Somewhere in the region of 10^{7-8} lactobacilli are thought to be shed per gram of human faeces (Rinttila *et al.*, 2004). Due to the relatively high numbers found there, most studies on lactobacilli in the human GI tract have focused on the intestine. However, it is worth remembering that the lactobacilli here represent a relatively small proportion of the overall bacterial population, probably less then 1% of the total (Sghir *et al.*, 2000; Harmsen *et al.*, 2002). In all, 17 species of lactobacilli are putative inhabitants of the human gut (Walter J, 2005). The most dominant colonizers are thought to be *Lactobacillus gasseri*, *L. crispatus*, *L. reuteri*, *L.casei* and *L. salivarius* but it is thought that some of these are transient colonizers that do not remain in the host for prolonged periods of time (Walter J, 2005; Tannock *et al.*, 2000; Kimura *et al.*, 1997).

The function of the intestinal flora is myriad including competitive interaction with pathogens, reduction in bacterial translocation, production of antimicrobial products, and anti-inflammatory signalling.

Lactobacilli are one of the most frequently used bacterial strains by microbiologists and food scientists today especially as a potential probiotic. Probiotics are described as 'living micro-organisms which upon ingestion in certain numbers, exert health affects beyond inherent basic nutrition' (Guarner and Schaafsma, 1998). This concept is not new as over a century ago Nobel Laureate Eli Metchnikoff reported that fermented milk with lactobacilli might be responsible for the increased longevity seen in Bulgarian peasants. This was the birth of what became the probiotic era and since then there has been a continual expansion of products available to consumers with purported probiotic properties. However, the requirement that the bacteria be intact and live may be too restrictive as recent data suggest that bacterial constituents, such as DNA or biologically active secreted metabolites called bacteriocins also account for some of the anti-inflammatory effects of probiotics (Isolauri *et al.*, 2004; Berg, 1996).

Criteria for designating a bacterial strain as a probiotic include human origin, acid and bile stability, absence of demonstrable pathogenicity, gastrointestinal transit and survival, production of antimicrobial substances, and modulation of the immune response (Dunne *et al.*, 2001). Since then, the terminology has evolved and the term 'pharmabiotics' now embraces all forms of therapeutic manipulation of the intestinal flora with anti-, pro-, pre-, or synbiotics (Tannock, 2004). Prebiotics are non-digestible food products, generally oligosaccharides, which on ingestion stimulate the growth of probiotic bacteria, and the term synbiotics describes the combination of a pre- and probiotic.

The last decade has seen the introduction of probiotic drinks and yoghurts on the market and this market continues to grow rapidly, with an estimated value of $US2 billion per annum year to the European market (Young, 1998).

Many of these strains lack good evidence base and need further validation in well designed randomized controlled trials. In this chapter we look at the role of the *Lactobacillus* strain in the gastrointestinal tract and the evidence for its role in conditions such as lactose intolerance, infectious diarrhoea, antibiotic-associated diarrhoea, inflammatory bowel disease, irritable bowel syndrome and in the prevention of colon cancer.

Lactobacillus and lactose intolerance

Lactose intolerance is the most common disorder of intestinal carbohydrate digestion with a prevalence of 7–20% in adult Caucasians in Europe and America and as high as 80–95% among Native Americans, 65–75% among Africans and African Americans (Scrimshaw and Murray, 1988). It is due to the loss of the enzyme lactase in the intestine resulting in diarrhoea, abdominal pain, bloating and flatulence after the ingestion of milk or milk-containing products. Traditionally it was treated with lactase replacement and dietary lactose restriction. Some probiotics such as *Lactobacillus* have a marked intestinal adherence and high beta-galactosidase activity, which may benefit lactose intolerant individuals. Several studies have looked at the role of various probiotic strains in the treatment of lactose intolerance with mixed results. A systematic review by Levri and coworkers found inconsistent results but did come to the conclusion that some strains warranted further investigation (Levri *et al.*, 2005). Further research is needed in this area to clarify the role if any that probiotics might have and which strains to use.

Lactobacillus and diarrhoeal illness

The role for probiotics in diarrhoeal illnesses such as infectious diarrhoea or antibiotic-associated diarrhoea has been extensively studied with a good evidence base. Infectious diarrhoea contributes substantially to paediatric morbidity and mortality worldwide, with an estimated 21–37 million episodes occurring in 16.5 million children less than 5 years of age annually in the USA (Glass *et al.*, 1991). This represents a huge burden on families and the healthcare system. Most gastrointestinal infections affecting children are due to rotavirus infection, which is characterized by vomiting and diarrhoea. Other commonly encountered pathogens include *Escherichia coli*, *Salmonella*, *Shigella*, *Campylobacter*, *Vibrio cholera* and *Giardia*. The role for probiotics in infectious diarrhoea is well validated as demonstrated by the recent Cochrane systematic review which showed probiotic feeding reduced the risk and duration of diarrhoea in adults and children (Allen *et al.*, 2004).

Antibiotic-associated diarrhoea is a common clinical problem, occurring in up to 25% of patients, and a quarter of cases are due to *Clostridium difficile* infection (Katz, 2006). With such an impact on healthcare resources, an inexpensive, safe and effective therapy would be desirable. Current treatment options include oral rehydration therapy, antibiotics, and gut motility agents such as loperamide. The rationale for using probiotics in infectious diarrhoea is based on evidence to show stabilization of the indigenous microbiota, reduction of rotavirus shedding, decreased intestinal permeability caused by rotavirus shedding and an increase in IgA-secreting cells to rotavirus (Isolauri *et al.*, 2004).

Naaber *et al.* has previously shown, using an *in vitro* model of 23 different *C. difficile* strains the antagonistic activity of various different lactobacilli. They found five strains (*L. paracasei* and *L. plantarum* species) to be antagonistic to all the *C. difficile* strains and concluded that the antagonistic activity was strain specific and correlated with hydrogen peroxide and lactic acid production (Naaber *et al.*, 2004). However, the few uncontrolled studies in humans did suggest a role for *Lactobacillus* GG in recurrent *C. difficile*-associated diarrhoea but these trials were small and will need larger controlled trials to support this preliminary data (Biller *et al.*, 1995; Gorbach *et al.*, 1987). The results of many of the trials using *Lactobacillus* in gastrointestinal infections including *Helicobacter* infection are summarized in Table 9.1.

Lactobacillus in inflammatory bowel disease

Inflammatory bowel disease (IBD) comprises Crohn's disease (CD) and ulcerative colitis (UC) and represents a significant healthcare burden with between 0.1% and 0.2% of developed countries affected (Shanahan, 2000b) The aetiology of IBD is complex but the current consensus is that it is due to an exaggerated immune response to enteric microflora in genetically susceptible individuals (Shanahan, 2000a; Elson, 2000). There

Table 9.1 Summary of studies of *Lactobacillus* therapy in GI infections

Diarrhoea	*Lactobacillus* strain	Trial outcome	Reference
Infantile diarrhoea	*Lactobacillus reuteri*	Reduced duration of diarrhoea	Shornikova *et al.*, 1997a
	L. reuteri	Reduced duration of rotavirus-associated diarrhoea	Shornikova *et al.*, 1997b
	L. rhamnosus and *L. reuteri*	Reduced duration of diarrhoea in non-hospitalised patients	Rosenfeldt *et al.*, 2002
	L. GG	Reduced duration of rotavirus-associated diarrhoea and enhanced IgA response	Kaila *et al.*, 1992
	L. GG	Reduced vomiting and diarrhoea in a cohort of hospitalized patients	Raza *et al.*, 1995
	L. GG	Reduced duration of diarrhoea and shortened hospital admission	Guandalini *et al.*, 2000
Traveller's diarrhoea	*L.* GG	Reduction in diarrhoea by a maximum of 39.5%	Oksanen *et al.*, 1990
	L. GG	Modest reduction in diarrhoea	Hilton *et al.*, 1997
	L. acidophilus and *L. bulgaricus*	No reduction in diarrhoea	de dios Pozo-Olano *et al.*, 1978
Antibiotic-associated diarrhoea	*L.* GG	Significant reduction in GI side-effects in triple therapy group	Armuzzi *et al.*, 2001
	L. GG	No reduction in diarrhoea in adults receiving antibiotics	Thomas *et al.*, 2001
	L. GG	Reduced incidence of diarrhoea in children receiving antibiotics for acute respiratory infection	Arvola *et al.*, 1999
	L. GG	Reduced diarrhoea in antibiotic-treated children	Vanderhoof *et al.*, 1999
	L. GG	Decreased duration of erythromycin-induced diarrhoea	Siitonen *et al.*, 1990
	L. acidophilus and *L. bulgaricus*	Decrease in ampicillin-induced diarrhoea in adults	Gotz *et al.*, 1979
Helicobacter infection	*L. gasseri* LG21	Improved [^{13}C]urea breath test and decreased gastric mucosal inflammation	Sakamoto *et al.*, 2001
	L. acidophilus	Inactivated culture increased the eradication rate of standard therapy	Canducci *et al.*, 2000
	L. johnsonii La1	Reduction in helicobacter gastritis and inflammation. No improvement in eradication	Felley *et al.*, 2001
	L. johnsonii La1 supernatant	Decreased breath test values in patients but persistence of *H. pylori* infection on biopsy	Michetti *et al.*, 1999
	L. johnsonii La1	Decreased gastritis score and *H. pylori* density after 16 weeks' treatment	Pantoflickova *et al.*, 2003
	Lactobacillus spp. (three strains)	Urea breath test remained positive after 30 days probiotic	Wendakoon *et al.*, 2002

is a large body of evidence to suggest the enteric microflora is a key component of the inflammatory process in IBD:

- It has long been known that the areas in the gut with the highest bacterial concentrations are the sites most frequently affected by IBD (Simon and Gorbach, 1984).
- The diversion of the faecal stream is associated with distal improvement in CD (Shanahan, 2002).
- Immune reactivity against enteric bacteria can be seen in CD patients, thus suggesting a loss of tolerance to commensal bacteria (Duchmann *et al.*, 1995; Macpherson *et al.*, 1996).
- Therapeutic efficacy is seen with antibiotics (Ursing *et al.*, 1982).
- There has been reports of increased numbers of bacteria within the mucosa of IBD patients versus healthy controls (Swidsinski *et al.*, 2002; Schultsz *et al.*, 1999).

The above evidence has led to a surge in interest in the therapeutic modification of the enteric flora with probiotics in particular the *Lactobacillus* strain in the management of the IBD patient.

Crohn's disease

There are several studies involving different strains of lactobacilli in the treatment of Crohn's disease and these are summarized in Table 9.2. The results are varied with many small studies, the results of which are inconclusive. *L. rhamnosus* GG has been used in two open-label studies in paediatric CD patients. One study demonstrated the immunostimulatory effects of L. GG by its ability to increase gut IgA levels but no clinical improvement was noted while the other study reported improved intestinal permeability and clinical scores in four children with mildly active CD (Malin *et al.*, 1996; Gupta *et al.*, 2000). More recently there has been several randomized controlled trials using *L. rhamnosus* GG and *L. johnsonii* LA1 in the induction and maintenance of remission of CD. These studies showed no statistical difference in the probiotic arms versus the placebo. The one study that did demonstrate a beneficial effect in the maintenance of a surgically induced remission was the probiotic mixture VSL#3 in addition to the antibiotic rifamixin compared to mesalazine alone but this was in abstract form only (Campieri *et al.*, 2000).

Ulcerative colitis

The results of published trials with pharmabiotics in UC are shown in Table 9.3. Two open label studies have demonstrated the efficacy of the probiotic combination VSL#3 in the induction and maintenance of remission in UC and also confirmed the presence of significant concentrations of probiotics in the faecal cultures (Bibiloni *et al.*, 2005; Venturi *et al.*, 1999). *L. rhamnosus* GG has also been shown to be equally efficacious as mesalazine in the maintenance of remission of UC in a large open label study (Zocco *et al.*, 2006).

Pouchitis

Pouchitis is estimated to occur in between 24% and 46% of patients after ileal pouch anal anastomosis and approximately 10% develop chronic pouchitis (Sandborn *et al.*, 1994; Fazio *et al.*, 1995). Its aetiology remains unknown but an imbalance in the bacterial flora in the pouch including reduced *Bifidobacteria* and lactobacilli and increased counts of *Clostridia perfringens* may be contributory (Ruseler-van Embden *et al.*, 1994). The human studies to date are summarized in Table 9.4. *VSL#3* has been well studied in three separate randomized controlled trials and has demonstrated good efficacy in the maintenance of remission after antibiotic induced remission and in the postoperative prevention of pouchitis (Gionchetti *et al.*, 2000; Gionchetti *et al.*, 2003; Mimura *et al.*, 2004). However, more recently Shen and coworkers published an open label study in 31 patients with antibiotic-dependent pouchitis using *VSL#3* and found only six patients remained on the product for the 8-month duration of follow-up with no difference in the Pouchitis Disease Activity Index (PDAI) from baseline (Shen *et al.*, 2005). There were some inherent problems with this study in that patients had to purchase the probiotic themselves and bear the cost, and the medication was self-administered with no measure of adherence.

In terms of lactobacilli alone, Kuisma and coworkers did show a change in the pouch flora

Table 9.2 Summary of human studies of probiotics in Crohn's disease (CD)

Study type	Organism used and duration	Trial outcome	Reference
Open labelled trial in paediatric CD	*Lactobacillus rhamnosus* GG for 10 days ($n=14$)	Increase in gut IgA response	Malin *et al.*, 1996
Open label trial in children with mildly active CD	*Lactobacillus rhamnosus* GG for 6 months. ($n = 4$)	Improved intestinal permeability and CDAI	Gupta *et al.*, 2000
RCT in maintenance of surgical induced remission CD	*VSL#3* with rifamixin versus mesalazine for 12 months ($n = 40$)	Endoscopic remission of 80% in probiotic group versus 60% in mesalazine group	Campieri *et al.*, 2000
Open label trial in mild/moderately active CD	*Lactobacillus salivarius* UCC118 for 3 months ($n = 25$)	Reduction in CDAI and reduced steroid use in the probiotic group	McCarthy *et al.*, 2001
RCT in maintenance of surgical induced remission	*Lactobacillus rhamnosus* GG for 12 months ($n = 45$)	No difference between two groups at 1 year	Prantera *et al.*, 2002
RCT in induction and maintenance of remission	*Lactobacillus rhamnosus* GG for 6 months ($n = 11$)	No benefit in induction or maintenance of remission	Schultz *et al.*, 2004
RCT in maintenance of remission in children	*Lactobacillus rhamnosus* GG for 2 years ($n=39$)	No difference in the median time to relapse compared to placebo	Bousvaros *et al.*, 2005
RCT in maintenance of remission	*Lactobacillus johnsonii* LA1 for 6 months ($n=48$)	No benefit in preventing endoscopic recurrence compared to placebo.	Marteau *et al.*, 2006
RCT in prevention of post operative endoscopic recurrence.	*Lactobacillus johnsonii LA1* for 12 weeks (n=70).	No difference in early endoscopic recurrence compared to placebo.	Van *et al.*, 2006

RCT = randomized control trial; n = total number of patients; CDAI = Crohn's disease activity index; VSL#3 comprises *Lactobacillus casei*, *L. plantarum*, *L. acidophilus*, *L. delbrueckii*, *Bifidobacterium infantis*, *B. breve*, *B. longum*, and *Streptococcus salivarius* subsp.

Table 9.3 Summary of human trials of probiotics in ulcerative colitis

Study type	Organism used	Trial outcome	Reference
Open label study in maintenance of remission	*VSL*#3 for 12 months (n = 20)	75% in remission	Venturi *et al.*, 1999
Open label study in induction of remission	*VSL*#3 for 6 weeks (n = 34)	53% remission rate	Bibiloni *et al.*, 2005
RCT in maintenance of remission	*Bifidobacteria*, Lactobacilli, Enterococci combination (BIFICO) for 8 weeks (n = 30).	20% relapse in treatment arm vs. 93% placebo group ($P < 0.01$)	Cui *et al.*, 2004
Open label study in the maintenance of remission.	*Lactobacillus rhamnosus* GG for 12 months (n = 187).	No difference in relapse rates compared to mesalazine alone	Zocco *et al.*, 2006
RCT in maintenance of a steroid induced remission	*Lactobacillus salivarius UCC118, Bifidobacterium infantis* 35624, or placebo for 12 months (n = 157)	No significant benefit with either probiotic over placebo	Shanahan *et al.*, 2006

Table 9.4 Summary of clinical trials of probiotics in pouchitis

Study type	Organism used	Trial outcome	Reference
Open label trial	*Lactobacillus* GG and prebiotic with antibiotic (n = 10)	Effective in inducing remission	Friedman and George, 2000
RCT to maintain remission after antibiotic induced remission	*VSL*#3 (n = 40)	15% relapse in probiotic arm vs. 100% placebo arm.	Gionchetti *et al.*, 2000
RCT in post-operative prevention	*VSL*#3 (n = 40)	Acute pouchitis in 10% probiotic arm vs. 40% placebo group ($P < 0.01$)	Gionchetti *et al.*, 2003
RCT to maintain remission in relapsing/ recurrent pouchitis after antibiotic-induced remission	*VSL*#3 (n = 36)	Remission in 85% probiotic group vs. 6% placebo ($P < 0.0001$)	Mimura *et al.*, 2004
Open label study in antibiotic-dependent pouchitis	*VSL*#3 (n = 31)	After 8 months only six remained on probiotic. No difference in PDAI.	Shen *et al.*, 2005
RCT in acutely active pouchitis	*Lactobacillus* GG (n = 20)	No benefit in clinical or endoscopic response.	Kuisma *et al.*, 2003
Case–control study in postoperative prevention	*Lactobacillus* GG (n = 117)	Decreased risk of pouchitis. Cumulative risk at 3 years: 7% vs. 29% (P = 0.011)	Gosselink *et al.*, 2004
Open label study in prevention of acute pouchitis	*Lactobacillus acidophilus* and *Bif lactis* (n = 51)	Improvement in PDAI, but not histology	Laake *et al.*, 2004

RCT, randomized control trial; n, total number of patients; PDAI, = Pouchitis Disease Activity Index.

with *L.* GG but no significant effect on clinical or endoscopic response (Kuisma *et al.*, 2003). However, Gosselink and coworkers did show it to be efficacious in the prevention of pouchitis after 3 years (Gosselink *et al.*, 2004).

It is apparent from the above studies that the issue of probiotics in IBD is far from confirmed. Further large well-powered studies are needed to clarify issues such as strains, dose, patient subtypes and whether the use of probiotics should be preceeded by a course of antibiotics. Also the issue of genetically engineered or 'turbo' probiotics has now been realised and proof of principle has been established in animal models of IBD. Steidlerand coworkers have created a genetically modified *Lactococcus lactis* producing the anti-inflammatory cytokine interleukin-10 which has resulted in significant reduction in inflammation when fed to the murine model of IBD (Steidler *et al.*, 2000). Whilst this is not a true *Lactobacillus* it gives us a glimpse at what may be to come in terms of tailor made probiotics.

Lactobacillus in irritable bowel syndrome

Irritable bowel syndrome (IBS) is the most frequent gastrointestinal diagnosis made by gastroenterologists and is estimated to affect 11–14% of the North American population and accounts for 30% of GP consultations. It is characterized by abdominal pain or discomfort relieved by defecation and alteration in bowel habit either constipation or diarrhoea with associated symptoms of flatulence, abdominal distension and mucous per rectum (Drossman *et al.*, 2002). The aetiology is unknown but one of the purported mechanisms is a disturbance in the gut microflora. Balsari and coworkers demonstrated significantly lower numbers of coliforms, lactobacilli and bifidobacteria in IBS patients compared with healthy control subjects (Balsari *et al.*, 1982). This has led to increased interest in therapeutic manipulation of gut microflora with various probiotics strains, especially lactobacilli.

Halpern and coworkers demonstrated a 50% improvement in overall GI function in 18 IBS patients using capsules of heat-killed *L. acidophilus* compared with placebo (Halpern *et al.*, 1996). There have been several trials using *L. plantarum* 299v with varying results. Niedzielin and coworkers compared *L. plantarum* in liquid form versus placebo in 40 patients and found a significant improvement in terms of abdominal pain and in the overall IBS symptom scores in the probiotic arm (Niedzielin *et al.*, 2001). Another study looking at *L. plantarum* in 60 IBS patients over a 4-week period demonstrated a significant decrease in abdominal pain and flatulence (Nobaek *et al.*, 2000). In contrast, a third placebo-controlled trial by Sen and coworkers in 12 patients failed to show any benefit or did not alter colonic fermentation (Sen *et al.*, 2002). Likewise, a randomized controlled trial of *L. casei* GG in 25 patients did not show any significant difference in symptoms but there was a trend toward a reduction in the number of unformed stools in the diarrhoea predominant subgroup (O'Sullivan and O'Morain, 2000).

The case for probiotics in IBS remains to be further validated with larger randomized placebo-controlled trials with a longer duration of treatment and a well-defined patient profile.

Lactobacillus and colorectal cancer

Colorectal cancer (CRC) is the second leading cause of cancer death and shows no sign of decreasing despite newer diagnostic tools and genetic testing. The vast majority of these cancers are the sporadic type and diet is thought to be a contributory factor in its aetiology. Many intestinal bacteria have been implicated in the conversion of dietary constituents into toxic or carcinogenic compounds via the secretion of certain enzymes like β-glucuronidase, nitroreductase, azoreductase and nitrate/nitrite reductase. Several probiotic bacteria especially the lactic acid bacteria have been shown to reduce this faecal enzyme activity (Burns and Rowland, 2000). Multiple animal studies have looked at various strains of lactic acid bacteria in their ability to reduce colonic tumours. Pool-Zobel and coworkers reported the ability of several strains including *L. acidophilus*, *L. gasseri*, *L. confusus*, *Streptococcus thermophilus*, *Bifidobacterium breve* and *B. longum*, to dose dependently reduce DNA damage in rats (Pool-Zobel *et al.*, 1996). Goldin and coworkers also showed the ability of *L.*

rhamnosus GG to interfere with the initiation of 1,2 diethylhydrazine-induced intestinal tumorigenesis in rats (Goldin *et al.*, 1996).

Whilst the evidence for an anti-cancer effect in animal models shows much promise, we await human trials like the EU sponsored SYNCAN randomized controlled trial which will look at a synbiotic preparation in a high risk group (polypectomized subjects) and a control group of volunteers (Van *et al.*, 2005). There is some evidence already to demonstrate that *L. acidophilus* reduced the excretion of faecal mutagens in healthy volunteers on a fried meat diet which is known to increase faecal mutagenicity (Lidbeck A *et al.*, 1992). Certainly the use of lactic acid bacteria as a possible anticarcinogenic agent would be revolutionary but needs further validation.

References

Allen, S.J., Okoko, B., Martinez, E., Gregorio, G., and Dans, L.F. (2004). Probiotics for treating infectious diarrhoea. Cochrane Database Syst. Rev. CD003048.

Armuzzi, A., Cremonini, F., Bartolozzi, F., Canducci, F., Candelli, M., Ojetti, V., Cammarota, G., Anti, M., De, L.A., Pola, P., Gasbarrini, G., and Gasbarrini, A. (2001). The effect of oral administration of *Lactobacillus* GG on antibiotic-associated gastrointestinal side-effects during *Helicobacter pylori* eradication therapy. Aliment. Pharmacol. Ther. *15*, 163–169.

Arvola, T., Laiho, K., Torkkeli, S., Mykkanen, H., Salminen, S., Maunula, L., and Isolauri, E. (1999). Prophylactic *Lactobacillus* GG reduces antibiotic-associated diarrhea in children with respiratory infections: a randomized study. Pediatrics *104*, e64.

Backhed, F., Ding, H., Wang, T., Hooper, L.V., Koh, G.Y., Nagy, A., Semenkovich, C.F., and Gordon, J.I. (2004). The gut microbiota as an environmental factor that regulates fat storage. Proc. Natl. Acad. Sci. USA *101*, 15718–15723.

Backhed, F., Ley, R.E., Sonnenburg, J.L., Peterson, D.A., and Gordon, J.I. (2005). Host-bacterial mutualism in the human intestine. Science *307*, 1915–1920.

Balsari, A., Ceccarelli, A., Dubini, F., Fesce, E., and Poli, G. (1982). The fecal microbial population in the irritable bowel syndrome. Microbiologica *5*, 185–194.

Berg, R.D. (1996). The indigenous gastrointestinal microflora. Trends Microbiol. *4*, 430–435.

Bernardeau, M., Guguen, M., and Vernoux, J.P. (2006). Beneficial lactobacilli in food and feed: long-term use, biodiversity and proposals for specific and realistic safety assessments. FEMS Microbiol. Rev. *30*, 487–513.

Bibiloni, R., Fedorak, R.N., Tannock, G.W., Madsen, K.L., Gionchetti, P., Campieri, M., De, S.C., and Sartor, R.B. (2005). VSL#3 probiotic-mixture induces remission in patients with active ulcerative colitis. Am. J. Gastroenterol. *100*, 1539–1546.

Biller, J.A., Katz, A.J., Flores, A.F., Buie, T.M., and Gorbach, S.L. (1995). Treatment of recurrent *Clostridium difficile* colitis with *Lactobacillus* GG. J. Pediatr. Gastroenterol. Nutr. *21*, 224–226.

Bousvaros, A., Guandalini, S., Baldassano, R.N., Botelho, C., Evans, J., Ferry, G.D., Goldin, B., Hartigan, L., Kugathasan, S., Levy, J., Murray, K.F., Oliva-Hemker, M., Rosh, J.R., Tolia, V., Zholudev, A., Vanderhoof, J.A., and Hibberd, P.L. (2005). A randomized, double-blind trial of *Lactobacillus* GG versus placebo in addition to standard maintenance therapy for children with Crohn's disease. Inflamm. Bowel. Dis. *11*, 833–839.

Burns, A.J. and Rowland, I.R. (2000). Anti-carcinogenicity of probiotics and prebiotics. Curr. Issues Intest. Microbiol. *1*, 13–24.

Campieri M., Rizzello F., Venturi A., Poggioli G., Ugolini F., Helwig U., Amadini C., Romboli E., Gionchetti P. (2000). Combination of antibiotic and probiotic treatment is efficacious in prophylaxis of post-operative recurrence of Crohn's disease: a randomised controlled study vs mesalazine. Gastroenterology *118*, A781.

Canchaya, C., Claesson, M.J., Fitzgerald, G.F., van, S.D., and O'Toole, P.W. (2006). Diversity of the genus *Lactobacillus* revealed by comparative genomics of five species. Microbiology *152*, 3185–3196.

Canducci, F., Armuzzi, A., Cremonini, F., Cammarota, G., Bartolozzi, F., Pola, P., Gasbarrini, G., and Gasbarrini, A. (2000). A lyophilized and inactivated culture of *Lactobacillus acidophilus* increases *Helicobacter pylori* eradication rates. Aliment. Pharmacol. Ther. *14*, 1625–1629.

Cui, H.H., Chen, C.L., Wang, J.D., Yang, Y.J., Cun, Y., Wu, J.B., Liu, Y.H., Dan, H.L., Jian, Y.T., and Chen, X.Q. (2004). Effects of probiotic on intestinal mucosa of patients with ulcerative colitis. World J. Gastroenterol. *10*, 1521–1525.

de dios Pozo-Olano J., Warram, J.H., Jr., Gomez, R.G., and Cavazos, M.G. (1978). Effect of a lactobacilli preparation on traveler's diarrhea. A randomized, double blind clinical trial. Gastroenterology *74*, 829–830.

Drossman, D.A., Camilleri, M., Mayer, E.A., and Whitehead, W.E. (2002). AGA technical review on irritable bowel syndrome. Gastroenterology *123*, 2108–2131.

Duchmann, R., Kaiser, I., Hermann, E., Mayet, W., Ewe, K., and Meyer zum Buschenfelde, K.H. (1995). Tolerance exists towards resident intestinal flora but is broken in active inflammatory bowel disease (IBD). Clin. Exp. Immunol. *102*, 448–455.

Dunne, C., O'Mahony, L., Murphy, L., Thornton, G., Morrissey, D., O'Halloran, S., Feeney, M., Flynn, S., Fitzgerald, G., Daly, C., Kiely, B., O'Sullivan, G.C., Shanahan, F., and Collins, J.K. (2001). In vitro selection criteria for probiotic bacteria of human origin: correlation with in vivo findings. Am. J. Clin. Nutr. *73*, 386S-392S.

Eckburg, P.B., Bik, E.M., Bernstein, C.N., Purdom, E., Dethlefsen, L., Sargent, M., Gill, S.R., Nelson, K.E., and Relman, D.A. (2005). Diversity of the human intestinal microbial flora. Science *308*, 1635–1638.

Edwards, C.A. and Parrett, A.M. (2002). Intestinal flora during the first months of life: new perspectives. Br. J. Nutr. *88* pl 1, Suppl 1, S11–S18.

Elson, C.O. (2000). Commensal bacteria as targets in Crohn's disease. Gastroenterology *119*, 254–257.

Fazio, V.W., Ziv, Y., Church, J.M., Oakley, J.R., Lavery, I.C., Milsom, J.W., and Schroeder, T.K. (1995). Ileal pouch-anal anastomoses complications and function in 1005 patients. Ann. Surg. *222*, 120–127.

Felley, C.P., Corthesy-Theulaz, I., Rivero, J.L., Sipponen, P., Kaufmann, M., Bauerfeind, P., Wiesel, P.H., Brassart, D., Pfeifer, A., Blum, A.L., and Michetti, P. (2001). Favourable effect of an acidified milk (LC-1) on *Helicobacter pylori* gastritis in man. Eur. J. Gastroenterol. Hepatol. *13*, 25–29.

Friedman, G. and George, J. (2000). Treatment of refractory 'pouchitis' with prebiotic and probiotic therapy. Gastroenterology. *118*, A4167.

Gionchetti, P., Rizzello, F., Helwig, U., Venturi, A., Lammers, K.M., Brigidi, P., Vitali, B., Poggioli, G., Miglioli, M., and Campieri, M. (2003). Prophylaxis of pouchitis onset with probiotic therapy: a double-blind, placebo-controlled trial. Gastroenterology *124*, 1202–1209.

Gionchetti, P., Rizzello, F., Venturi, A., Brigidi, P., Matteuzzi, D., Bazzocchi, G., Poggioli, G., Miglioli, M., and Campieri, M. (2000). Oral bacteriotherapy as maintenance treatment in patients with chronic pouchitis: a double-blind, placebo-controlled trial. Gastroenterology *119*, 305–309.

Glass, R.I., Lew, J.F., Gangarosa, R.E., Lebaron, C.W., and Ho, M.S. (1991). Estimates of morbidity and mortality rates for diarrheal diseases in American children. J. Pediatr. *118*, S27–S33.

Goldin, B.R., Gualtieri, L.J., and Moore, R.P. (1996). The effect of *Lactobacillus* GG on the initiation and promotion of DMH-induced intestinal tumors in the rat. Nutr. Cancer 25, 197–204.

Gorbach, S.L., Chang, T.W., and Goldin, B. (1987). Successful treatment of relapsing *Clostridium difficile* colitis with *Lactobacillus* GG. Lancet 330, 1519.

Gosselink, M.P., Schouten, W.R., van Lieshout, L.M., Hop, W.C., Laman, J.D., and Ruseler-van Embden, J.G. (2004). Delay of the first onset of pouchitis by oral intake of the probiotic strain *Lactobacillus rhamnosus* GG. Dis. Colon Rectum *47*, 876–884.

Gotz, V., Romankiewicz, J.A., Moss, J., and Murray, H.W. (1979). Prophylaxis against ampicillin-associated diarrhea with a lactobacillus preparation. Am. J. Hosp. Pharm. *36*, 754–757.

Guandalini, S., Pensabene, L., Zikri, M.A., Dias, J.A., Casali, L.G., Hoekstra, H., Kolacek, S., Massar, K., Micetic-Turk, D., Papadopoulou, A., de Sousa, J.S., Sandhu, B., Szajewska, H., and Weizman, Z. (2000). *Lactobacillus* GG administered in oral rehydration solution to children with acute diarrhea: a multicenter European trial. J. Pediatr. Gastroenterol. Nutr. *30*, 54–60.

Guarner, F. and Schaafsma, G.J. (1998). Probiotics. Int. J. Food Microbiol. 39, 237–238.

Gupta, P., Andrew, H., Kirschner, B.S., and Guandalini, S. (2000). Is lactobacillus GG helpful in children with Crohn's disease? Results of a preliminary, open-label study. J. Pediatr. Gastroenterol. Nutr. *31*, 453–457.

Halpern, G.M., Prindiville, T., Blankenburg, M., Hsia, T., and Gershwin, M.E. (1996). Treatment of irritable bowel syndrome with Lacteol Fort: a randomized, double-blind, cross-over trial. Am. J. Gastroenterol. *91*, 1579–1585.

Harmsen, H.J., Raangs, G.C., He, T., Degener, J.E., and Welling, G.W. (2002). Extensive set of 16S rRNA-based probes for detection of bacteria in human feces. Appl. Environ. Microbiol. *68*, 2982–2990.

Hilton, E., Kolakowski, P., Singer, C., and Smith, M. (1997). Efficacy of *Lactobacillus* GG as a diarrheal preventive in travelers. J. Travel Med. *4*, 41–43.

Isolauri, E., Salminen, S., and Ouwehand, A.C. (2004). Microbial-gut interactions in health and disease. Probiotics. Best Pract. Res. Clin. Gastroenterol. *18*, 299–313.

Kaila, M., Isolauri, E., Soppi, E., Virtanen, E., Laine, S., and Arvilommi, H. (1992). Enhancement of the circulating antibody secreting cell response in human diarrhea by a human *Lactobacillus* strain. Pediatr. Res. *32*, 141–144.

Kassen, R. and Rainey, P.B. (2004). The ecology and genetics of microbial diversity. Ann. Rev. Microbiol. *58*, 207–231.

Katz, J.A. (2006). Probiotics for the prevention of antibiotic-associated diarrhea and *Clostridium difficile* diarrhea. J. Clin. Gastroenterol. *40*, 249–255.

Kimura, K., McCartney, A.L., McConnell, M.A., and Tannock, G.W. (1997). Analysis of fecal populations of bifidobacteria and lactobacilli and investigation of the immunological responses of their human hosts to the predominant strains. Appl. Environ. Microbiol. *63*, 3394–3398.

Kuisma, J., Mentula, S., Jarvinen, H., Kahri, A., Saxelin, M., and Farkkila, M. (2003). Effect of *Lactobacillus rhamnosus* GG on ileal pouch inflammation and microbial flora. Aliment. Pharmacol. Ther. *17*, 509–515.

Laake, K.O., Line, P.D., Grzyb, K., Aamodt, G., Aabakken, L., Roset, A., Hvinden, A.B., Bakka, A., Eide, J., Bjorneklett, A., and Vatn, M.H. (2004). Assessment of mucosal inflammation and blood flow in response to four weeks' intervention with probiotics in patients operated with a J-configurated ileal-pouch-anal-anastomosis (IPAA). Scand. J. Gastroenterol. 39, 1228–1235.

Levri, K.M., Ketvertis, K., Deramo, M., Merenstein, J.H., and D'Amico, F. (2005). Do probiotics reduce adult lactose intolerance? A systematic review. J. Fam. Pract. *54*, 613–620.

Lidbeck A., Overvik E., Rafter J., and et al (1992). Effect of *Lactobacillus acidophilus* supplements on mutagen excretion in feces and urine in humans. Microb. Ecol. Health Dis. 5, 59–67.

Macpherson, A., Khoo, U.Y., Forgacs, I., Philpott-Howard, J., and Bjarnason, I. (1996). Mucosal antibodies in inflammatory bowel disease are directed against intestinal bacteria. Gut 38, 365–375.

Malin, M., Suomalainen, H., Saxelin, M., and Isolauri, E. (1996). Promotion of IgA immune response in patients with Crohn's disease by oral bacteriotherapy with *Lactobacillus* GG. Ann. Nutr. Metab *40*, 137–145.

Manichanh, C., Rigottier-Gois, L., Bonnaud, E., Gloux, K., Pelletier, E., Frangeul, L., Nalin, R., Jarrin, C., Chardon, P., Marteau, P., Roca, J., and Dore, J. (2006). Reduced diversity of faecal microbiota in Crohn's disease revealed by a metagenomic approach. Gut 55, 205–211.

Marteau, P., Lemann, M., Seksik, P., Laharie, D., Colombel, J.F., Bouhnik, Y., Cadiot, G., Soule, J.C., Bourreille, A., Metman, E., Lerebours, E., Carbonnel, F., Dupas, J.L., Veyrac, M., Coffin, B., Moreau, J., Abitbol, V., Blum-Sperisen, S., and Mary, J.Y. (2006). Ineffectiveness of *Lactobacillus johnsonii* LA1 for prophylaxis of postoperative recurrence in Crohn's disease: a randomised, double blind, placebo controlled GETAID trial. Gut 55, 842–847.

Mazmanian, S.K., Liu, C.H., Tzianabos, A.O., and Kasper, D.L. (2005). An immunomodulatory molecule of symbiotic bacteria directs maturation of the host immune system. Cell *122*, 107–118.

McCarthy J., O'Mahony L., Dunne C., Kelly P., Feenay M., Kiely B., O'Sullivan G., Collins JK, Shanahan F. (2001). An open label of a novel probiotic as alternative to steroid in mild/moderately active Crohn's disease. Gut *49(3)*: A2447.

Michetti, P., Dorta, G., Wiesel, P.H., Brassart, D., Verdu, E., Herranz, M., Felley, C., Porta, N., Rouvet, M., Blum, A.L., and Corthesy-Theulaz, I. (1999). Effect of whey-based culture supernatant of *Lactobacillus acidophilus* (johnsonii) La1 on *Helicobacter pylori* infection in humans. Digestion *60*, 203–209.

Mimura, T., Rizzello, F., Helwig, U., Poggioli, G., Schreiber, S., Talbot, I.C., Nicholls, R.J., Gionchetti, P., Campieri, M., and Kamm, M.A. (2004). Once daily high dose probiotic therapy (VSL#3) for maintaining remission in recurrent or refractory pouchitis. Gut 53, 108–114.

Naaber, P., Smidt, I., Stsepetova, J., Brilene, T., Annuk, H., and Mikelsaar, M. (2004). Inhibition of *Clostridium difficile* strains by intestinal *Lactobacillus* species. J. Med. Microbiol. 53, 551–554.

Niedzielin, K., Kordecki, H., and Birkenfeld, B. (2001). A controlled, double-blind, randomized study on the efficacy of *Lactobacillus plantarum* 299V in patients with irritable bowel syndrome. Eur. J. Gastroenterol. Hepatol. *13*, 1143–1147.

Nobaek, S., Johansson, M.L., Molin, G., Ahrné, S., and Jeppsson, B. (2000). Alteration of intestinal microflora is associated with reduction in abdominal bloating and pain in patients with irritable bowel syndrome. Am. J. Gastroenterol. 95, 1231–1238.

O'Hara, A.M. and Shanahan, F. (2006). The gut flora as a forgotten organ. EMBO Rep. 7, 688–693.

O'Sullivan, G.C., Kelly, P., O'Halloran, S., Collins, C., Collins, J.K., Dunne, C., and Shanahan, F. (2005). Probiotics: an emerging therapy. Curr. Pharm. Des. *11*, 3–10.

O'Sullivan, M.A. and O'Morain, C.A. (2000). Bacterial supplementation in the irritable bowel syndrome. A randomised double-blind placebo-controlled crossover study. Dig. Liver Dis. 32, 294–301.

Oksanen, P.J., Salminen, S., Saxelin, M., Hamalainen, P., Ihantola-Vormisto, A., Muurasniemi-Isoviita, L., Nikkari, S., Oksanen, T., Porsti, I., Salminen, E., and. (1990). Prevention of travellers' diarrhoea by *Lactobacillus* GG. Ann. Med. 22, 53–56.

Pantoflickova, D., Corthesy-Theulaz, I., Dorta, G., Stolte, M., Isler, P., Rochat, F., Enslen, M., and Blum, A.L. (2003). Favourable effect of regular intake of fermented milk containing *Lactobacillus johnsonii* on *Helicobacter pylori* associated gastritis. Aliment. Pharmacol. Ther. *18*, 805–813.

Pool-Zobel, B.L., Neudecker, C., Domizlaff, I., Ji, S., Schillinger, U., Rumney, C., Moretti, M., Vilarini, I., Scassellati-Sforzolini, R., and Rowland, I. (1996). *Lactobacillus*- and bifidobacterium-mediated antigenotoxicity in the colon of rats. Nutr. Cancer *26*, 365–380.

Prantera, C., Scribano, M.L., Falasco, G., Andreoli, A., and Luzi, C. (2002). Ineffectiveness of probiotics in preventing recurrence after curative resection for Crohn's disease: a randomised controlled trial with *Lactobacillus* GG. Gut *51*, 405–409.

Raza, S., Graham, S.M., Allen, S.J., Sultana, S., Cuevas, L., and Hart, C.A. (1995). *Lactobacillus* GG promotes recovery from acute nonbloody diarrhea in Pakistan. Pediatr. Infect. Dis. J. *14*, 107–111.

Rinttila, T., Kassinen, A., Malinen, E., Krogius, L., and Palva, A. (2004). Development of an extensive set of 16S rDNA-targeted primers for quantification of pathogenic and indigenous bacteria in faecal samples by real-time PCR. J. Appl. Microbiol. *97*, 1166–1177.

Rosenfeldt, V., Michaelsen, K.F., Jakobsen, M., Larsen, C.N., Moller, P.L., Tvede, M., Weyrehter, H., Valerius, N.H., and Paerregaard, A. (2002). Effect of probiotic *Lactobacillus* strains on acute diarrhea in a cohort of nonhospitalized children attending day-care centers. Pediatr. Infect. Dis. J. *21*, 417–419.

Ruseler-van Embden, J.G., Schouten, W.R., and van Lieshout, L.M. (1994). Pouchitis: result of microbial imbalance? Gut 35, 658–664.

Sakamoto, I., Igarashi, M., Kimura, K., Takagi, A., Miwa, T., and Koga, Y. (2001). Suppressive effect of *Lactobacillus gasseri* OLL 2716 (LG21) on *Helicobacter pylori* infection in humans. J. Antimicrob. Chemother. *47*, 709–710.

Sandborn, W.J., Tremaine, W.J., Batts, K.P., Pemberton, J.H., and Phillips, S.F. (1994). Pouchitis after ileal pouch-anal anastomosis: a Pouchitis Disease Activity Index. Mayo Clin. Proc. *69*, 409–415.

Satokari, R.M., Vaughan, E.E., Smidt, H., Saarela, M., Matto, J., and de Vos, W.M. (2003). Molecular approaches for the detection and identification of bifidobacteria and lactobacilli in the human gastrointestinal tract. Syst. Appl. Microbiol. *26*, 572–584.

Schultsz, C., Van Den Berg, F.M., Ten Kate, F.W., Tytgat, G.N., and Dankert, J. (1999). The intestinal mucus layer from patients with inflammatory bowel disease harbors high numbers of bacteria compared with controls. Gastroenterology *117*, 1089–1097.

Schultz, M., Timmer, A., Herfarth, H.H., Sartor, R.B., Vanderhoof, J.A., and Rath, H.C. (2004). *Lactobacillus*

GG in inducing and maintaining remission of Crohn's disease. BMC. Gastroenterol *4*, 5.

Scrimshaw, N.S. and Murray, E.B. (1988). The acceptability of milk and milk products in populations with a high prevalence of lactose intolerance. Am. J. Clin. Nutr. *48*, 1079–1159.

Sen, S., Mullan, M.M., Parker, T.J., Woolner, J.T., Tarry, S.A., and Hunter, J.O. (2002). Effect of *Lactobacillus plantarum* 299v on colonic fermentation and symptoms of irritable bowel syndrome. Dig. Dis. Sci. *47*, 2615–2620.

Sghir, A., Gramet, G., Suau, A., Rochet, V., Pochart, P., and Dore, J. (2000). Quantification of bacterial groups within human fecal flora by oligonucleotide probe hybridization. Appl. Environ. Microbiol. *66*, 2263–2266.

Shanahan, F. (2000b). Immunology. Therapeutic manipulation of gut flora. Science *289*, 1311–1312.

Shanahan, F. (2002). Crohn's disease. Lancet *359*, 62–69.

Shanahan, F. (2000a). Probiotics and inflammatory bowel disease: is there a scientific rationale? Inflamm. Bowel. Dis. *6*, 107–115.

Shanahan, F., Guarner, F., Von Wright A., Vilpponen-Salmela T., O'Donoughue D., Kiely B., and Progid Investigators. (2006). A one year, randomised, double-blind, placebo controlled trial of a *Lactobacillus* or a Bifidobacterium probiotic for maintenance of steroid-induced remission of Ulcerative Colitis. Gastroenterology *130*, A–44.

Shen, B., Brzezinski, A., Fazio, V.W., Remzi, F.H., Achkar, J.P., Bennett, A.E., Sherman, K., and Lashner, B.A. (2005). Maintenance therapy with a probiotic in antibiotic-dependent pouchitis: experience in clinical practice. Aliment. Pharmacol. Ther. *22*, 721–728.

Shornikova, A.V., Casas, I.A., Isolauri, E., Mykkanen, H., and Vesikari, T. (1997a). *Lactobacillus reuteri* as a therapeutic agent in acute diarrhea in young children. J. Pediatr. Gastroenterol Nutr. *24*, 399–404.

Shornikova, A.V., Casas, I.A., Mykkanen, H., Salo, E., and Vesikari, T. (1997b). Bacteriotherapy with *Lactobacillus reuteri* in rotavirus gastroenteritis. Pediatr. Infect. Dis. J. *16*, 1103–1107.

Siitonen, S., Vapaatalo, H., Salminen, S., Gordin, A., Saxelin, M., Wikberg, R., and Kirkkola, A.L. (1990). Effect of *Lactobacillus* GG yoghurt in prevention of antibiotic associated diarrhoea. Ann. Med. *22*, 57–59.

Simon, G.L. and Gorbach, S.L. (1984). Intestinal flora in health and disease. Gastroenterology *86*, 174–193.

Steidler, L., Hans, W., Schotte, L., Neirynck, S., Obermeier, F., Falk, W., Fiers, W., and Remaut, E. (2000). Treatment of murine colitis by *Lactococcus lactis* secreting interleukin-10. Science *289*, 1352–1355.

Swidsinski, A., Ladhoff, A., Pernthaler, A., Swidsinski, S., Loening-Baucke, V., Ortner, M., Weber, J., Hoffmann, U., Schreiber, S., Dietel, M., and Lochs, H. (2002). Mucosal flora in inflammatory bowel disease. Gastroenterology *122*, 44–54.

Tannock, G.W. (2004). A special fondness for lactobacilli. Appl. Environ. Microbiol. *70*, 3189–3194.

Tannock, G.W. (1995). Normal Microbiota: An Introduction to Microbes Inhabiting the Human Body. London: Chapman & Hall.

Tannock, G.W., Munro, K., Harmsen, H.J., Welling, G.W., Smart, J., and Gopal, P.K. (2000). Analysis of the fecal microflora of human subjects consuming a probiotic product containing *Lactobacillus rhamnosus* DR20. Appl. Environ. Microbiol. *66*, 2578–2588.

Thomas, M.R., Litin, S.C., Osmon, D.R., Corr, A.P., Weaver, A.L., and Lohse, C.M. (2001). Lack of effect of *Lactobacillus* GG on antibiotic-associated diarrhea: a randomized, placebo-controlled trial. Mayo Clin. Proc. *76*, 883–889.

Ursing, B., Alm, T., Barany, F., Bergelin, I., Ganrot-Norlin, K., Hoevels, J., Huitfeldt, B., Jarnerot, G., Krause, U., Krook, A., Lindstrom, B., Nordle, O., and Rosen, A. (1982). A comparative study of metronidazole and sulfasalazine for active Crohn's disease: the cooperative Crohn's disease study in Sweden. II. Result. Gastroenterology *83*, 550–562.

Van, G.A., Dewit, O., Louis, E., de, H.G., Baert, F., Fontaine, F., Devos, M., Enslen, M., Paintin, M., and Franchimont, D. (2007). Multicenter randomized-controlled clinical trial of probiotics (*Lactobacillus johnsonii*, LA1) on early endoscopic recurrence of Crohn's disease after ileo-caecal resection. Inflamm. Bowel. Dis. *13*, 135–42.

Van, L.J., Clune, Y., Bennett, M., and Collins, J.K. (2005). The SYNCAN project: goals, set-up, first results and settings of the human intervention study. Br. J. Nutr. *93* (Suppl.), S91–S98.

Vanderhoof, J.A., Whitney, D.B., Antonson, D.L., Hanner, T.L., Lupo, J.V., and Young, R.J. (1999). *Lactobacillus* GG in the prevention of antibiotic-associated diarrhea in children. J. Pediatr. *135*, 564–568.

Vanhoutte, T., De, P., V., De, B.E., Verbeke, K., Swings, J., and Huys, G. (2006). Molecular monitoring of the fecal microbiota of healthy human subjects during administration of lactulose and *Saccharomyces boulardii*. Appl. Environ. Microbiol. *72*, 5990–5997.

Venturi, A., Gionchetti, P., Rizzello, F., Johansson, R., Zucconi, E., Brigidi, P., Matteuzzi, D., and Campieri, M. (1999). Impact on the composition of the faecal flora by a new probiotic preparation: preliminary data on maintenance treatment of patients with ulcerative colitis. Aliment. Pharmacol. Ther. *13*, 1103–1108.

Walter, J.- (2005). The Microecology of Lactobacilli in the Gastrointestinal Tract. In Probiotics and Prebiotics: Scientific Aspects, G.W. Tannock, ed. (Norfolk, UK: Caister Academic Press).

Wendakoon, C.N., Thomson, A.B., and Ozimek, L. (2002). Lack of therapeutic effect of a specially designed yogurt for the eradication of Helicobacter pylori infection. Digestion *65*, 16–20.

Xu, J. and Gordon, J.I. (2003). Inaugural Article: Honor thy symbionts. Proc. Natl. Acad. Sci. USA *100*, 10452–10459.

Young, J. (1998). European market developments in prebiotic- and probiotic-containing foodstuffs. Br. J. Nutr. *80*, S231–S233.

Zocco, M.A., dal Verme, L.Z., Cremonini, F., Piscaglia, A.C., Nista, E.C., Candelli, M., Novi, M., Rigante, D.,

Cazzato, I.A., Ojetti, V., Armuzzi, A., Gasbarrini, G., and Gasbarrini, A. (2006). Efficacy of *Lactobacillus* GG in maintaining remission of ulcerative colitis. Aliment. Pharmacol. Ther. 23, 1567–1574.

Zoetendal, E.G., Akkermans, A.D., and de Vos, W.M. (1998). Temperature gradient gel electrophoresis analysis of 16S rRNA from human fecal samples reveals stable and host-specific communities of active bacteria. Appl. Environ. Microbiol. 64, 3854–3859.

Zoetendal, E.G., Cheng, B., Koike, S., and Mackie, R.I. (2004). Molecular microbial ecology of the gastrointestinal tract: from phylogeny to function. Curr. Issues Intest. Microbiol. 5, 31–47.

Lactobacillus in the Vagina: Why, How, Which Ones, and What Do They Do?

10

Gregor Reid

Abstract

Of the 50 known species of *Lactobacillus*, at most 20 are able to colonize the intestine, and *L. iners* and *L. crispatus* appear to be the most commonly isolated in women in various countries around the globe. This consistency of isolation is all the more intriguing given the disparity of societal cultures and diet. For the most part, the origin of the lactobacilli is the woman's own intestinal microbiota, with passive transfer occurring along the skin from the anus to the perineum, vulva and vagina. The microbiota is disrupted by hormone levels, douching, sexual practices, the types of organisms ascending from the anus, and other factors such as alterations in innate immunity. The exogenous administration of certain strains of lactobacilli (probiotics) has been shown to reduce the risk of infection and help eradicate bacterial vaginosis. The mechanisms of action of the most documented probiotics, *Lactobacillus rhamnosus* GR-1 and *L. reuteri* RC-14, appear to consist of physicochemical displacement ability, as well as anti-infective compounds released from the cells, and immune-modulatory factors. Further development of these and other strains will lead to new approaches to help women retain and regain their vaginal health, and reduce their risk of various problematic, and in some cases, life threatening, conditions.

The origin of the vaginal microbiota

The female anatomy is such that the anus is approximately 4 cm from the vaginal opening, and is linked by the perineal skin (Fig. 10.1). This is not a sterile area, and bacteria transfer along the perineum onto the labia and into the vagina, on a daily basis that is natural and not because of poor hygiene. The urethra opens into the vagina and this short urinary drainage tube is only 4 cm from the bladder. Thus, bacteria such as *Escherichia coli* that infect the bladder actually come from the person's own intestine via the anus, perineum, vagina and urethra. Likewise, for bacteria such as *Gardnerella*, *Prevotella*, *Mobiluncus* and *Atopobium* that cause bacterial vaginosis (BV), and *Candida* that cause yeast vaginitis, the process is one of ascension along these surfaces from the host herself (Mardh *et al.*, 2003; Antonio *et al.*, 2005). There is evidence that sexual intercourse can transfer these organisms, and in some cases this might re-colonize the female following eradication of the organisms by use of antimicrobial agents (al-Wali *et al.*, 1989; Foxman *et al.*, 1992; Spinillo *et al.* 2002).

In a novel series of studies, a set of 12 new genetic regions of *E. coli* were identified that may encode factors that help certain strains transfer from the rectum to the vagina and bladder (Xie *et al.* 2006). This is particularly interesting and may explain why many *E. coli* strains found readily in the intestine, never make it to the vagina. In exactly the same manner, normal, beneficial bacteria such as *Lactobacillus* species, and probiotic lactobacilli that make it to the anus after oral intake, also passively transfer from the anus to the perineum, vulva and then vagina. They can colonize the urethra but they do not infect either the vagina or the bladder. It would be interesting to look for expression factors in lactobacilli that

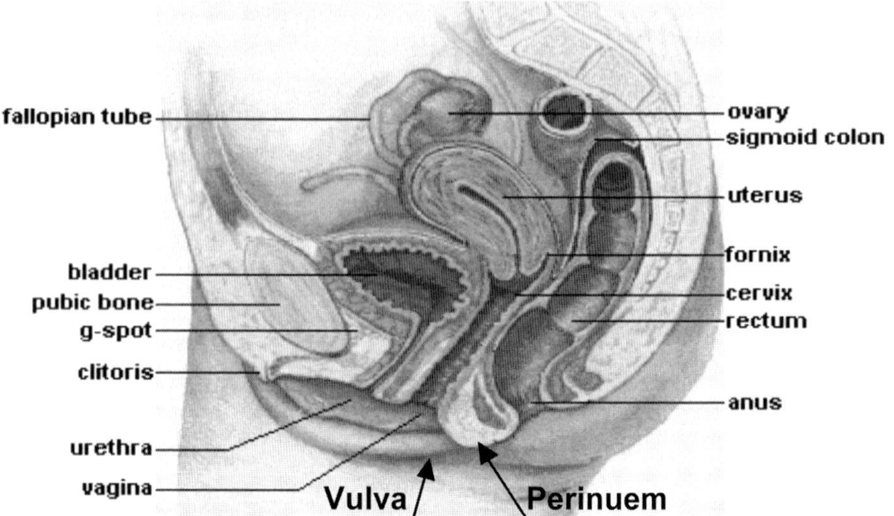

Figure 10.1 The female anatomy showing the close proximity of anus to vagina, the route taken by lactobacilli and urogenital pathogens. Adapted from original – copyright sought from Creative Commons Licenses.

aid in this transfer, as it may identify more efficacious probiotics. Likewise, uropathogenic *E. coli* have genes yceP, yqgA, ygiD, and aaeX that aid in biofilm formation (Hancock and Klemm, 2007), and therefore retention at mucosal sites. A study proposed that 114 genes were important for biofilm development and maintenance, especially encoding cell structures, small-molecule transport, energy metabolism and regulatory functions (Junker *et al.* 2006). No such genes have been identified in lactobacilli, and it is possible that they are not needed in the vagina because the host is tolerant to the indigenous organisms and some strains produce biosurfactants that affect the area surrounding the cell, making it less receptive to competitor organisms (Millsap *et al.*, 1994).

The *Lactobacillus* composition of the vaginal microbiota

The *Lactobacillus* genus are a phenotypically heterogeneous group of facultatively anaerobic, catalase-negative, non-spore-forming, Gram-positive, rod-shaped organisms. They are renowned for producing lactic acid as a major metabolic end-product and over 50 species have been identified (Falsen *et al.*, 1999). The question is how many and which types transfer through the intestine and ascend to survive and populate the vagina?

Prior to the late 1990s, culture techniques had been used to determine the composition of the normal microbes in the healthy vagina. A biochemical system termed API 50 CH (BioMérieux, Inc., Marcy l'Etoile, France), was used as the primary means to identify and classify *Lactobacillus* species. However, this method has been shown to incorrectly identify some species of lactobacilli as *Lactobacillus acidophilus*, thereby putting into question the belief that *L. acidophilus* were the most frequently isolated *Lactobacillus* species in the vagina. In a study comparing this API 50 CH carbohydrate fermentation method with molecular techniques, the former was found to misidentify *Lactobacillus* species and to produce uninterpretable results (Boyd *et al.* 2005).

In 1998, a molecular method using ribotyping was described as a more accurate means of identifying and grouping *Lactobacillus* species, and some vaginal isolates were re-classified (Zhong *et al.*, 1998). This sensitive but rapid method for species differentiation combines Southern hybridization of chromosomal DNA fingerprints with the use of *E. coli* rRNA probes, thereby discriminating between species and individual strains. The method involves the separation and identification of the rRNA genes present within the bacterial genome, which exist in various copy numbers. Thus, it is possible to delineate species based on differences in the restriction fragment length polymorphisms of the rRNA genes. The ribotyping study showed that *L. acidophilus* RC-14 previously classified by the API 50 CH method, fitted more closely into the *L. fermentum* grouping. Thus, it was re-named. On reflection, had a type strain of *L. reuteri* been included in this study, the RC-14 strain would likely have been re-named *L. reuteri* at that time, rather than in 2006 when DNA–DNA hybridization and PCR-denaturing gradient gel electrophoresis (DGGE) were performed on it.

The utilization of molecular methodologies per se did not resolve the identity of the vaginal microbiota right away, and while genotypic methods were performed on vaginal samples in the United States (Antonio *et al.* 2005) and Japan (Song *et al.*, 1995) and reported that *L. crispatus*, *L. jensenii*, *L. gasseri*, and *L. vaginalis* were the most common species of lactobacilli, a major flaw existed. The samples were first cultured, thus only those recoverable on agar plates were identified. It was only when samples were immediately tested for DNA content, that a more comprehensive picture arose, with a new species *L. iners* being identified and found to the most common *Lactobacillus* species in women in Sweden, US, Canada, Russia and Nigeria (Falsen *et al.*, 1999; Vasquez *et al.* 2002; Burton *et al.* 2003; Anukam *et al.* 2006; Ferris *et al.* 2007).

Even with the use of specific 16S genes, there is a lot of diversity amongst strains, such that at the 97% cut off, multiple types of organisms still arise. This microdiversity will require a sub-analysis of genome composition or content, or other means of classifying organisms more precisely (Doolittle, 2006). Thus, it will be no surprise if *L. reuteri* RC-14 one day is again reclassified, perhaps to a new species, or a non-reuterin producing species, or some other nomenclature (Cadieux *et al.*, 2008).

The constant reclassification of bacteria emphasizes that any attempts to re-populate the vagina with probiotic lactobacilli, will require that the strains possess properties suitable for such a task, and be selected for these rather than what nomenclature they belong to. For now, an approximation is presented in Fig. 10.2a and b of the *Lactobacillus* species most likely to be recovered from the vagina.

Disruptions and fluctuations within the microbiota

Several papers have provided valuable insight into the rapidity with which the vaginal microbiota can change in some subjects, yet the stability it retains in others. The reasons for the change is often not clear, although vaginal douching and sexual activity can increase the risk (Mbizvo *et al.* 2004). Keane and others (1997) followed a group of women over the menstrual cycle. Daily samples showed fluctuations in 5 of 6 women that ranged from normal to BV (Fig. 10.3).

The findings were supported by a study of 51 women during the menstrual cycle, in which significant, transient changes in the microbiota were found (Schwebke and Weiss, 2001). During

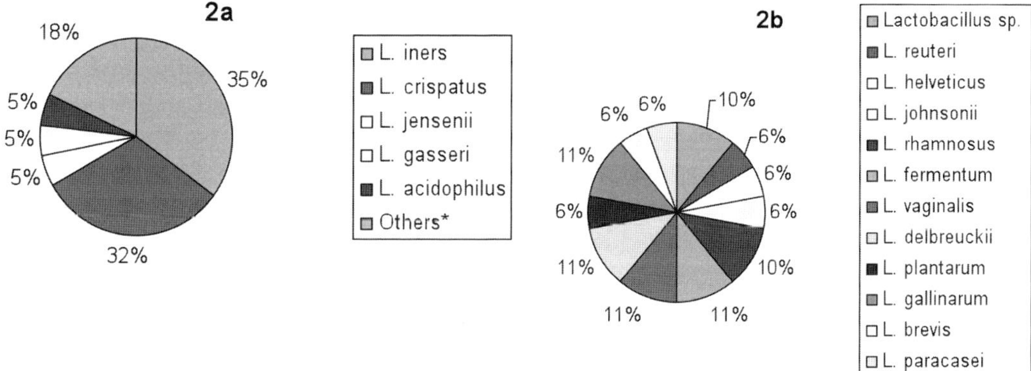

Figure 10.2 (a) An approximation and generalization of the percentage distribution of *Lactobacillus* species found in the healthy vagina or pre-menopausal women. The 'other' 18% comprises the organisms shown in (b).

menses, the most significant changes occurred, and overall only 11 (22%) of the 51 women maintained a *Lactobacillus*-predominant flora throughout the cycle.

The extent to which the vaginal microbiota can change is shown by the finding of 29% prevalence of BV in US women aged 14–49 years (Allsworth and Peipert, 2007). Such a rate may seem surprisingly high, but it agrees with our previous studies (Reid *et al.*, 2003) and emphasizes the need to improve the diagnosis, prevention and treatment of this condition, particularly since it is associated with increased risk of sexually transmitted infections and preterm labour (Cherpes *et al.*, 2003; Myer *et al.*, 2005).

Fredricks *et al.* (2005) used a broad-range PCR assay of 16S rDNA to examine the vaginal microbiota, and showed a change from a *L. crispatus* and *L. iners* colonization to a very diverse BV microbiota of *G. vaginalis, Prevotella, Atopobium* and other species. The change was noted after two months, as that is when sampling occurred, so it is not clear how rapid the events occurred. Nevertheless, the extent of the change was dramatic.

In premenopausal women, two interesting findings have been made:

1 Contrary to their not being any lactobacilli because of drops in oestrogen, lactobacilli can be recovered from the vagina.
2 The fluctuations are also seen between samples taken one week apart. This is illustrated in Fig. 10.4, which shows examples of the types of changes that were detected: from BV to normal, normal or BV throughout, and BV to normal.

A more close examination of the microbiota was carried out on a post-menopausal woman who had a long history of BV. Interestingly, even although lactobacilli were detected by PCR-DGGE, the Nugent scores showed BV in every sample. Of note, the types of pathogens detected showed quite a range and included some well-known pathogenic types (Table 10.1).

The influence of hormone levels on vaginal lactobacilli have been known for some time. *In vitro* experiments have shown that uroepithelial cell receptivity to *L. rhamnosus* GR-1 appears to peak in mid menstrual cycle, suggesting that prior to, and immediately following, menses, the host is more likely to be colonized by species other than lactobacilli (Fig. 10.5) (Chan *et al.*, 1984).

Thus, in theory, the host may be at greater risk of BV or urinary tract infection at the end and beginning of her menstrual cycle. This is borne out by studies of post-menopausal women whereby recovery of lactobacilli is associated with significantly reduced infection rates (Raz and Stamm, 1993). The disruption of the

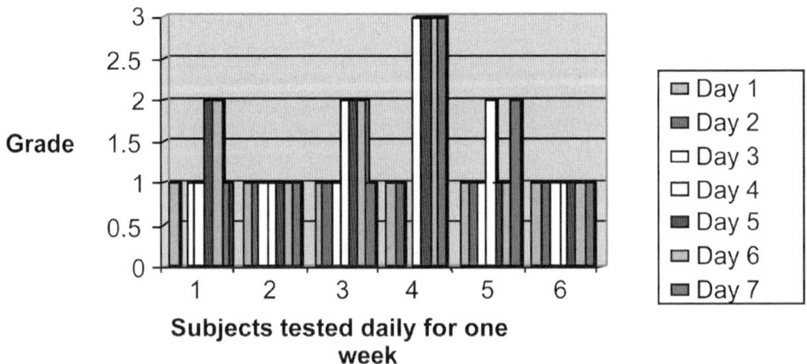

Figure 10.3 Created from data published by Keane *et al.* (1997). Grade 1 represents Normal microbiota (*Lactobacillus* dominated), Grade 2 is Intermediate (mixed *Lactobacillus* and other bacterial morphotypes) and Grade 3 represents BV (few or no lactobacilli and large numbers of *G. vaginalis* and other morphotypes indicative of BV).

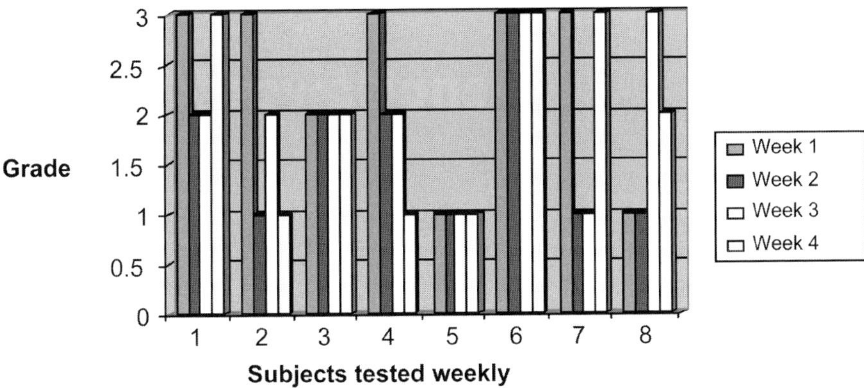

Figure 10.4 Created from data published by Burton and Reid (2002). Grade 1 represents Normal microbiota (*Lactobacillus* dominated), Grade 2 is Intermediate (mixed *Lactobacillus* and other bacterial morphotypes) and Grade 3 represents BV (few or no lactobacilli and large numbers of *G. vaginalis* and other morphotypes indicative of BV).

Table 10.1 Vaginal bacteria recovered from a woman with a history of BV. The number represent the number of strains of that species recovered on the samples taken weekly for 6 weeks. The four lactobacilli species detected were *L. johnsonii, L. gasseri, L. delbrueckii* and *L. reuteri*

	Week 1	Week 2	Week 3	Week 4	Week 5	Week 6
Lactobacillus	4	4	2	3	1	3
Klebsiella	1	0	1	1	1	1
Serratia	1	0	0	0	0	0
Citrobacter	1	0	0	0	0	0
Morganella	0	1	1	1	1	1
Kluyvera	0	0	1	0	0	0
E. coli	0	0	0	1	1	1
S. epidermidis	0	0	0	1	1	1

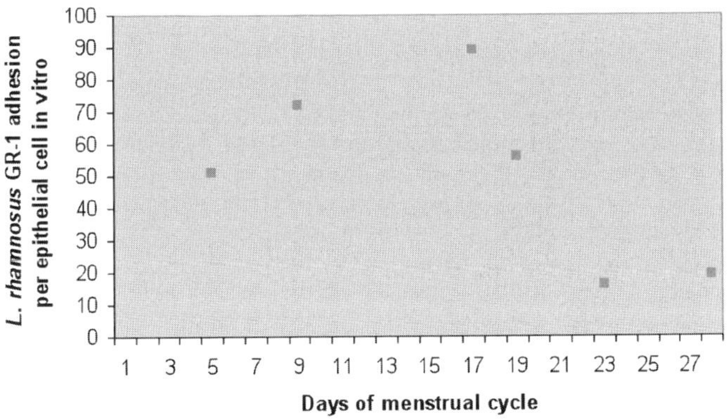

Figure 10.5 *In vitro* adhesion per cell values for uroepithelial cells recovered on days 5, 9, 17, 19, 23 and 28 of the menstrual cycle (Chan *et al.,* 1984).

microbiota after menopause not only reduces the lactobacilli count, but it results in a more diverse, multi-speciated composition (Heinemann and Reid, 2005), from which more infections can ensue.

Recently, another proposal was put forward for the rapid transition from a *Lactobacillus*-dominated vaginal microbiota to BV, namely due to alterations in innate immunity (Witkin *et al.*, 2007). The theory is that a microbial-induced inhibition of Toll-like receptor expression and/or activity may block induction of proinflammatory immunity and lead to the proliferation of atypical vaginal bacteria. A lack of 70-kDa heat-shock protein production and release in response to foreign microbes would cause failure to activate antimicrobial immune responses. In line with this concept, perhaps the role of lactobacilli is to trigger the host's immune response and displace the pathogens.

Replenishment of the lactobacilli and evidence of effects

The clinical observation that lactobacilli may be associated with a healthy urogenital tract (Bruce *et al.*, 1973) led to the concept that replenishment of the vagina with exogenous lactobacilli might reduce the risk of infection. This concept is called probiotics, defined as 'live microorganisms which when administered in adequate amounts confer a health benefit on the host' (FAO/WHO, 2001).

As stated above, the characteristics of the organism(s) are more important that the species. The question is which characteristics reflect the probiotic potential in the host? In 1980, when our group first began to look for suitable probiotic lactobacilli, the ability of bacteria to adhere to epithelial cells was viewed as a key component in colonization and competition with other inhabitants of a given ecological niche. In addition, it was believed that in order to change the microbial environment back to normal, the pathogens had to be killed or their growth inhibited. Thus, these types of assays were employed to find suitable probiotic strains. From these came *L. rhamnosus* GR-1 and *L. reuteri* RC-14. The former was documented for its ability to inhibit *E. coli* and other Gram negative pathogens (Reid *et al.*, 1987), while the latter was chosen for its ability to interfere with Gram positive organisms (Velraeds *et al.*, 1998). In hindsight, these *in vitro* assays do not necessarily either predict or verify probiotic activity, but at least provide information about the strains. As it happens, these two strains have gone on to be effective probiotics, and a human study has shown a degree of correlation with the adhesion ability of *L. rhamnosus* GR-1 (Reid *et al.*, 1995).

The guidelines for what constitutes a probiotic clearly state the need for studies that prove benefits to the host. These have been particularly well established for *L. rhamnosus* GR-1 and *L. reuteri* RC-14, but not yet for other strains, even those already commercialized, as illustrated in Table 10.2. Some fail the Guidelines for multiple reasons (no strain designation, no end of shelf-life guarantee, no clinical safety of efficacy data published, no mechanism of action reported) (Reid, 2005).

Mechanisms of interference by *Lactobacillus* over pathogens in the vagina

It was long thought that probiotic strains had to colonize the vagina or other sites in order to confer their effects. In more recent times, it has been recognized that probiotic strains do not tend to stay for long periods in the host, and thus their activity as they pass through or are temporarily retained, becomes even more critical. Some of these mechanisms will now be discussed.

Displacement

Of note, the regular intake of probiotic *Lactobacillus rhamnosus* GR-1 and *Lactobacillus reuteri* RC-14 not only results in transfer of these organisms to the vagina in a natural way, but it also does two other important things that benefit the vaginal health of the host. These are reducing the number of pathogenic bacteria and yeast that transfer from the anus to the vagina, as shown in Fig. 10.6, and allowing the indigenous lactobacilli of the host to increase in number.

The ability of lactobacilli to displace urogenital pathogens has been shown *in vitro* (Chan *et al.*, 1985; Hawthorn *et al.*, 1990; Velreads *et al.*, 1996, 1998; Saunders *et al.*, 2007) and in humans (Burton *et al.*, 2003). Biosurfactant substances were shown to be capable of induc-

Table 10.2 There are clearly products for vaginal health on the market or under examination, but few have any clinical data supporting their claims. The consumer should ask the supplier to provide published human trials on their products (as they appear in shops), on the chance that they were done but are not easy to find on English literature libraries.

Strains@	Availability Commercially*	Documentation as per PubMed** citation
L. rhamnosus GR-1 and *L. reuteri* RC-14	Lacibios, Poland; Ombe Austria; Ecoflora India, Fem-Dophilus, USA; Bion Flore, France; Lactogyn, Croatia	Vaginal colonization, and efficacy in prevention and treatment of infection (Reid *et al*., 1995; Cadieux *et al.* 2002; Gardiner *et al.* 2002 ; Reid et al 2001a; 2001b; 2003a; 2003b; 2004. Anukam *et al.* 2006a; 2006b).
L. crispatus CTV05	Not commercialized	Shown to populate the vagina safely, and has potential to be beneficial (Marrazzo *et al.* 2006)
L. rhamnosus, L. acidophilus, S. thermophilus and *L. bulgaricus.*	Femina Flora, Roots Herbal, USA	Dubious claims have been seen. No studies or rationale for the strain selection. No published human trial data found on PubMed**. Not a probiotic.
L. acidophilus NAS	GyNatren, Natren USA	Issued a warning letter from FDA in 2006 for misleading information. No clinical studies proving claims**.
L acidophilus, L casei, L lactis, B bifidum	Symbiofem Plus, Symbio Pharm, Germany	No strain designations, no published human trial data found on PubMed**. Not a probiotic.
L acidophilus, L. plantarum, L. rhamnosus, L. casei and L. delbruekii	Vagilac, Italy, USA, other countries.	No strain designations, no published human efficacy data on PubMed**. Not a probiotic.
L. acidophilus, L. rhamnosus, S. thermophilus	Probaclac Vaginal, various distributors, USA	No strain designations, no published human efficacy data**. Not a probiotic.
L. acidophilus	Gynoflor Vaginal capsule, Medinova, Switzerland	Contains estrogen which may cause side-effects including cancer (WHO 1981; Rossouw *et al.* 2002). Has been used in one study to treat BV (Parent *et al.*, 1996).
L. rhamnosus, L. acidophilus, S. thermophilus and *L. bulgaricus.*	Fermalac, Rosell-Lallemand and various distributors.	No effect in preventing vulvovaginal candidiasis (Pirotta *et al.* 2004), but used by some women who have few other options.
L. rhamnosus	Lactovaginal, Poland	No strain designations, no published human trial data found on PubMed**. Not a probiotic

@ refers to strains proven or claiming to be of use for vaginal health, and known to the author.

*Commercial products known to the author. No comment is made as to manufacturing quality of the products or the extent to which they may have benefited people who have used them. In addition, no claim is made as to whether the products listed are any more or any less useful than products no listed here. When selecting probiotic products for disease prevention or treatment, a physician should be consulted where appropriate.

**Following a PubMed search using 'probiotics', 'vagina', 'efficacy', 'Lactobacillus' and 'product name' search words.

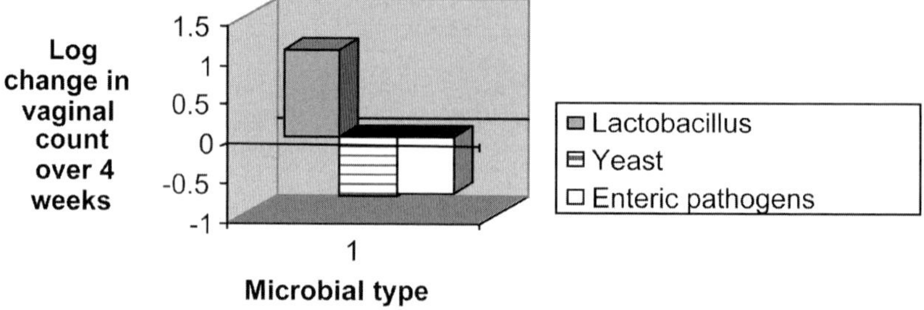

Figure 10.6 Adapted from Reid *et al.* (2003a), this shows the average change in viable count of lactobacilli, yeast and enteric pathogen (e.g. *E. coli*) in vaginal samples following four weeks of daily use of *L. rhamnosus* GR-1 and *L. reuteri* RC-14. The change is compared to the counts in the 32 subjects treated with placebo.

ing this effect, at least from *in vitro* studies (Velraeds *et al.*, 1998). Strain *L. reuteri* RC-14 is particularly effective at producing these surface-tension lowering biosurfactant molecules, within which collagen-binding proteins and peptides of unknown function exist (Velraeds *et al.*, 1996; Heinemann *et al.*, 2000; Howard *et al.*, 2000; Reid *et al.*, 2002). Thermodynamic principals could explain part of this displacement phenomenon, in that alteration in surface tension of the surrounding areas, by substances that produce a surfactant environment, can be a major factor in changing the adhesiveness of bacteria (Absolom *et al.*, 1983). This could certainly be a main functional factor on biomaterial surfaces, but evidence from animal experiments on wound infections indicated more of a multi-factorial effect taking place.

A foreign body was placed under the skin of rats and infected with *S. aureus*. The result was infection and inflammation at the site within three days (Gan *et al.*, 2002). However, when challenged with *L. reuteri* RC-14 or biosurfactants and their components, not only was the spread of the pathogen halted, but significantly less inflammation was noted (Gan *et al.*, 2002, 2003). An examination of the site showed that the *S. aureus* were not eradicated, but the infectious nidus was clearly not as severe as controls. This led to a series of studies investigating the mechanisms of interference by the lactobacilli.

Anti-infective signalling

A first hypothesis was that the lactobacilli were secreting some sort of signalling substance that affected the pathogen. A two chamber system was then developed and used to examine this phenomenon (Fig. 10.7). The bacterial types were separated by a membrane so that only proteins or peptides or other small molecules could pass between them.

The results showed that *L. reuteri* RC-14 secreted cell-cell communication signals that inhibited the expression of both staphylococcal superantigen-like protein 11, a putative staphylococcal exotoxin, and RNAIII, the effector molecule of the *agr* locus, a quorum sensing system that up-regulates the expression of many secreted proteins upon entering late-exponential phase, and also represses the expression of many cell wall-associated proteins (Laughton *et al.* 2006). This was the first evidence of anti-infective signalling between lactobacilli and pathogens. This then explained, at least in part, the animal study findings, and indicated the presence of a system that could potentially keep infection in-check, without necessarily having to rapidly kill the offending pathogen. Nevertheless, these studies did not examine the anti-inflammatory effects noted in the rats. Two studies were undertaken to investigate this possibility.

Anti-inflammatory activity

Potential anti-inflammatory mechanisms of lactobacilli have been reported to function at the level of mucosal epithelial cells by modulating production of inflammatory cytokines through inhibiting or regulating a key signalling molecule, nuclear factor κB (NF-κB) (Neish *et al.* 2000; Kelly *et al.* 2004) and producing heat shock proteins through inhibiting proteasome (Petrof *et al.* 2004). Animal studies have also shown that

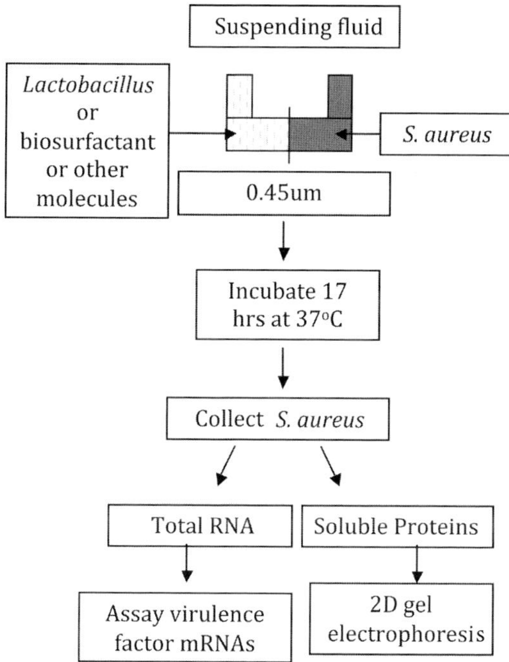

Figure 10.7 Two-chamber system used to study signalling exchanges between *L. reuteri* RC-14 whole cells or extracts, and *S. aureus*.

lactobacilli can attenuate inflammation (Doron *et al.*, 2005), possibly through autocrine and/or paracrine routes (Madsen *et al.*, 1999; Sheil *et al.* 2004, Pena *et al.* 2003). *L. reuteri* has been shown to inhibit colitis in interleukin-10 (IL-10)-deficient mice, in part through up-regulation of nerve growth factor (NGF), inhibition of translocation of NF-kappaB to the cell nuclei, prevention of degradation of IkappaB, and inhibition of IL-8 synthesis induced by pathogens (Ma *et al.* 2004).

To further examine a global effect on inflammation, *L. rhamnosus* GR-1 and *L. reuteri* RC-14 (the latter in very low numbers due to poor survival in yogurt) were supplemented into yoghurt and ingested daily by patients with inflammatory bowel disease (IBD) for 30 days (Baroja *et al.*, 2007). The proportion of CD4+CD25^{high} T-cells increased significantly with treatment, and the basal proportion of TNF-α+/IL-12+ monocytes and myeloid dendritic cells (DC) decreased. Serum IL-12 concentrations and proportions of IL-2+ and CD69+ T-cells from stimulated cells decreased in the IBD patients. The increase in CD4+CD25^{high} T-cells correlated with the decrease in the percentage of TNF-α- or IL-12-producing monocytes and DC, and the effects were confirmed by a follow-up study in which subjects consumed unsupplemented yoghurt. This study showed again the anti-inflammatory effects of the probiotics *in vivo*.

As macrophages are well populated in wound and intestinal tissues and are the major source of various cytokines, they provide a good system to study immune responses and cytokine release in response to probiotics. Preliminary studies showed *L. rhamnosus* GR-1 to be superior to *L. reuteri* RC-14 in the induction of anti-inflammatory cytokines in macrophages, so strains *L. rhamnosus* GR-1 and GG were examined further. It was discovered that these strains induce production of anti-inflammatory cytokines including IL-10 and granulocyte-colony stimulating factor (G-CSF), with the latter playing a crucial role in suppressing key inflammatory cytokines TNF-α (TNF) and IL-12 through activating STAT3 and subsequently inhibiting c-Jun N-terminal kinases (JNKs) in macrophages (Kim *et al.* 2006).

A third study in which *L. rhamnosus* GR-1 was inserted vaginally in four subjects and the host gene responses tested by human microarrays, showed that anti-microbial defenses were turned on by the treatment (Kirjavainen *et al.*, 2008). Thus, the probiotic strains do not simply turn off all inflammatory processes. Rather, they appear to modulate the host's response in such a way as to help cope with any infecting organisms, without necessarily inducing a symptomatic, inflammatory reaction. More studies will further clarify this phenomenon.

Antimicrobial activity

Other factors, such as the production of antimicrobial substances hydrogen peroxide (H_2O_2), bacteriocins and lactic acid, are believed to influence the effectiveness of lactobacilli in the vagina (McGroarty and Reid, 1988; Reid *et al.*, 1988; Aroutcheva *et al.* 2001; Ocana *et al.* 2004; Tomas *et al.* 2004; Voravuthikunchai *et al.* 2006). The case for H_2O_2 is due to the high incidence of H_2O_2-producing strains being isolated from the vagina of healthy women (Klebanoff *et al.*, 1991; Hillier *et al.*, 1993), and the ability of this enzyme to inhibit or kill other organisms. On the other hand, BV has been shown to develop despite the presence of H_2O_2-producing strains (Rosenstein *et al.*, 1997), and strains such as *L. rhamnosus* GR-1 that do not produce much H_2O_2 or only produce low amounts such as *L. reuteri* RC-14, can be effective vaginal probiotics.

Bacteriocins have long been identified in lactobacilli, although few have been commercialized per se. It is difficult to know their effectiveness *in vivo*, and mutants need to be produced and tested, at least in a system that mimics the human condition, to assess if they protect the host. Based upon studies in the oral cavity with bacteriocin-producing *S. salivarius* K12 and a mouse study using *L. salivarius* (Corr *et al.*, 2007), a case can be made that these compounds play a role in protecting the host from infection (Burton *et al.* 2006).

Lactic acid is clearly a by-product of lactobacilli, and this substance helps maintain a low pH in the vagina. Indeed, the level of lactic acid produced by strains has been used in some cases to select for probiotic lactobacilli (Aslim and Kilic, 2006), while lactic acid per se has been used as a therapeutic agent for the vagina (Tansupasiri *et al.* 2005). Parallel flow cell studies have shown that 0.19 mmol/L lactic acid can detach 30% fimbriated uropathogenic *E. coli* 67 biofilms at pH 3 and 15% at pH 4 (Reid *et al.* 2005). Further studies have shown that lactic acid and H_2O_2 cause an up-regulation of uropathogen outer membrane proteins, similar to a stress reaction, and this may make the pathogens more susceptible to clearance (Cadieux et a. unpublished). Similarly, in studies of *Candida albicans*, both *L. rhamnosus* GR-1 and lactic acid have been shown to significantly reduce biofilm formation and kill the yeast (Koehler *et al.* 2006). This latter finding is potentially important for probiotics designed to prevent vulvovaginal candidiasis.

A final functionality worthy of further investigation is the augmentation of antimicrobial activity with the conjoint use of probiotics. A clinical trial of 106 women has shown that the use of *L. rhamnosus* GR-1 and *L. reuteri* RC-14 for 30 days in addition to 7 days' metronidazole, results in significantly improved cure of BV (Anukam *et al.* 2006a). Although the replenishment of the vaginal vault with lactobacilli in part explains this outcome, other attributes of the probiotic appear to be occurring. A study has shown that while concentrations $\geq$ 5000 mg/ml *in vitro* completely suppressed the growth of *Lactobacillus*, those between 128 and 256 mg/ml actually stimulated the growth (Simoes *et al.* 2001), and we confirmed this finding (Anukam and Reid, 2008). Even though the probiotics were not administered at the same time as the metronidazole in the human study, it is feasible that the drug enhanced the transfer of the lactobacilli from the anal area to vagina, thereby helping displace and kill the BV pathogens. Whether or not the lactobacilli also produced substances that adversely affected the BV organisms, remain to be found.

Summary

The development of molecular tools has allowed a clearer understanding of the microbes that inhabit the vagina, their source and mechanisms of interfering with infectious processes. The ability of exogenous (probiotic) lactobacilli to populate the host and induce the return of indigenous lactobacilli and a healthy vaginal tract, is now possible, although further studies are needed to fully understand the strengths and limitations of this intervention.

Acknowledgements

Dr Reid's lab is funded by the Natural Sciences and Engineering Research Council, AFMnet and the Ontario Ministry of Agriculture and Food.

Conflict statement

Dr Reid holds patents on some aspects of the use of *Lactobacillus* strains GR-1 and RC-14 mentioned in this chapter.

References

Absolom, D.R., Lamberti, F.V., Policova, Z., Zingg, W., van Oss, C.J., and Neumann, A.W. (1983). Surface thermodynamics of bacterial adhesion. Appl. Environ. Microbiol. *46*, 90–97.

Allsworth, J.E., and Peipert, J.F. (2007). Prevalence of bacterial vaginosis: 2001–2004 national health and nutrition examination survey data. Obstet. Gynecol. *109*, 114–120.

al-Wali, W., Hamilton-Miller, J.M., Joshi, S., and Brumfitt, W. (1989). A case of recurrently sexually transmitted urinary tract infection. Genitourin. Med. *65*, 397–8.

Antonio, M.A. D., Rabe, L.K., and Hillier, S.L. (2005). Colonization of the rectum by *Lactobacillus* species and decreased risk of bacterial vaginosis. J. Infect. Dis. *192*, 394–398.

Anukam, K.C., Osazuwa, E., Ahonkhai, I., Ngwu, M., Osemene, G., Bruce, A.W., and Reid, G. (2006a). Double blind placebo-controlled treatment of bacterial vaginosis with metronidazole and probiotic *Lactobacillus rhamnosus* GR-1 and *Lactobacillus reuteri* RC-14. Microbes Infect. *8*, 1450–1454.

Anukam, K.C., Osazuwa, E.O., Ahonkhai, I., and Reid G. (2006b). *Lactobacillus* vaginal microbiota of women attending a reproductive health care service in Benin city, Nigeria. Sex. Transm. Dis. *33*, 59–62.

Anukam, K.C., Osazuwa, E., Osemene, G.I., Ehigiagbe, F., Bruce, A.W., and Reid, G. (2006c) Clinical study comparing probiotic *Lactobacillus* GR-1 and RC-14 with metronidazole vaginal gel to treat symptomatic bacterial vaginosis. Microbes Infect. *8*, 2772–2776.

Anukam, K.C., and Reid, G. (2008). Growth promotion of probiotic *Lactobacillus rhamnosus* GR-1 and *Lactobacillus plantarum* KCA in metronidazole. Microbiol. Ecol. Health Dis. *20*, 48–52.

Aroutcheva, A., Gariti, D., Simon, M., Shott, S., Faro, J., Simoes, J.A., Gurguis, A., and Faro, S. (2001). Defense factors of vaginal lactobacilli. Am. J. Obstet. Gynecol. *185*, 375–379.

Aslim, B., and Kilic, E. (2006). Some probiotic properties of vaginal lactobacilli isolated from healthy women. Jpn. J. Infect. Dis. *59*, 249–253.

Baroja, M.L., Kirjavainen, P.V., Hekmat, S. and Reid, G. (2007). Anti-inflammatory effects of probiotic-yogurt in inflammatory bowel disease patients. Clin Experimental Immunol. in press.

Brown, C.J., Wong, M., Davis, C.C., Kanti, A., Zhou, X., and Forney, L.J. (2007). Preliminary characterization of the normal microbiota of the human vulva using cultivation-independent methods. J. Med. Microbiol. *56*(Pt 2), 271–276.

Boyd, M.A., Antonio, M.A., and Hillier, S.L. (2005). Comparison of API 50 CH strips to whole-chromosomal DNA probes for identification of *Lactobacillus* species. J. Clin. Microbiol. *43*, 5309–5311.

Bruce, A.W., Chadwick, P., Hassan, A., and VanCott, G.F. (1973). Recurrent urethritis in women. Can. Med. Assoc. J. *108*, 973–976.

Burton, J.P., Cadieux, P.A., and Reid, G. (2003). Improved understanding of the bacterial vaginal microbiota of women before and after probiotic instillation. Appl. Environ. Microbiol. *69*, 97–101.

Burton, J.P., Chilcott, C.N., Moore, C.J., Speiser, G., and Tagg, J.R. (2006). A preliminary study of the effect of probiotic *Streptococcus salivarius* K12 on oral malodour parameters. J. Appl. Microbiol. *100*, 754–764.

Burton, J.P., and Reid, G. (2003). Evaluation of the bacterial vaginal flora of twenty postmenopausal women by direct (Nugent Score) and molecular (polymerase chain reaction and denaturing gradient gel electrophoresis) techniques. J. Infect. Dis. *186*, 777–780.

Cadieux, P., Burton, J., Kang, C.Y., Gardiner, G., Braunstein, I., Bruce, A.W., and Reid, G. (2002). *Lactobacillus* strains and vaginal ecology. JAMA *287*, 1940–1941.

Cadieux, P., Wind, A., Sommer, P, Schaefer, L., Crowley, K., Britton, R.A. and Reid, G. (2008). Evaluation of reuterin production in urogenital probiotic *Lactobacillus reuteri* RC-14. Appl. Environ. Microbiol.

Chan, R.C. Y., Bruce, A.W., and Reid G. (1984). Adherence of cervical, vaginal and distal urethral normal microbial flora to human uroepithelial cells and the inhibition of adherence of uropathogens by competitive exclusion. J. Urol. *131*, 596–601.

Chan, R.C. Y., Reid, G., Irvin, R.T., Bruce, A.W. and Costerton, J.W. (1985). Competitive exclusion of uropathogens from uroepithelial cells by *Lactobacillus* whole cells and cell wall fragments. Infect. Immun. *47*, 84–89.

Cherpes, T.L., Meyn, L.A., Krohn, M.A., and Hillier, S.L. (2003). Risk factors for infection with herpes simplex virus type 2: role of smoking, douching, uncircumcised males, and vaginal flora. Sex. Transm. Dis. *30*, 405–410.

Corr, S.C., Li, Y., Riedel, C.U., O'Toole, P.W., Hill, C., and Gahan, C.G. (2007). Bacteriocin production as a mechanism for the antiinfective activity of *Lactobacillus salivarius* UCC 118. Proc. Natl Acad. Sci. USA. *104*, 7617–7621.

Doolittle, W.F. (2006). What is a bacterial species? Microbiology Today 33, 148–151.

Doron, S., Snydman, D.R., and Gorbach, S.L. (2005) *Lactobacillus* GG: bacteriology and clinical applications. Gastroenterol. Clin. North Am. *34*, 483–498.

Falsen, E., Pascual, C., Sjödén, B., Ohlén, M., and Collins, M. (1999). Phenotypic and phylogenetic characterization of a novel *Lactobacillus* species from human sources: description of *Lactobacillus iners* sp. nov. Int. J. Syst. Bacteriol. *49*, 217–221.

FAO/WHO. (2001). Evaluation of health and nutritional properties of powder milk and live lactic acid bacteria. Food and Agriculture Organization of the United Nations and World Health Organization Expert Consultation Report. http://www.fao.org/es/ESN/Probio/probio.htm.

Ferris MJ, Norori J., Zozaya-Hinchliffe M., Martin DH. Cultivation-independent analysis of changes in bacterial vaginosis flora following metronidazole treatment. J. Clin. Microbiol. 2007 Jan 3; [Epub ahead of print]

Foxman, B., Manning, S.D., Tallman, P., Bauer, R., Zhang, L., Koopman, J.S., Gillespie, B., Sobel, J.D., and Marrs, C.F. (2002). Uropathogenic *Escherichia coli* are more likely than commensal *E. coli* to be shared between heterosexual sex partners. Am. J. Epidemiol. *156*, 1133–1140.

Fredricks, D.N., Fiedler, T.L., and Marrazzo, J.M. (2005). Molecular identification of bacteria associated with bacterial vaginosis. N. Engl. J. Med. 353, 1899–1911.

Gan, B.S., Kim, J., Reid, G., Cadieux, P., and Howard, J.C. (2002). *Lactobacillus fermentum* RC-14 inhibits *Staphylococcus aureus* infection of surgical implants in rats. J. Infect. Dis. *185*, 1369–1372.

Gan, B.S., Kim, J., Reid, G., Cadieux, P., and Howard, J.C. (2003). Probiotic *Lactobacillus* RC-14 and its biosurfactant prevent *Staph. aureus* infection: A new approach in the prevention of implant infection? Plastic Surgery Forum *25*, 164–167.

Gardiner, G., Heinemann, C., Beuerman, D., Bruce, A.W., and Reid, G. (2002). Persistence of *Lactobacillus fermentum* RC-14 and *L. rhamnosus* GR-1, but not *L. rhamnosus* GG in the human vagina as demonstrated by randomly amplified polymorphic DNA (RAPD). Clin. Diag. Lab. Immunol. *9*, 92–96.

Hancock, V., and Klemm, P. (2007). Global gene expression profiling of asymptomatic bacteriuria *Escherichia coli* during biofilm growth in human urine. Infect. Immun. *75*, 966–976.

Hawthorn, L.A., and Reid, G. (1990). Exclusion of uropathogen adhesion to polymer surfaces by *Lactobacillus acidophilus*. J. Biomed. Mater. Res. *24*, 39–46.

Heinemann, C., Van Hylckama Vlieg, J.E.T., Janssen, D.B., Busscher, H.J., van der Mei, H.C., and Reid, G. (2000). Purification and characterization of a surface-binding protein from *Lactobacillus fermentum* RC-14 inhibiting *Enterococcus faecalis* 1131 adhesion. FEMS Microbiol. Lett. *190*, 177–180.

Heinemann C., and Reid G., (2005). Vaginal microbial diversity among postmenopausal women with and without hormone replacement therapy. Can. J. Microbiol. *51*, 777–81.

Hillier, S.L., Krohn, M.A., Rabe, L.K., Klebanoff, S.J., and Eschenbach, D.A. (1993). The normal vaginal flora, H2O2-producing lactobacilli, and bacterial vaginosis in pregnant women. Clin. Infect. Dis. *16* (Suppl. 4), S273–S81.

Howard, J., Heinemann, C., Thatcher, B.J., Martin, B., Gan, B.S., and Reid, G. (2000). Identification of collagen-binding proteins in *Lactobacillus* spp. With surface-enhanced laser desorption/ionization-time of flight ProteinChip technology. Appl. Environ. Microbiol. 66, 4396–4400.

Junker, L.M., Peters, J.E., and Hay, A.G. (2006). Global analysis of candidate genes important for fitness in a competitive biofilm using DNA-array-based transposon mapping. Microbiology *152*(Pt 8), 2233–2245.

Keane, F.E., Ison, C.A., and Taylor-Robinson, D. (1997) A longitudinal study of the vaginal flora over a menstrual cycle. Int. J.S.T.D. AIDS. 8(8), 489–94.

Kelly, D., Campbell, J.I., King, T.P., Grant, G., Jansson, E.A Coutts, A.G., Pettersson, S., and Conway, S. (2004). Commensal anaerobic gut bac teria attenuate inflammation by regulating nuclearcytoplasmic shuttling of PPAR-gamma and RelA. Nat. Immunol. 5: 104–112.

Kim, S.O., Sheikh, H.I., Ha, S-D., Martins, A., and Reid G., (2006). G-CSF mediated inhibition of JNK is a key mechanism for *Lactobacillus rhamnosus*-induced suppression of TNF production in macrophages. Cell. Microbiol. *8*, 1958–1971.

Kirjavainen, P.K., Laine, R.M., Carter, D., Hammond, J-A., and Reid, G. (2008). Expression of antimicrobial defense factors in vaginal mucosa following exposure to *Lactobacillus rhamnosus* GR-1. Int. J. Probiotics. In press.

Klebanoff, S.J., Hillier, S.L., Eschenbach, D.A., and Waltersdorph, A.M. (1991). Control of the microbial flora of the vagina by H_2O_2-generating lactobacilli. J. Infect. Dis. *164*, 94–100.

Koehler, G. and Reid, G. (2006). Mechanisms of probiotic interference with *Candida albicans*. ASM Conference on Candida and Candidiasis. Published abstract.

Laughton, J., Devillard, E., Heinrichs, D., Reid, G., and McCormick, J. (2006). Inhibition of expression of a staphylococcal superantigen-like protein by a soluble factor from *Lactobacillus reuteri*. Microbiology *152*, 1155–1167.

Ma, D., Forsythe, P., and Bienenstock, J. (2004). Live *Lactobacillus reuteri* is essential for the inhibitory effect on tumor necrosis factor alpha-induced interleukin-8 expression. Infect. Immun. 72, 5308–5314.

Madsen, K.L., Doyle, J.S., Jewell, L.D., Tavernini, M.M., and Fedorak, R.N. (1999). *Lactobacillus* species prevents colitis in interleukin 10 gene-deficient mice. Gastroenterology 116: 1107–1114.

Mardh, P.A., Novikova, N., and Stukalova, E. (2003). Colonisation of extragenital sites by *Candida* in women with recurrent vulvovaginal candidosis. Br. J. Obstet. Gynaecol. *110*, 934–937.

Marrazzo, J.M., Cook, R.L., Wiesenfeld, H.C., Murray, P.J., Busse, B., Krohn, M., and Hillier, S.L. (2006). Women's satisfaction with an intravaginal *Lactobacillus* capsule for the treatment of bacterial vaginosis. J. Women's Health (Larchmt). *15*, 1053–1060.

Mbizvo, M.E., Musya, S.E., Stray-Pedersen, B., Chirenje, Z., and Hussain, A. (2004). Bacterial vaginosis and intravaginal practices: association with HIV. Cent. Afr. J. Med. *50*(5–6), 41–46.

McGroarty, J.A. and Reid, G. (1988). Detection of a *Lactobacillus* substance which inhibits *Escherichia coli* Can. J. Microbiol. *34*, 974–978.

Myer, L., Denny, L., Telerant, R., Souza, M., Wright, T.C. Jr, and Kuhn, L. (2005). Bacterial vaginosis and susceptibility to HIV infection in South African women: a nested case-control study. J. Infect. Dis. *192*(8), 1372–80.

Neish, A.S., Gewirtz, A.T., Zeng, H., Young, A.N., Hobert, M.E., Karmali, V., Rao, A.S., and Madara, J.L. (2000). Prokaryotic regulation of epithelial responses by inhibition of IkappaB-alpha ubiquitination. Science *289*, 1560–1563.

Ocana, V.S., and Elena Nader-Macias M. (2004). Production of antimicrobial substances by lactic acid bacteria II: screening bacteriocin-producing strains with probiotic purposes and characterization of a *Lactobacillus* bacteriocin. Methods Mol. Biol. *268*, 347–53.

Parent, D., Bossens, M., Bayot, D., Kirkpatrick, C., Graf, F., Wilkinson, F.E., and Kaiser, R.R. (1996). Therapy of bacterial vaginosis using exogenously-applied *Lactobacilli acidophili* and a low dose of estriol: a placebo-controlled multicentric clinical trial. Arzneimittelforschung. *46*(1), 68–73.

Pena, J.A., Rogers, A.B., Ge, Z., Ng, V., Li, S.Y., Fox, J.G., and Versalovic, J. (2005). Probiotic *Lactobacillus* spp. diminish *Helicobacter hepaticus*-induced inflammatory bowel disease in interleukin-10-deficient mice. Infect. Immun. 73**,** 912–920.

Petrof, E.O., Kojima, K., Ropeleski, M.J., Musch, M.W., Tao, Y., De Simone, C., and Chang, E.B. (2004). Probiotics inhibit nuclear factor-kappaB and induce heat shock proteins in colonic epithelial cells through proteasome inhibition. Gastroenterology 127**,** 1474–1487.

Pirotta, M., Gunn, J., Chondros, P., Grover, S., O'Malley, P., Hurley, S., and Garland, S. (2004). Effect of *Lactobacillus* in preventing post-antibiotic vulvovaginal candidiasis: a randomised controlled trial. BMJ *329*, 548.

Raz, R., and Stamm, W.E. (1993). A controlled trial of intravaginal estriol in postmenopausal women with recurrent urinary tract infections. N. Engl. J. Med. *329*, 753–6.

Reid, G. (2005). The importance of guidelines in the development and application of probiotics. Curr. Pharma. Design *11*, 11–16.

Reid, G., Anukam, K., James, V.L., van der Mei, H.C., Heinemann, C., Busscher, H.J. and Bruce, A.W. (2005). Oral probiotics for maternal and newborn health. J. Clin. Gastroenterol. *39*, 353–354.

Reid, G., Beuerman, D., Heineman, C., and Bruce, A.W. (2001). Probiotic *Lactobacillus* dose required to restore and maintain a normal vaginal flora. FEMS Immun. Med. Microbiol. *32*, 37–41.

Reid, G., Bruce, A.W., and Taylor, M. (1995). Instillation of *Lactobacillus* and stimulation of indigenous organisms to prevent recurrence of urinary tract infections. Microecol. Ther. 23, 32–45.

Reid, G., Burton, J., Hammond, J.-A., and Bruce, A.W. (2004). Nucleic acid-based diagnosis of bacterial vaginosis and improved management using probiotic lactobacilli. J. Medicinal Food 7, 223–228.

Reid, G., Charbonneau, D., Erb, J., Kochanowski, B., Beuerman, D., Poehner, R., and Bruce, A.W. (2003a). Oral use of *Lactobacillus rhamnosus* GR-1 and *L. fermentum* RC-14 significantly alters vaginal flora: randomized, placebo-controlled trial in 64 healthy women. FEMS Immunol. Med. Microbiol. 35: 131–134.

Reid, G., Cook, R.L., and Bruce, A.W. (1987). Examination of strains of lactobacilli for properties which may influence bacterial interference in the urinary tract. J. Urol. 138: 330–335.

Reid, G., Gan, B.S., She, Y-M., Ens, W., Weinberger, S., and Howard, J.C. (2002). Rapid identification of probiotic lactobacilli biosurfactant proteins by ProteinChip MS/MS tryptic peptide sequencing. Appl. Environ. Microbiol. 68(2) 977–980.

Reid, G., Hammond, J-A., and Bruce, A.W. (2003b). Effect of lactobacilli oral supplement on the vaginal microflora of antibiotic treated patients: randomized, placebo-controlled study. Nutraceut. Food. 8: 145–148.

Reid, G., McGroarty, J.A., Angotti, R., and Cook, R.L. (1988). *Lactobacillus* inhibitor production against *E. coli* and coaggregation ability with uropathogens. Can. J. Microbiol. 34: 344–351.

Reid, G., Millsap, K., and Bruce, A.W. (1994). Implantation of *Lactobacillus casei* var *rhamnosus* into the vagina. The Lancet 344: 1229.

Rossouw, J.E., Anderson, G.L., Prentice, R.L., LaCroix, A.Z., Kooperberg, C., Stefanick, M.L., Jackson, R.D., Beresford, S.A., Howard, B.V., Johnson, K.C., Kotchen, J.M., and Ockene J. (2002). Writing Group for the Women's Health Initiative Investigators. Risks and benefits of estrogen plus progestin in healthy postmenopausal women: principal results from the Women's Health Initiative randomized controlled trial. JAMA 288, 321–333.

Rosenstein, I.J., Fontaine, E.A., Morgan, D.J., Sheehan, M., Lamont, R.F., and Taylor-Robinson, D. (1997). Relationship between hydrogen peroxide-producing strains of lactobacilli and vaginosis-associated bacterial species in pregnant women. Eur. J. Clin. Microbiol. Infect. Dis. 16, 517–522.

Saunders, S., Bocking, A.Challis, J.P., and Reid, G. (2007). Effect of *Lactobacillus* challenge on *Gardnerella vaginalis* biofilms. Coll. Surfaces B: Biointerfaces (in press).

Sheil, B., McCarthy, J., O'Mahony, L., Bennett, M.W., Ryan, P., Fitzgibbon, J.J., Kiely, B., Collins, J.K., and Shanahan, F. (2004). Is the mucosal route of administration essential for probiotic function? Subcutaneous administration is associated with attenuation of murine colitis and arthritis. Gut 53, 694–700.

Simoes, J.A., Aroutcheva, A.A., Shott, S., and Faro, S. (2001). Effect of metronidazole on the growth of vaginal lactobacilli *in vitro*. Infect. Dis. Obstet. Gynecol. 9(1), 41–5.

Song, Y., Kato, N., Matsumiya, Y., Liu, C., Kato, H., and Watanabe, K. (1999). Identification of and hydrogen peroxide production by fecal and vaginal lactobacilli isolated from Japanese women and newborn infants. J. Clin. Microbiol. 37, 3062–3064.

Spinillo, A., Carratta, L., Pizzoli, G., Lombardi, G., Cavanna, C., Michelone, G., and Guaschino, S. (1992). Recurrent vaginal candidiasis. Results of a cohort study of sexual transmission and intestinal reservoir. J. Reprod. Med. 37, 343–7.

Schwebke, J.R, and Weiss, H. (2001). Influence of the normal menstrual cycle on vaginal microflora. Clin. Infect. Dis. 32, 325.

Tansupasiri, A., Puangsricharern, A., Itti-arwachakul, A., and Asavapiriyanont, S. (2005). Satisfaction and tolerability of combination of lactoserum and lactic acid on the external genitalia in Thai women. J. Med. Assoc. Thai. *88*, 1753–1757.

Tomas, M.S., Claudia Otero, M., Ocana, V., and Elena Nader-Macias, M. (2004). Production of antimicrobial substances by lactic acid bacteria I: determination of hydrogen peroxide. Methods Mol. Biol. *268*, 337–346.

Vasquez, A., Jakobsson, T., Ahrne, S., Forsum, U., and Molin, G. (2002). Vaginal lactobacillus flora of healthy Swedish women. J. Clin. Microbiol. *40*, 2746–2749.

Velraeds, M.C., van der Belt, B., van der Mei, H.C., Reid, G., and Busscher, H.J. (1998). Interference in initial adhesion of uropathogenic bacteria and yeasts silicone rubber by *a Lactobacillus acidophilus* biosurfactant. J. Med. Microbiol. *49*, 790–794.

Velraeds, M.C., van der Mei, H.C., Reid, G., and Busscher, H.J. (1996). Physicochemical and biochemical characterization of biosurfactants released from *Lactobacillus* strains. Coll. Surf. B: Biointerfaces 8, 51–61.

Voravuthikunchai, S.P., Bilasoi, S., and Supamala, O. (2006). Antagonistic activity against pathogenic bacteria by human vaginal lactobacilli. Anaerobe. *12*, 221–226.

Witkin, S.S., Linhares, I.M., Giraldo, P., amd Ledger, W.J. (2007). An altered immunity hypothesis for the development of symptomatic bacterial vaginosis. Clin. Infect. Dis. *44*, 554–557.

World Health Organization. (1981) Report of a WHO Study Group on Recommended Health-Based Limits in Occupational Exposure to Selected Organic Solvents. ISBN 9241206640.

Xie, J., Foxman, B., Zhang, L., and Marrs, C.F. (2006). Molecular epidemiologic identification of Escherichia coli genes that are potentially involved in movement of the organism from the intestinal tract to the vagina and bladder. J. Clin. Microbiol. *44*, 2434–2441.

Zhong, W., Millsap, K., Bialkowska-Hobrzanska, H., and Reid, G. (1998). Differentiation of *Lactobacillus* species by molecular typing. Appl. Environ. Microbiol. *64*, 2418–2423.

Zhou, X., Bent, S.J., Schneider, M.G., Davis, C.C., Islam, M.R., and Forney, L.J. (2004). Characterization of vaginal microbial communities in adult healthy women using cultivation-independent methods. Microbiology *150*(Pt 8), 2565–2673.

From Probiotics, Prebiotics and Synbiotics to 'Living Drugs'

11

Åsa Ljungh and Torkel Wadström

The human intestine harbours an immense collection of microbes which have co-evolved with us. Recent studies indicate that the gut microbes regulate energy harvest from the diet and participate in the peripheral body metabolism. Elie Metchnikoff understood that a gut microbial dysbiosis severely affects many body functions, including a complex interplay of gut-brain interactions, now under intense study. Most probiotic strains belong to the genus *Lactobacillus*. The promising results of a first generation of probiotic microbes, evaluated in animal models as well as natural infections in animals and humans indicate a promising future for coming generations of probiotics. Antibiotic-associated, travellers' and pediatric diarrhea have been most studied, and more recently, inflammatory bowel disease and irritable bowel syndrome. The probiotic strains should be thoroughly characterized. Probably, future probiotics will often contain mixes of strains with complementary characteristics, tailormade for different gastrointestinal diseases, vaginosis or as delivery systems for vaccines, immunoglobulins and other protein based therapies.

The development of many new molecular biology based methods to study the great complexity of the indigenous microflora of the skin as well as of mucosal surfaces only in the last decade has opened up new frontiers to study microbe interactions and crosstalks with the host in health and disease (Chapter 4). During this decade we have observed a continuously increasing problem with antibiotic resistance, and the appearance of new 'superbugs', such as community-acquired methicillin-resistant *Staphylococcus aureus* (CA-MRSA), extended-spectrum-β lactamase producing *Escherichia coli*, *Klebsiella* spp, *Proteus* spp and *Salmonella* spp (ESBL), and other nosocomial pathogens in all parts of the world (Andersson, 2003). The immense costs to develop new antimicrobials should hopefully now increase our interest in 'the non-antibiotic approach', including bacteriophage and bacteriotherapy. To escalate the recruitment of scientists in academia *and* industry to study the normal microbiota, how it is affected by broad and narrow spectrum antibiotic therapy, and effects on pathogens is one possible way.

The increasing problem today with antibiotic-associated diarrhoea (AAD) and nosocomial spread of new 'superstrains' of *Clostridium difficile* clearly indicates that we need a better understanding of the indigenous gut microflora to combat AAD by an optimal pro- and prebiotic approach. This includes our understanding of best candidate species and strains from *in vitro* screening studies in animal models and human clinical studies (Ljungh and Wadström, 2006).

Saccharomyces boulardii, a probiotic yeast is made into 'living drug' (Ultra-Levura®, Biocodex® and others), and now used to treat AAD and infectious diarrhoea (Vandenplas *et al.* 2007). However, studies in intensive care unit (ICU) identified this fungal 'drug'-strain in patients with fungaemia (Boyle *et al.*, 2006). It is thus contraindicated in severely ill and/or immunocompromised patients. Studies in specific mouse models, such as the $IL10^{-/-}$ mouse are promising tools to evaluate *new* probiotic drug candidates to prevent and also to treat bacterial, viral and parasitic infections in immunocompromised patients (Chapter 8). These patient groups include children with various immunodeficiencies and cancer patients on cytostatic drug treatment. For these patients it is important that selected probiotics, prebiotics, fibers and mixes of lactic acid bacteria (LAB) can stimulate the gut indigenous microflora and improve the gut barrier function (Chapter 7). Other mouse and rat models were developed in recent years to simulate the decay of this barrier, including the epithelial tight junctions (TJ) which opens up in several microbe-induced diarrhoeas, such as those caused by *Salmonella enterica*, enteropathogenic *Escherichia coli* (EPEC, EAEC, EaggEC) and *Shigella* sp. (Tien M-T *et al.*, 2006). A promising first study with a probiotic LAB product (*Lactobacillus plantarum*) indicates that further studies in polarized epithelial models of $CaCO_2$ and T84 cell lines, mouse models and human volunteer studies should be encouraged (Chapter 5).

Despite the great progress with oral rehydration therapy (ORT) for acute diarrhoea in paediatrics during the last decades, few well-designed studies for travellers' diarrhoea have been published. A recent collaborative study showed that probiotics was a useful adjunct to ORT to treat infectious diarrhoea in children (Canani *et al.*, 2007). However, the great variations in the study design, including probiotic products studied and patient groups and subgroups, gave nonconclusive results. This included the *L. rhamnosus* strain GG with various delivery systems, such as yoghurt, pills and capsules.

Thus, today antibiotic prophylaxis and treatment is not commonly recommended by specialists in travel medicine. This should certainly encourage more research on basic antidiarrhoeal effects induced by pro-, pre- and synbiotics, as well as combinations with ORT and other natural remedies such as *nutraceuticals* (Penner *et al.*, 2005). In the traditional Indian medicine and in other Asian countries we have a number of candidates, now appearing in Western 'alternative' medicine. Studies on these compounds should be encouraged to combat these diseases in hunger epidemics as in the Dalfour province of Sudan today, also suggested by calls by Doctors Without Borders.

Recent studies clearly show that probiotic strains should be systematically studied in order to combat, i.e. inhibit, displace and compete with gastrointestinal (GI) pathogens, like *C. difficile*, salmonellae and various pathogenic *E. coli* (Vandenplas *et al.*, 2007). Altogether, this approach is most timely with a rapid global increase in travelling as well as the fact that diarrhoea is a major killer of children in the Third World. Moreover, such strain mixes may also be promising as an adjunct treatment in respiratory diseases in children in combination with antibiotics.

More recently, several clinical studies in patients with *Helicobacter pylori* gastritis, peptic ulcer disease (PUD) and dyspepsia using specific probiotic yoghurts as well as tablets show promising results to suppress the infection and reduce the symptoms, especially the acid reflux induced by eradication therapy (Franceschi *et al.*, 2007). In Italy, a commercial *Bacillus* probiotic product has been used to treat dyspeptic patients (Nista *et al.*, 2004). The rapid increase in patients with GERD (gastro-oesophageal reflux disease), now treated with proton pump inhibitor and other acid-suppressing therapies, may well be better explained soon by favourable probiotic cell wall, flagellae, nucleic acid and Toll-like receptor (TLR) interactions, down-regulating *H. pylori* bile and acid stress-induced gastric, oesophageal and gut inflammation. A number of animal models are now well developed and standardized for studies of such living drug/probiotic mixtures to be delivered to these mucosal surfaces by various homing procedures. However, some LAB cell lysis products seem to suppress *H. pylori* infection, just like the normal LAB flora of the murine fore stomach (Wang *et al.*, 2000). A similar natural probiotic stomach–gut barrier exists in a number of other species: pig, horse, and ruminants. The stomach LAB and other microflora of primates, including humans, has been poorly studied (Monstein *et al.*, 2000). Thus, the old dogma that the normal gastric acidity is a good gut barrier and prevents GI infections should be better analysed as well as the gastric microflora of our aging population in the whole industrialized world. Overgrowth of the small bowel microflora with enterococci and other potential pathogens, some with carcinogenic – cocarcinogenic metabolites should be analysed in *H. pylori* gastritis, mild as well as chronic, with mucosal inflammation, atrophy and a high risk to absorb food and microbial carcinogenic substances (Quigly and Quera, 2003). An optimal prophylaxis and long time treatment strategy seems to be of great demand considering the rapid decline in many modern societies of the natural indigenous gastric colonization of humans and closely related primates with a common African origin (Linz *et al.*, 2007). Our recent understanding indicates that this vanishing gastric infection closely followed us through the evolution for more than 58,000 years. A link between the disappearance of the gastric *H pylori* and the rapid rise in modern western societies of asthma, skin and food allergies as well as Crohn's disease (CD) and ulcerative colitis (UC) has also been suggested (Blaser *et al.*, 2008).

A first generation of a probiotic drug strategy to treat CD and UC by a specific *E coli* strain has been developed in Germany, and gave similar treatment results as a classical UC drug, i.e. Salazopyrin® (Jones and Fox-Orenstein, 2007). Interestingly, a specific eight-strain mix of LAB, VSL#3, developed in Bologna, also had good treatment effects of UC patients with pouchitis, i.e. severe inflammation in the peroperatively created pouch in the lower ileum (Gionchetti *et al.*, 2007; Katz, 2003). However, the great technical problems to produce and stabilize such a strain mix product are important for manufacturing of living drug products. Several patents on how to culture and lyophilize LAB strains exist in a rapidly expanding biotech industry today producing LAB, also as yoghurt starter cultures, for ensilage and food preservation. In Chapter 8, aspects that the chronic inflammation in UC predisposes to DNA damage, mutations and colonic tumours should stimulate studies in well defined animal models and in similar inflammatory bowel disease (IBD) conditions in dogs (and other pet animals) before VSL#3 alternative products are studied in expensive clinical trials. Also, well designed *in vitro* models to study inactivation of food and other mutagens as well as bile acid derived mutagens seem most attractive. The high exposure of many non-western diet consuming populations to food pathogens and various mutagens, such as aflatoxin and other mycotoxins should give priority to such research areas, including GMO (gene modified organisms) to define new 'living drugs' to combat gut, biliary tract and liver diseases (Ahmed, 2003). These can also stabilize the gut barrier function which becomes damaged in acute and chronic gut infections (O'Hara *et al.*, 2006; Saulnier *et al.*2006). In this way, we may find alternatives to antibiotic treatment in gastrointestinal surgery as well as in small bowel overgrowth syndromes, often associated with irritable bowel syndrome (IBS) (Creed, 2007; Spiller *et al.*, 2007), and maybe with autism and associated gut-brain disorders (Horvath and Perman, 2002).

Such 'living drugs' to target the stomach, including the cardia region, to treat reflux disease (GERD) and suppress oesophagitis and gastritis is also a promising area to develop new therapy to avoid overuse of antibiotics, and to combat antibiotic resistance and 'superbugs', including intestinal colonization with multiple resistant hospital pathogens, like MRSA, vancomycin-resistant enterococci (VRE) and extended spectrum β-lactamase-producing enterics (ESBL) (Andersson, 2003).

Another product commonly produced by LAB are antibacterial peptides, bacteriocins, which suppress closely related strains but some also suppress an important food pathogen, *Listeria monocytogenes* and others (Ljungh and Wadström, 2006). These antibacterial peptides of LAB form a new expanding research area, also as candidates to combat *H pylori* and other gut pathogens, as well as skin pathogens, such as MRSA and other gram positive pathogens with several different immunostimulatory and proinflammatory properties (Kimbrell *et al.*,2008). It should be remembered that Shinefield already in the early 1970s developed a skin interference therapy for treatment of severe staphylococcal skin infections such as the scalded skin syndrome by a specific attenuated *S aureus* strain, 524A. The development of new antibiotics up to now has delayed further studies on topical bacteriotherapy. Interestingly, the great interest today in new toxin producing 'superbugs', such as strains of *S aureus* and group A streptococci, inducing toxic shock-like syndromes (TSS) opens up new fields for research on experimental bacteriophage as well as bacteriotherapy. A number of good animal models are available for studies of prototype products (Casey *et al.*, 2007).

Since bacteriotherapy has been used for thousands of years in Asian traditional medicine, including yoghurt preparations to treat vaginal discharge it seems appropriate to activate a number of research lines to study the skin, oral, urogenital and gut mucosa microflora by modern real-time PCR and PCR-DGGE methods (Chapter 4) to analyse the culturable and unculturable parts of (a) normal individuals before and after weanling, (b) before and after antibiotic therapy, and (c) in immunocompromised patients, and (d) in the ageing populations. These studies should permit a better selection of strains as 'living drugs'. A thorough knowledge of LAB metabolomics and other probiotic candidates will allow a better design of LAB mix products and similar products with nutraceuticals to combat chronic inflammation which may develop into cancer of the skin and various mucosal surfaces, including IBD-associated colon cancer (Chapter 8).

So far, research on pro- and prebiotics has been driven primarily by interests and lobbying by big dairy industries (i.e. Nestlé, Danone, Valio, Yakult) and not by the 'big Pharma' (i.e. pharmaceutical industries). This has created a special situation in Japan with the FOSHU guidelines and now calls in the EU community (without a US equivalent to FDA) on how to create rules to make appropriate health claims for pro- and prebiotic based products, such as fermented milk products, cheese, fruit juices and vegetables (Chapter 9).

Our knowledge today in mechanisms of action of the probiotic cross-talk with mucosal surfaces, normal as well as damaged epithelial layers in gut infections, post-infection IBS and IBD is certainly too incomplete to design optimal probiotics, prebiotics and synbiotics to combat these various diseases, and for prevention of enteric infections, IBS and IBD (Jones and Fox-Orenstein, 2007; O'Hara *et al.*, 2006). A breakthrough in this research area might be the finding that animals harbouring *Bacteroides fragilis* expressing polysaccharide A were protected from colitis induced by the enterohepatic *Helicobacter hepaticus* (Mazmanian *et al.*, 2008).

To better understand the mode of action of various pro-, pre- and synbiotics, the IL-$10^{-/-}$ mouse is today probably the best model to simulate chronic gastric and intestinal inflammation (Hjerrild Zeuthen *et al.*, 2006; Chapter 8). However, to compare experiments in germ-free IL-$10^{-/-}$ and mice with a defined gut flora, diet and housing conditions is important to avoid a 'silent' chronic gut inflammation induced by enterohepatic *Helicobacter* species (*H. hepaticus, H. bilis*, etc.) and other agents causing chronic gut inflammation in rodents (Solnick *et al.*, 2006). The great interest today in the gut-liver link in IBD and primary sclerosing cholangitis as well as diet-gut microflora, obesity and type II diabetes indicates that we will get other well-defined mouse models to study these links as well as related diseases, such as chronic cholangitis and gallstone disease (Fox *et al.*, 2003).

We do hope that medical research foundations will put a high priority on basic research to explore mechanisms of action for probiotics, including metabolomics and proteomics to optimize future combinations of strains *and* to tailor-make prebiotics. The rapid increase of new prebiotics beside some classical ones, i.e. inulin, β-glucans, fructo- and galactooligo-saccharides will now include studies on chitosan, other natural glycoconjugates and compounds, such as tannins, degraded by probiotics in laboratory studies (Bruggencate *et al.*, 2006). Thus, the borderline area of prebiotic and nutraceutical research will probably disappear in the near future. However, new *in vitro* findings have to be evaluated in rodent models of comparative medicine as well as in pigs, dogs, cats and humans to confirm *in vivo* effects to prevent (and treat?) diarrhoea, *H. pylori* gastritis, IBD, IBS etc, and other health effects (Creed, 2007). This is most important also in very young animals to create a better basis for pro-, pre- and synbiotics to prevent and treat necrotizing enterocolitis in preterm infants. The possibility to suppress an abnormal Th1/Th2 immune response by such products in children, prone to develop chronic disease, UC and chronic liver disease is another area to be explored, including new animal and organ culture models. The first promising clinical studies with a synbiotic product in severely ill surgical patients in London and Berlin (Bengmark, 2004) should encourage other studies to compare treatments with probiotics and antibiotics in well-designed studies.

The complex cell surface proteins (CSPs) of LAB and *Bifidobacteria* (Chapter 5) may allow targeting of vaccine components as well as HIV cell entry inhibitors to the vaginal mucosa to combat HIV/Aids (Liu *et al.*, 2007). A better understanding of the pro- and prebiotic interplay in the vagina will permit development of new vaginosis prophylaxis and therapy (Chapter 10). Similar mixes may also be considered to treat various chronic skin inflammations, such as atopic dermatitis (Callard and Harper, 2007). Several recent patents in such 'new' applications propose cosmetic as well as medical ointments and other products to combat various skin disorders and STD. Such studies could also encourage research in possible use of pro- and prebiotics to combat other allergic disorders (food allergy, asthma, etc.), a very controversial area in paediatric research today.

Finally, TLR-signalling pathways and the specific mechanisms which govern the interaction between lactobacillus and the surface epithelium as well as gut allergen processing (Rizzello *et al.*, 2007) will certainly prove to be a fruitful research area (Chapter 7; Strober *et al.*, 2007). While pathogenic bacteria evoke phosphorylation of the ERK pathway and activate AP-1, probiotic strains generally modulate the NF-κB pathway (Iyer *et al.*, 2008). Considering that the collective gut microbial genome is >200 times bigger than the human genome it is not surprising that the gut microbiota governs so many functions – the gut–liver link, the gut–brain link, probably allergic diseases, and even obesity, based on that 'obesity-causing' bacteria release more calories from food than those harboured by lean people (Mullard, 2008). Results of the on-going EC project, MetaHIT, on the role of the gut microbiota in obesity and IBD are awaited with great interest.

References

Ahmed, F.E. (2003). Genetically modified probiotics in foods. Trends Biotechnol. *21*, 491–497.

Andersson, D.I. (2003). Persistence of antibiotic resistant bacteria. Curr. Opin. Microbiol. *6*, 452–456.

Bengmark, S. (2004). Synbiotics to strengthen gut barrier function and reduce morbidity in critically ill patients. Clin. Nutr. *23*, 441–445.

Blaser, M.J., Chen, Y., and Reibman, J. (2008). Does *Helicobacter pylori* protect against asthma and allergy? Gut *57*, 561–562?

Boyle, R.J., Robins-Browne, R.M., and Tang, M.L.K. (2006). Probiotic use in clinical practice: what are the risks? Am. J.Clin. Nutr. *83*, 1256–1264.

Bruggencate, S.J.M.T., Bovee-Oudenhoven, I.M.J., Lettink-Wissink, M.L.G., Katan, M.B., van der Meer, R. (2006) Dietary fructooligosaccharides affect intestinal barrier function in healthy men. J. Nutr. *136*, 70–74.

Callard, R.E. and Harper, J.I. (2007) The skin barrier, atopic dermatitis and allergy: a role for Langerhans cells? Trends Immunol. *28*: 294–298.

Canani, R.B., Cirillo, P., Terrin, G., Co, L., Spagnuloto, M.I., De Vicenzo, A., Albano, F., Passariello, A. De Marco, G., Manguso, F., Guarino, A. (2007). Probiotics for treatment of acute diarrhoea in children: randomised clinical trial of five different preparations. Br. Med. J. 335, 340–346.

Casey, P.G., Gardiner, G.E., Casey, G., Bradshaw, B., Lawlor, P.G., Lynch, P.B., Leonard, F.C., Stanton, C., Ross, R.P., Fitzgerald, G.F., and Hill, C. (2007). A five-strain probiotic combination reduces pathogen shedding and alleviates disease signs in pigs challenged with *Salmonella enterica* serovar typhimurium. Appl. Environ. Microbiol. *73*, 18558–1863.

Creed, F. (2007). Cognitive behavioural model of irritable bowel syndrome. Gut *56*, 1039–1041.

Fox, J., Dewhirst, F., Shen, Z., Feng, Y., Taylor, N., Paster, B., Ericson, R., Lau, C., Correa, P., and Araya, J. (2003). Hepatic species identified in bile and gallbladder tissue from Chileans with chronic cholecystitis. Gastroenterol. *114*, 755–763.

Franceshi, F., Cazzato, A., Nista, E.C., Scarpellini, E., Roccarina, D., Gigante, G., Gasbarrini, G., and Gasbarrini, A. (2007). Role of probiotics in patients with *Helicobacter pylori* infection. Helicobacter *12*, 59–63.

Gionchetti, P., Rizzello, F., Morselli, C., Poggioli, G., Tambasco, R., Calabrese, C., Brigidi, P., Vitali, B., Straforini, G., and Campieri, M. (2007). High-dose probiotics for the treatment of active pouchitis, Dis. Colon Rectum *50*, 2075–2082.

Hjerrild Zeuthen, L., Risager Christensen, H., and Frøkiær, H. (2006). Lactic Acid Bacteria inducing a weak interleukin-12 and tumor necrosis factor alpha response in human dendritic cells inhibit strongly stimulating Lactic Acid Bacteria but act synergistically with gram-negative bacteria. Clin. Vaccine Immunol. *13*, 365–375.

Horvath, K. and Perman, J.A. (2002). Autism and gastrointestinal symptoms. Curr. Gastroenterol. Rep. *4*, 251–258.

Iyer, C., Kosters, A., Sethi, G., Kunnumakkara, A.B., Aggarwal, B.B., and Versalovic, J. (2008). Probiotic *Lactobacillus reuteri* promotes TNF-induced apoptosis in human myeloid leukemia-derived cells by modulation of NF-κB and MAPK signaling. Cell. Microbiol. *10*, 1442–1452.

Jones, J.L., Fox-Orenstein, A.E. (2007). The role of probiotics in inflammatory bowel disease. Dig. Dis. Sci. 52, 607–11.

Katz, J. (2003). Prevention is the best defense: probiotic prophylaxis of pouchitis. Gastroenterology *124*, 1535–1538.

Kimbrell, M.R., Warshakoon, H., Cromer, J.R., Malladi, S., Hood, J.D., Balakrishna, R., Scholdberg, T.A. and David, S.A. (2008). Comparison of the immunostimulatory and proinflammatory activities of candidate gram-positive endotoxins, lipoteichoic acid, peptidoglycan, and lipopeptides, in murine and human cells. Immunol. Lett. *118*, 132–141.

Linz, B., Balloux, F., Moodley, Y., Manica, A., Liu, H., Roumagnac, P., Falush, D., Stamer, C., Prugnolle, F., van der Merwe, S.W., Yamaoka, Y., Graham, D.Y., Perez-Trallero, E., Wadström, T., Suerbaum, S., Achtman, M. (2007). An African origin for the intimate association between humans and *Helicobacter pylori*. Nature 445, 915–918.

Liu, J.J., Reid, G., Jiang, Y., Turner, M.S., Tsai, C.-C. (2007) Activity of HIV entry and fusion inhibitors expressed by the human vaginal colonizing probiotic *Lactobacillus reuteri* RC-14. Cell. Microbiol. *9*, 120–130.

Ljungh, Å. and Wadström, T. (2006). Lactic acid bacteria as probiotics. Curr. Iss. Intest. Microbiol. *7*, 73–89.

Mazmanian, S.K., Round, J.L., and Kasper, D.L. (2008). A microbial symbiosis factor prevents intestinal inflammatory disease. Nature *453*, 620–625.

Monstein, H.-J., Tiveljung, A., Kraft, C.H., Borch, K., and Jonasson, J. (2000). Profiling of bacterial flora in gastric biopsies from patients with *Helicobacter pylori*-associated gastritis and histologically normal control individuals by temperature gradient gel electrophoresis and 16S rDNA sequence analysis. J. Med. Microbiol. *49, 817–822.*

Mullard, A. (2008). The inside story. Nature *453*, 578–580.

O'Hara, A.M., O'Reagan, P., Fanning, A., O'Mahony, C., Macsharry, J., Lyons, A., Bienenstock, J., O'Mahony, L., Shanahan, F. (2006) Functional modulation of human intestinal epithelial cell responses by *Bifidobacterium infantis* and *Lactobacillus salivarius*. Immunol. *118*, 202–215.

Penner, R., Fedorak, R.N., and Madsen, K.L. (2005). Probiotics and nutraceuticals: non-medicinal treatments of gastrointestinal diseases. Curr. Opin. Pharmacol. 5, 596–603.

Quigley, E. and Quera, R. (2003). Small intestinal bacterial overgrowth: Roles of antibiotics, prebiotics, and probiotics. Gastroenterology *130*, S78–S90.

Rizzello, C.G., De Angelis, M., Di Cagno, R., Camarca, A., Silano, M., Losito, I., De Vincenzi, M., De Bari, M.D., Palmisano, F., Maurano, F., Gianfrani, C., Gobbetti, M. (2007) Highly efficient gluten degradation by Lactobacilli and fungal proteases during food processing: New perspectives for celiac disease. Appl. Environm. Microbiol. *73*, 4499–507.

Saulnier, N., Zocco, M.A., Di Caro S., Gasbarrini, G., and Gasbarrini, A. (2006). Probiotics and small bowel flora: Molecular aspects of their interactions. Genes Nutr. *1*, 107–116.

Solnick, J.V., Franceschi, F., Roccarina, D., and Gasbarrini, A. (2006). Extragastric manifestations of *Helicobacter pylori* infection – other *Helicobacter* species. Helicobacter *11*, 46–51.

Spiller, R., Aziz, Q., Creed, F., Emmanuel, A., Houghton, L., Jones, R., Kumar, D., Rubin, G., Trudgill, N., Whorwell, P. (2007). Guidelines on the irritable bowel syndrome: mechanisms and practical management. Gut 56, 1770–1798.

Strober, W., Fuss, I., Mannon, P. (2007). The fundamental basis of inflammatory bowel disease. J. Clin. Invest. *117*, 514–21.

Tien, M.T., Girardin, S.E., Regnault, B., Le Bourhis, L., Dillies, M.A., Coppée, J.Y., Bourdet-Sicard, R., Sansonetti, P.J., and Pédron, T. (2006). Anti-inflammatory effect of *Lactobacillus casei* on Shigella-infected human intestinal epithelial cells. J. Immunol. *176*, 1228–1237.

Vandenplas, Y., Nel, E., Waltermeyer, G.A., Walele, A., Wittenberg, D., and Zuckerman, M. (2007). Probiotics in infectious diarrhoea: are they indicated? A review focusing on *Saccharomyces boulardii*. SA J. Child Health *1*, 116–119.

Wang X., Willén, R., and Wadström, T. (2000). Astaxanthin-rich algal meal and vitamin C inhibit *Helicobacter pylori* infection in BALB/cA mice. Antimicrob. Agents Chemother. *44*, 2452–2457.

Index